Ira H. Bernstein ▪ Nancy A. Rowe

Statistical Data Analysis Using Your Personal Computer

Sage Publications
International Educational and Professional Publisher
Thousand Oaks ▪ London ▪ New Delhi

For information:

 Sage Publications, Inc.
2455 Teller Road
Thousand Oaks, California 91320
E-mail: order@sagepub.com

Sage Publications Ltd.
6 Bonhill Street
London EC2A 4PU
United Kingdom

Sage Publications India Pvt. Ltd.
M-32 Market
Greater Kailash I
New Delhi 110 048 India

Printed in the United States of America

Library of Congress Cataloging-in-Publication Data

Bernstein, Ira H.
  Statistical data analysis for the personal computer / by Ira H.
Bernstein and Nancy A. Rowe.
     p. cm.
  ISBN 0-7619-1780-2 ó ISBN 0-7619-1781-0 (pbk.)
  1. Statisticsó Data processing. 2. Microcomputers. I. Rowe, Nancy A.
II. Title.
  QA276.4 .B48526 2001
  001.4í22í02856 dc21   2001000422

01   02   03   04   05   06   07   7   6   5   4   3   2   1

Acquiring Editor:	C. Deborah Laughton
Editorial Assistant:	Veronica Novak
Production Editor:	Diane S. Foster
Editorial Assistant:	Cindy Bear
Typesetter/Designer:	Technical Typesetting Inc.
Cover Designer:	Jane Quaney

Contents

Preface

Like so many who have taught statistics to students who are intelligent yet not highly gifted at thinking algebraically, we have struggled with how to make key concepts clear. On the one hand, we do not wish to reduce matters to a "cookbook" because we have seen how poorly this equips students to make appropriate decisions about their data, which is one of the major reasons the courses are taught in the first place. On the other hand, we feel a need to provide something more tangible than is taught in the typical text. The approach we took to this supplement basically revives an approach to teaching that was attempted many years ago—the use of demonstrations.

Demonstrations go back at least 125 years in the teaching of statistics, as noted by a reviewer. At that time, Galton developed a device he called a "quincunx," which he used to demonstrate the binomial distribution (Stigler, 1986, pp. 275–280). Any number of those of us who are senior can recall flipping a series of pennies to generate binomial distributions. Unfortunately, these were very time consuming and the point of the demonstration had long been lost to those who needed it most by the time the demonstration was completed. Various companies marketed devices to speed up the process, but these met with limited success.

As computers became more and more available, their use as didactic devices was quickly seized upon and exploited. At first, this required rather extensive programming skills. However, as more and more students become familiar with the essentials of such programs as SAS and SPSS, it is possible to write programs that they can easily understand and that do not require much effort to prepare. That is the spirit in which this book was written.

The authors are grateful to Joan Ellason and William Schweinle III for their interest in the project.

Finally, to quote Prof. Stanley Cohen: "You never complete a book, they just take it from you."

We dedicate this book to Linda, Cari, and Dina Bernstein and to Kathy Rowe.

Part I

1

Introduction

CHAPTER OBJECTIVES

The general goal of this book is to consider the process of analyzing the data one encounters in the behavioral sciences. The particular purpose of this chapter is to serve as an overview. Although we will be dealing with quantitative matters in the remaining chapters, we do not in any way wish to suggest that this is all that is involved in conducting research. Consequently, we will approach research from a broader perspective than will be true later in the book.

▣ INTRODUCTION

This is a book about the **practicalities** of data analyses. As such, it contrasts with the many fine books devoted to statistical **theory**, particularly those concerned with multivariate statistics and psychometric theory. In no way do we suggest that this book can replace knowledge of the statistical foundations of what is to be done. Similarly, these books often fail to provide basic "how to" information about understanding the meaning of results obtained from computer packages. This book thus complements the bulk of material written about data analysis. We assume that you have at least some familiarity with the language and various topics taught in graduate statistics, multivariate analysis, and psychometric theory courses, but do not assume that you are experts. In particular, we seek to relate two specific issues:

3

1. We will construct data with properties of theoretical interest. For example, many empirical problems revolve around the question of whether a particular set of measures fits a single factor-model. Such data, first described by Spearman (1904), consist of a single latent variable common to a series of observables plus the random error specific to each observable. We suggest that an important part in being able to recognize an outcome of this form in "real" data is to have seen it simulated.

2. We will present the results of such simulations in terms of the output of the two most popular statistical packages, SAS and SPSS. In this way, the reader can see how specific properties of the data relate to the various items of output. For example, one can note how changing the factor structure changes the scree (relative magnitudes of successive eigenvalues). Thus, we will "walk you through" the results of the most common procedures.

If you run into problems with terminology, you may wish to consult an Internet glossary. These include

> www.stats.gla.ac.uk/steps/glossary
> www.cas.lancs.ac.uk/glossary_v1.1/main.html
> www.animatedsoftware.com/ statglos/statglos.htm
> www.statsoftinc.com/textbook/ieglfra.html
> odwin.ucsd.edu/glossary
> www.cbs.nl/isi/glossary.htm
> www.ruf.rice.edu/~lane/hyperstat/glossary.html

The following may lead to other glossaries

> www.oswego.edu/~kane/econometrics/glossaries.htm
> www.grad.cgs.edu/wise/glossaryf.shtml

◉ WHAT IS NEEDED FOR EFFECTIVE BEHAVIORAL SCIENCE RESEARCH

Given a particular problem, an effective investigator must know (or work with someone who knows):

a. The general question to be asked, leading to the **goals** of the research, which jointly involve **hypotheses** and **strategic issues**

b. The **substantive literature** on the problem at hand

c. **General statistical theory**

d. How to employ relevant **computer programs**

e. Information from apparently **dissimilar areas** that can contribute to the problem because of common points

f. How to **implement** these goals

g. How to **communicate** the findings

This is not a rank ordering and it deals with the order of events only in a general sense, but it does point out the breadth and difficulty of doing research. It also explicates the importance of computer literacy. Although the goals are quite separate from the **means** used to achieve these goals and how they are **communicated**, all contribute to the final result.

Setting Research Goals

Every study has to begin with some form of **question**. The question may be stated quite broadly, as all research is in its initial stages, or it may be narrowed considerably by previous work. The question provides an **objective** to be aimed for. Sometimes the question is one you decide to ask, as in a thesis or dissertation; other times it is one someone else has asked but needs your expertise in answering, as in much consulting. Questions, in turn, lead to **hypotheses** as to what the answer might be and to a **strategy**. These are the general features that define your **goals**.

A question can therefore be satisfied by expressions such as "What causes antisocial behavior in adolescents?" "What affects memory for pictures?" and in thousands of other ways. The need for an objective may seem obvious, but large numbers of studies arise simply because a database is available, and many hours are spent fruitlessly on "file drawer" studies that are concerned with making sense out of a haphazard aggregation of data. Some investigators argue that objectives should follow from a highly structured theory; others are more pragmatic and note the many areas in which there is no well-developed theory. The very important pros and cons of this issue are discussed in a variety of sources. For example, see Greenwald, Pratkanis, Lieppe, and Baumgardner (1986) and succeeding articles in *Psychological Review* among many other interchanges. However, this topic is beyond the scope of this book. Nonetheless, if you do pursue a highly structured theory in a context where it is expensive to gather new data, do not conceive of your problem so narrowly that the neglect of a relevant variable makes the database useless in the future. The life span of very explicit theories in the behavioral sciences is not that great though the general concepts that underlie them can be.

The more time you spend planning any form of study, the better, in general, as long as you maintain sight of your objectives. It is, of course, possible to spend so much time planning that you never get the study done. First, there is a certain background, which consists of the knowledge of the research team and the available equipment, including computers. The main steps involved are as follows:

1. Be sure to address the questions you originally asked. This is not to say that your main conclusions may not reflect part of the wonderful serendipity of research. Indeed, much successful research eventuates far from its original target. However, regardless of what originally stimulated you, you do need to answer those original questions. For example, if you are doing a thesis or dissertation, your original proposal may be viewed as a contract ("If I do what I proposed to do, appropriately communicate the results, and complete the other requirements, I should obtain my degree"). You need to fulfill the basic terms of that contract. Likewise, if you are doing research on a consulting basis, you need to answer the questions that you were paid to answer.

2. You want others using appropriate methods to draw the same conclusions you did. One of the guiding, and perhaps controversial, theses of this book is that situations in which one and only one statistical procedure can supply "truth" are the exception, if they exist at all. For example, there has been vigorous debate on what numbers to place in the diagonals of a correlation matrix. Although one might argue that unities should be placed there, by definition, it is common practice to use the term "correlation matrix" when other numbers appear in these positions. In fact, unities are placed there in a component analysis, but numbers less than unity are placed there in common factor analysis. Much ink has been spilled regarding differences between these two methods. In fact, when either there is a modestly large number of variables (say, a dozen or so or more) or the variables correlate highly with at least one other (as

should be the case anyway), your results will hardly differ (Nunnally & Bernstein, 1994). Likewise, one commonly computes the sum of a series of variables and then correlates each of the underlying variables with that equally weighted sum. This result will scarcely differ from looking at the structure of the first factor in a factor analysis, which generally provides an optimally weighted sum. Indeed, if you are performing an analysis you are not extremely familiar with, you should consider looking also at a simpler analysis even if you will eventually report the more sophisticated one.

3. You want your conclusions to reflect the questions you asked and not be an artifact of method. Investigators using a method that they are not familiar with can easily confuse what is an artifact of method with an empirical finding. For example, various people have argued that multidimensional scaling of a matrix of correlations is preferable to factor analysis because it leads to one less dimension. Among those who point out why this is a pure artifact, Davison (1985) provides perhaps the clearest explanation.

4. You want your conclusions to reflect the questions you asked and not be an artifact of variables you did not take into account. It is easy to see this happen in retrospect. For example, differences attributable to ethnicity often reflect differences of an economic nature. By definition, all field studies involve variables that are naturally occurring rather than manipulated. Consequently, these variables may be, and usually are, confounded with other variables. There are some the literature or plain common sense dictates control over; others may only be known at a later date.

5. You want your audience to understand how you reached your conclusions. A structural equation or an item response model may be perfect for a dissertation, but neither is likely to be well understood when making a presentation to an advertising agency about a marketing strategy whose personnel have little or no background in statistics. Basically, if the client cannot visualize why tactic *A* is superior to tactic *B*, your efforts will fall on deaf ears. This is not to say you should not employ these sophisticated methods. Part of the art of data analysis is in knowing what kinds of evidence your audience will expect. Likewise, one of the dilemmas of data analysis is that you must often make things tangible at the same time you are appearing to use the "latest" methods.

Substantive Literature

Suppose you are studying the interrelationships among a set of personality measures. You must know what others working on similar problems have found (e.g., what factor structures have been previously reported). This point is, of course, nothing new because any type of research demands familiarity with the literature. There is little more we can add to this relatively obvious point because it is rare to encounter a topic worthy of inquiry that totally lacks prior relevant data.

Statistical Theory

Statistical theory means the formal mathematics underlying what you are doing. For example, factor analysis can be summarized by the equation $\hat{\mathbf{R}} = \hat{\mathbf{B}}'\hat{\boldsymbol{\Theta}}\hat{\mathbf{B}} + \hat{\mathbf{U}}$. This means that an estimated correlation matrix ($\hat{\mathbf{R}}$) can be represented as the sum of two parts. One is the product of a factor pattern that contains regression coefficients relating factors (latent variables) to observables ($\hat{\mathbf{B}}'$) and the correlations among these latent variables ($\hat{\boldsymbol{\Theta}}$). The second is the uniqueness matrix ($\hat{\mathbf{U}}$) that describes (a) error variance (measurement error) plus (b) specific variance that describes what is idiosyncratic about each variable. The equation has the further implication that you have some familiarity with the rules that govern how matrices are added, subtracted, and so on. You will further have to know that a primary factor pattern in exploratory factor analysis is usually obtained by a procedure known as eigenanalysis and that each column of $\hat{\mathbf{B}}$ is a normalized (unit length) eigenvector multiplied by the square root of its associated eigenvalue.

Over the years, two particular problems of a mathematical nature have been observed. We call the first of these "Charlie the Tuna statistics" in honor of the erstwhile but fictional hero (?) of any number of commercials of yesteryear. Poor Charlie always tried to show that he had good taste, but was continually rejected by the company that canned the tuna because all it wanted was fish that tasted good. Let us ignore the question of why someone's lifelong ambition should be to become someone's salad. Instead, consider the corresponding error made in statistical analysis—trying to show that you have good taste by performing the most sophisticated statistical analysis possible when the goal is to create the good taste of answering a question. Although specific studies will not be cited, consider the growing literature that employs structural equation modeling.

A structural equation model demands that the investigator have a well-developed theory of how variables relate before testing the theory. All too often, a preliminary model is tested and found wanting, leading to a revision which is then accepted. There is absolutely nothing wrong with this if the new model is then tested on a fresh sample to ensure the results are not due to capitalization upon chance. This, however, is the exception rather than the rule.

We call the second problem "Henny-Penny statistics," named in honor of the chicken in the fable that always saw the sky falling. The correlate here is the criticism that a study is invalid because a particular assumption has not been met (e.g., error is not normally distributed). In fact, you cannot possibly do a statistical analysis that fulfills all its mathematical assumptions. What you can do is know enough of the literature (often based on computer simulations) to know which violations are important and which are not. We suggest that a common example of Henny-Penny involves criticisms of those who aggregate items into scales using a Likert or similar format (Likert scales consist of attitudinal items denoting strength of agreement, e.g., using the format "strongly agree," "agree," "neutral," "disagree," and "strongly disagree"). When aggregating, one should ensure that the items used are related to one another as indexed by a statistic such as Cronbach's alpha measure of internal consistency reliability. However, one should not worry about the equality of intervals for the numbers used to denote successive points along the scale. Replacing "1," "2," ... with more "exact" numbers will make no difference to the conclusions you draw about the correlates of the items on a scale of even modest length.

Using Computer Packages

About 50 years ago, many now-common statistical models such as factor analysis were regarded as esoteric because of the enormous amounts of hand calculation they required. Various approximations had to be used, as, by their very nature, the eigenanalyses that were the basis of preferred methods were and are now too laborious to be performed by hand. About 25 years ago, relevant computer programs became available for mainframe computers, which most academic investigators had access to, but the public did not. Today, the vastly easier to use descendants of these programs are widely available on personal computers, and nearly everyone has access to them. Consequently, nearly all graduate students in the behavioral sciences must know at least something about factor analysis.

This availability of computer programs has, however, created a separate need from that involved in knowing the formal mathematics. You must know the syntax of a suitable computer language such as SAS or SPSS and the various options. Specifically, if you want to perform an exploratory maximum likelihood factor analysis on variables labeled X1 to X12, extract two factors, and rotate these to a varimax criterion using SAS, you would have to know how to define a data set, say A, that contains these variables and issue the command `Proc Factor Data=A Method=ML N=2 Rotate=V;VAR X1-X12;`, among perhaps other statements.

Knowing a computer language like SAS, which includes knowing how to interpret its results, is not a simple chore, but do not confuse this with the remaining things you need to know, especially the mathematics. In particular, beware of the person labeled as a "genius" who can implement any analysis in a package like SAS. Be sure that that is not all he or she knows! Likewise, people who work at university computer centers can be a godsend in catching "bugs" in programs, but it is not necessarily their job to know the mathematics and, if they do, know the applicability of the statistical model to your problem at hand.

Information From Other Areas

At the same time, the wider your experience the more likely it is that you have run across some approach that has not been tried in the context you are now working. This is where things can be really innovative— finding that a particular strategy has worked in a situation that appears different but is abstractly similar is where some of the most major contributions come from. In psychology, rather dramatic contributions have been made from a variety of areas in mathematics, physics, physiology, and engineering, such as the applicability of Fourier analysis. At a more mundane level, the more you learn about how to handle databases in general, the easier it is for you to work with a particular set.

Implementing Goals

In turn, implementation involves the following:

Identifying the relevant variables

Identifying the relevant population(s)

Defining the data input (how to get data inside the computer)

Doing a pilot study

Modifying previous efforts based on the results of the pilot study

Conducting the main study

Analyzing the data

Writing the report

1. Identify and define **relevant variables**. The minimal requirement of a theory is that it specifies at least some variables of interest. Your study might also need to include obvious ones or ones of general interest. For example, even though your theory may not deal with sex differences, it is generally a good idea to identify the gender of your subjects. Defining a variable is different from identifying it—definition includes the specific operations and units you will use. For example, a life-span developmental study and a study of infants generally both require that age be a variable. However, in a life-span study, you usually would define age in years, whereas a developmental study will typically require it be defined in months or weeks. Included in this heading is the development of scales or composites of individual items. For example, liberalism/conservatism may be an important theoretical variable. However, it is commonly the case that a single item (e.g., "Rate how liberal you are on a 7-point scale with 1 = most conservative and 7 = most liberal") does not have as desirable properties as the aggregate of a series of items dealing with specific political items. The latter allows psychometric analyses to infer how well the construct is defined.

2. Identify the **relevant population(s)**. Can you do your study with college students, generally, for the most part, the easiest to obtain, or do you require some specific population? If you require a special population, can you obtain the desired number? Equally important, can you get permission from relevant sources? Unfortunately, those controlling some of the most interesting data, such as those showing that a particular therapy is ineffective, forbid publication of often-interesting results for fear of legal repercussions.

3. Define the **data input**. One of the least interesting to discuss but vital steps is getting the data into a form that permits analysis. Try to avoid duplicating clerical efforts, as when protocols are first obtained in a noncomputerized form and then input as a second step. Often, you cannot physically enter data as they are being obtained (e.g., during an interview), but you can obtain most, if not

all, of what you need by using mark-sense forms that can be quickly scanned at a later time. Part of this definitional process is to write a **codebook** that contains the names, possible values, and input location of each variable in the study. For example, one entry might say that the fourth variable in the data set is age, which is coded as the 10th and 11th digits of each record. The purpose of the codebook is both to let someone else go back to your data and reconstruct all analyses and to let you go back after you have forgotten what you did. This forgetting will take place no matter how much time you have spent on the project.

4. Conduct a **pilot study**. Nothing is more essential than a trial run on a few subjects. In addition to the things that you notice yourself from a data analysis, you can often get subjects to tell you other things you have overlooked if you talk to them in an informal rather than a formal research environment, as Krieckhaus and Eriksen (1960) noted many years ago.

5. **Modify** steps 1 to 4 as needed. Typically, you need to go through these steps at least once.

6. Conduct the **real study**.

7. Analyze the **data**. Your analysis should begin with procedures that ensure your data have been properly entered. For example, standard statistical packages such as SAS allow you to get the mean, standard deviation, and, most important, minimum and maximum values on all quantitative variables. This helps identify at least some miscoding. If a variable falls into a small number of categories, such as gender or ethnicity, obtain a frequency distribution. This will allow you to see if there are coding errors or incorrectly defined categories. Specifically, make sure that you do not have observations in an unassigned category and the distribution of scores "makes sense." Next, you perform the planned analyses, but it is crucial that you not stop there. In addition to those analyses that are dictated by the initial theory, any study of merit provides data that suggest new analyses.

Communicating Results

Finally, you must communicate your results. At this point, you should know what is an appropriate level for your audience. You should always think in terms of first writing a preliminary report before writing the final report. Though beyond the scope of this book, writing a coherent report, whether designed for a journal or a corporation work group, is not an easy chore. Obtain feedback as to both substance and style. In this sense, find an "enemy" who will take the time and trouble to identify errors that a "friend" would overlook (Von Békésy, 1960).

◙ STRUCTURE OF THIS BOOK

The focus of this book is much narrower than this general structure. The next chapter, which constitutes Part I of the book along with this chapter, considers basic hardware and software. In particular, software needs fall into the follow two general categories:

1. General programs that need to be in every laboratory, for example, spreadsheets like Excel and general-purpose statistical packages like SAS or SPSS

2. Programs that perform more specialized types of analyses such as EQS or LISREL for structural equation modeling or Multilog for item response analysis.

Part II deals with **creating** data structures and consists of Chapters 3 to 6. In the course of performing various statistical analyses, you will be concerned with such issues as seeing if your data are normally distributed and if a series of measures defines a single factor, among others. To answer these questions, it is useful to explore where these data structures come from. For example, you can make decisions better about whether your data contain a single factor if you have run analyses in which the data truly reflected a single factor or reflected more than one factor. These structures will generally be created using a spreadsheet program (Excel), SAS' programming language, and SPSS' programming language. Note that the data created in Excel can easily be pasted into SPSS and, less easily, into SAS so the Excel programs will generally suffice. Chapter 3 deals with univariate simulations; Chapter 4 deals with basic bivariate simulations; Chapter 5 deals with multivariate simulations; and Chapter 6 deals with simulations of categorical data.

Finally, Part III involves using the simulations to understand common statistical algorithms, particularly their output, and consists of Chapters 7 to 11. Chapter 7 involves basic correlation analysis and exploratory factor analysis. Chapter 8 considers confirmatory factor analysis, clustering, and multidimensional scaling. Chapter 9 involves multiple regression and covariance structures. Chapter 10 considers discriminant analysis, classification analysis, and multivariate analysis of variance. Finally, Chapter 11 considers developments in the analysis of categorical data that are more recent, particularly those involving psychometric theory.

Note on Where to Find the Programs in This Book

Although it is not necessary to enter the SAS and SPSS programs we have provided by hand, you certainly can do so. However, we have made them, and, in the future, other, programs available at ftp.uta.edu/lhb/simulations. Other material will be made available there as well.

▣ PROBLEMS

1.1. Choose a problem area of interest that requires you to establish a database.

1.2. Using a suitable library search engine such as PsychLit (OVID), obtain references on the topic. A formal investigation would require your reading these sources rather than simply going through the abstracts, but use your data to identify relevant variables.

1.3. Choose a relevant recent journal article that incorporated a database of several variables. Read it in detail to see what variables were chosen for analysis. Which variables were omitted that might be incorporated in a replication to improve the study?

1.4. Choose a suitable statistics package such as SAS (your instructor may have made this choice for you). Install it on your own computer now if you plan to do so. In any event, open it up and become familiar with its major options. For example, learn how you can save results to disk.

2

Hardware and Software Needs

CHAPTER OBJECTIVES

This chapter is concerned with the more directly computational aspects of data analysis. In particular, this chapter will explore

1. The goals of statistical analysis in the most general form, specifically, the creation of a database.

2. The computer hardware you will need to effectively analyze a database.

3. The auxiliary equipment you will need to effectively analyze a database.

4. The software you will need. In particular, we will examine the roles played by (a) ASCII (text) editors, (b) word processors, and (c) general-purpose statistical packages such as SAS and SPSS. Selected specialized programs such as those used for structural equation modeling, item response analysis, and numerical estimation will be considered in a later chapter.

▣ GENERAL GOALS OF STATISTICAL ANALYSIS

Concept of a Database

There are many specific goals of a statistical analysis. One person may be interested in determining whether a proposed medication is superior to an existing treatment or not. A second may be interested in seeing whether a series of items can be viewed as comprising a scale. A third may be interested in developing an advertising strategy. What unites them is that all are gathering data to make **decisions**. The data they gather must be structured in some way so that they can be read for analysis.

A database is a specific collection of information with a two-dimensional or **matrix** structure. Each row typically represents a **record** (from the standpoint of a computer scientist) or **case** (from the standpoint of an empirically oriented investigator). In some situations, as when the data are in a text file, one or more columns (successive characters) are grouped into a **field** or **variable** (again from the respective standpoints of a computer scientist and investigator). How this grouping is accomplished reflects the **formatting** used by the programmer in defining the data. Thus, if the numbers 1234 are in columns 1 to 4 of a text file, they may represent the single number 1234, the pair of numbers 12 and 34, or various other arrangements, depending on the formatting (formatting applies to both data that are read in and data that are printed out). Conversely, in other situations, as when the data are in a spreadsheet, the concept of field and column are synonymous as varying numbers of characters may be placed in a given cell.

Depending on the application, the two-dimensional structure may be viewed as part of a more complex structure. For example, suppose the database consists of a survey taken before and after some experimental manipulation. A variable indicating whether the result was the first or second test might define one field and the individual survey items additional fields. Respondents would be the records. However, it is also useful to think of these more complex data as having a three-dimensional structure: fields (items), records (respondents), and occasions (pre- or posttest). This idea can be expanded to more complex designs having additional numbers of dimensions.

In principle, one could think of performing the calculations manually or perhaps with the benefit of a hand calculator, and that is the way things were done in the past when calculators required a desktop rather than one's hand. However, it is rare that an analysis of any complexity would now be done this way. The computer, more specifically, the personal computer (PC), is clearly the standard for any meaningful form of data analysis, so consideration will now be given to this basic tool.

Generating a Database

Computers excel at creating databases. That is, suitable programs readily break down (**parse**) input into records (cases) and fields (variables). One basic distinction is between a numeric and an alphanumeric or text field. As the name indicates, a numeric field must only contain one or more numbers; entering 1C2 in a numeric field will produce an error. In contrast, letters, numbers, or any other symbols, such as punctuation marks, may be entered into an alphanumeric field.

Freefield (Delimited) Versus Fixed-Width Input

Computer packages create databases in different ways, but there are only two basic ways to write the program defining the data. These strategies depend on whether the file to be read and used to create the database is freefield (delimited) or **fixed width**.

The fields (or variables) read in from a freefield (delimited) file are determined by their position relative to other fields in the file rather than by their exact column location. Each field is separated from the fields on either side of it by a designated symbol (or symbols)—the delimiter(s). Data values occupying the same field in different records can take up a different number of spaces and different column positions in the file. Any symbol can be used as a delimiter, but the most common are spaces, commas, and tabs. Two records in a comma-delimited freefield file might appear as follows:

```
6,Hispanic,female, 123,567,96
46,white,male, 133,456,27
```

Note that the value of the first variable occupies column 1 in the first record, but occupies columns 1 and 2 in the second record, `Hispanic` is eight characters long, but `white` is only five, and so on. A statement such as `INPUT AGE RACE $ SEX $ X Y Z;` would read these data in

SAS. From the first record this would result in AGE=6 RACE=Hispanic SEX=female X=123 Y=567 Z=96. The data from the second record would be assigned similarly. (The $ in the input syntax indicates that the field [or variable] being defined is alphanumeric rather than numeric.) A symbol other than the one delimiting the data fields may be used to denote the end of a record. For example, the Enter key generates symbols called "line feed" and "carriage return" that many programs use to separate records. This also causes the monitor to go to a new line in a word processor. Some programs (e.g., SAS) continue looking on successive records until they find each value they were told to look for, and this can involve different numbers of lines for different records (it is a common source of error when a value has been left off a line). Conversely, SAS and other packages may be programmed to look for a fixed number of fields per record and report an error if they fail to find these fields. The user can usually define the character (delimiter) to be used to separate fields, but SAS, like other packages, uses a space as the default delimiter.

A fixed-width file has no delimiters between fields. Each field must take up exactly the same number of spaces (columns) in every record, and it must occupy exactly the same column positions in each record. For example, the preceding two records would take the form

```
 6 hispanic female 12356796
46 white    male   13345627
```

The first field could consist of columns 1 and 2, the second field could consist of columns 3 to 10, and the third field could consist of columns 11 to 16. Blank spaces must appear in the columns of any field if the actual data value does not complete it (notice the three spaces after white in the second line of data). If the fields are to be the same in this fixed-width example as in the preceding freefield example, the fourth field fills columns 17 to 19, the fifth 20 to 22, and the sixth 23 and 24. However, from looking at the two records, unlike in the freefield example, you cannot tell what the fields are that occupy columns 17 to 24. The programmer must determine the lengths and positions of the fields by using **formats**. For example, the SAS statement INPUT AGE 1-2 RACE $ 3-10 SEX $ 11-16 X 17-19 Y 20-22 Z 23-24; would parallel the previous freefield example. In particular, each record would be parsed (broken down by the program) to define variable X using columns 17 to 19, Y using columns 20 to 22, and Z using columns 23 and 24. However, the SAS statement INPUT AGE 1-2 RACE $ 3-10 SEX $

11-16 X 17-18 Y 22-24; would define X as 12 and Y as 796 in the first record and X as 13 and Y as 627 in the second, skipping the contents in columns 19 to 21. Thus, changing formats causes different data to be read in, unlike the freefield example, even though the individual data values in the two examples are identical.

Formatting also allows such things as an implicit decimal point to be employed. For example, INPUT @17 X 3.2 +1 Y 4.0; reads data starting at column 17, putting $X = 1.23$, skipping a space, and defining Y as 6796.

Both of the preceding strategies assume that the underlying file to be read by a statistical software package is a flat ASCII text file. In prior times, data were almost always read initially into an ASCII file and programmers had to write what were often complicated format statements. Currently, however, initial data entry often employs a spreadsheet like Microsoft Excel or a database like dBase. This allows the database to be defined as the data are entered. Current statistical packages like SAS and, especially, SPSS allow spreadsheet and database data to be brought in (imported) directly. Also, SAS and SPSS allow one to enter data into the program in spreadsheet fashion. This is often more convenient and accurate than using a format statement. Still another option is to cut and paste data and/or program statements directly into the program window, which is useful with small amounts of information.

▣ COMPUTER HARDWARE

Basic Equipment

To begin, it will be assumed that analysis will be performed on a laptop or desktop computer. The examples given will be Windows-based rather than Macintosh-based, but there is not much difference should you prefer to work on a Macintosh. Following usual practice, the term "personal computer" will denote the former and "microcomputer" will denote either one. Both are to be contrasted with mainframe computers serving multiple users. The primary reason you will use a mainframe computer (if, in fact, you ever will) is to access data from a large database, as when you access networked sites on the World Wide Web. The data are then transferred to your personal computer for analysis. This downloading may or may not involve a network. For example, you may simply log on to your computer account from your home or office to a university mainframe. If you do need to work with another computer, networked or not, you will need to learn the appropriate transfer protocols. If the computers are networked, data transfer can be as simple as pointing and clicking on files in a Web browser such as Internet Explorer or Netscape or they can use highly idiosyncratic and arcane interfaces. If you have never worked with networking transfer protocols such as FTP and Telnet, it would be useful for you to do so because the principles are quite generally useful, but this book will not consider the many forms of data transfer between computers.

Mainframes were once necessary because they housed the computer packages that operated on the data. Unless you work with massive databases, there is less reason to use a mainframe to access such programs because companies such as SAS and SPSS have put major development efforts into their microcomputer software for several years now. However, there are charges to rent or purchase these programs, so an occasional user, especially in a university setting, may find it more cost efficient to use the mainframe program. If you do work with extremely large databases, you may be able to profit from what are known as **client–server protocols** that allow things to be done in a more efficient way than overwhelming a PC with massive data transfers.

We will be discussing microcomputers with occasional reference to mainframes without making reference to desktop workstations such as those manufactured by Spark or Sun. The latter were preferred for demanding analyses just a few years ago, but they are currently not that much faster than stock microcomputers sold for much less. Moreover, workstations are more limited as to available software.

It is difficult to think of a situation in which having more in the way of computer resources, specifically processor speed and capacity, is a disadvantage. However, the "law of diminishing return" is very applicable. Virtually any computer you purchase, including those in the sub-$1000 range, will perform nearly all of the behavioral science applications even rather advanced users will engage in. Generally, what taxes a computer most is not the need for analyses of the kind this book is concerned with but graphics applications as in video games and computer-assisted design (CAD) for architecture.

It is specifically assumed that you have access to the following relatively standard hardware items:

1. A computer running Windows or equivalent Macintosh. The most recent versions of these operating systems are not necessary though they generally offer conveniences not found in earlier versions. Linux has been gaining ground as an alternative operating system to Windows, and many, but not all, major programs are now or will be available soon for this platform.

2. A processor sufficient to run whatever version of software you plan to use.

3. A modem for logging onto a network or mainframe. If you are fortunate enough to have a direct connection at work or at a university, your use of a modem will diminish. However, you may still find it useful to connect from a home to an office computer.

4. A compact disc read-only memory (CD-ROM) drive. Nearly all programs are available for installation on CD-ROM, an act of unmitigated mercy for anyone who has ever had to insert dozens of floppy disks (and had a defect on the last one ruin the process).

5. A mass storage device, such as a tape drive, recordable or rewritable CD-ROM, LS-120 drive, Zip drive, Jaz drive, or, in the near future, a rewritable digital video disc (DVD). Although hard disk storage is now so cheap that much archival data can be kept on one without compromising your needs to store other data, inadvertent data loss is much less likely on these other devices. You should also use them to make periodic backups of your system. There are also archival locations on the Internet such as idrive (www.idrive.com) and ifloppy (www.ifloppy.net). Universities and businesses usually have archiving facilities for accounts on their own mainframes. However, it is usually a good idea to make your archives on your own as you may need to go back to a data set to perform additional analyses long after the original project was thought completed.

Typically, the newer the package, the more computing "horsepower" is needed. Conversely, some older DOS-based programs such as Thissen's (1988) Multilog run quite well on older machines. Specifically, you may find some DOS programs will run best on computers that do not even have Windows installed.

Bubble Scanners

Perhaps the single most useful item of auxiliary equipment is access to what is commonly called a "bubble" or "mark-sense" scanner. Basically, respondents answer questions by blackening in spaces on special answer sheets. The sheets are then read into the scanner and to an ASCII file. These are widely used by colleges for such administrative purposes as registration and grading. Although faculty members are generally quite used to them for scoring tests, mark-sense forms are not used nearly often enough to capture research data. Bubble scanners are quite different from the more widely available graphical scanners that are marketed for use with microcomputers although some bubble scanners are available for microcomputer use. Scanner output is typically downloaded from a mainframe, but you may be able to make arrangements with the site doing the scanning to e-mail the data or place it in an FTP site for you. Two of the larger companies that manufacture bubble scanners and their associated Web sites are National Computer Systems (NCS, www.ncs.com) and Scantron (www.scantron.com). In addition to marketing scanners, both sell a wide variety of tools, including software for analyzing surveys.

Bubble scanners can be programmed to accept a wide range of forms, including those with survey questions already written on them. However, the most common strategy, especially for students, is to use a standard answer sheet such as the NCS Trans-optic® 4521 ("blue") form, which dates back to 1977. This form provides the following items:

1. Name or other alphanumeric data in 20 fields

2. Sex, alphanumerically coded M or F in one field

3. Education level, numerically coded 0–16 in two fields

4. Birth month, numerically coded 1–12 in two fields

5. Birth day, coded 0–3 and 0–9 in two fields

6. Birth year, coded 0–99 in two fields

7. Special codes, 15 fields each coded 0–9

8. 200 items each coded from 1 to 5

Figure 2.1 shows that portion of the sheet containing the identifying data.

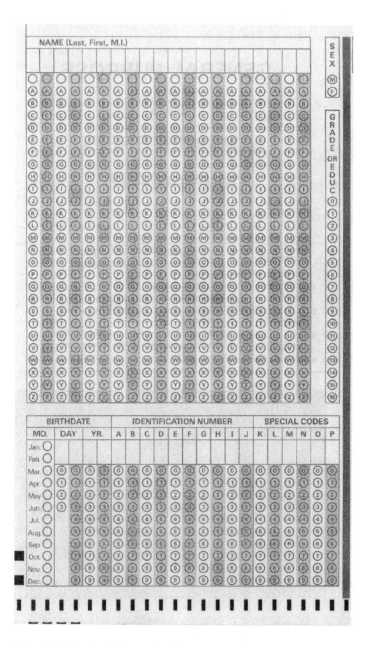

Figure 2.1. Identification information from a general-purpose NCS 4521 answer sheet.

A second common NCS form is the 6703 ("green"), which performs the same functions except that it encodes 120 items on a scale of 1 to 10.

If you have a large amount of data to gather, consider using one of these standard forms. Even a relatively old model scanner can quickly encode your data. There is no other item that you are not likely to own personally that is of more value to statistical analysis than a bubble scanner.

▣ GENERAL-PURPOJE JOFTWARE

ASCII (Text) Editors

A **text** editor creates files as a sequence of characters drawn from a set of 127 with a set of extended characters increasing the size of the character set to 256. Each character corresponds to a byte on a computer. There are many different character sets, but the most important of these is the American Standard Code for Information Interchange (ASCII), so we will follow common practice and use "text" editor and "ASCII" editor interchangably. These characters include numbers, letters, other symbols such as punctuation marks, and still other nonprinting characters that perform such actions as generating a line feed. The individual characters comprising an ASCII file do not have a format in the sense that you cannot italicize some characters and leave others in a Roman face. However, your text editor may let you print the file in a font of your choice, such as a bold. Moreover, you can open the file in a word processor, format it as you choose, and resave it in word processor (non-ASCII) format. The distinction between an ASCII editor and file and a word processor and its files is important. Programs and their associated data ordinarily must be in ASCII format. One important exception is when the data are saved in a database or spreadsheet format, which many statistical packages can read.

Windows comes with a text editor called Notepad. It is limited in various ways, most specifically as to the file size it can work with. This is normally not a problem with programs, as they are usually relatively short, but it can very definitely be a problem if one needs to edit a large ASCII database. For this and related reasons, a variety of other more sophisticated ASCII editors have been made available. Wordpad, also a word processor that first came with Windows 95, is more powerful. Norton

Desktop for Windows, which was a very popular shell prior to Windows 95, had an excellent ASCII editor. Various others are available as shareware or, in some cases such as NoteTab, freeware.

An ASCII (text) editor is perhaps also the most common and simplest way to create programs and small databases. Data in other formats are easily converted to ASCII, either freefield or formatted for input to programs. However, ASCII may be awkward when each record has many fields or is freefield because it may be difficult to locate what you are seeking to edit and to detect errors.

Word Processors

Word processors like Microsoft Word can be used to create ASCII files. They generally have even more extensive editing capabilities than the typical ASCII editor. However, be sure to save any programs or databases you are working on in ASCII (text) format. If you do not, they will be saved in the word processor's own format, which statistical programs typically cannot read. However, all is not lost if you make this common error because you can reopen the file and resave it as text. If you can develop the habit of saving material in the proper format, there is little reason you need to work with an ASCII editor.

Spreadsheets

Spreadsheet programs include Excel, Lotus 1-2-3, and Quattro-Pro. Their distinguishing characteristic is that you can enter information in cells of a two-dimensional matrix commonly called a flat file and perform computations on these cells to obtain needed results. For example, you can put the numbers 3, 5, and 10 in the first three rows of the first column, identified as A1, A2, and A3 in Excel's notation, which we will follow. You can then enter `=average(A1:A3)` in the cell immediately below (cell A4), which will return the value 6, and this value can be formatted as 6, 6., 6.0, and so on. Cells can also contain text, and the sheet can also contain graphics or sound files. When used as a database, a spreadsheet codes fields of data such as names in its columns and records, as of individuals, in its rows. Bernstein and Havig (1998) describe the general application of spreadsheet programs to statistical analysis. A relational

HINTS ON USING AN ASCII EDITOR
OR WORD PROCESSOR

▶ Word wrap automatically causes text to start on a new line when a certain cursor position is reached. You typically want this feature disabled when you are writing programs or creating a database. ASCII editors let you do this directly. If your word processor does not, change to a very small font size and large degree of magnification—the font size will not become part of the file when it is saved in ASCII. In some rare cases, you may also have to press Enter after each line to separate the lines.

▶ Always use a fixed-width (nonproportional) font such as Courier, Courier New, Prestige, or Monospace when looking at a database or program so that you can see if fields are properly aligned.

▶ If you are using a word processor, remember to save data and programs as ASCII rather than formatted text in the "Save as Type" box of the File Save menu. Failure to do so does not permanently lose what you have created though, as you can reopen it and then resave it.

▶ Be especially careful so as to avoid mistyping easily confused symbols such as "1" and "l." Some fonts make this task easier than others.

▶ A statistical program may have a default maximum length limit, typically 132 characters (bytes), which is a carryover from the days they ran on mainframe computers and output to line printers, which had this length as a maximum. This limit can usually be overridden with an appropriate option such as **LRECL=XXX**, where **XXX** is the desired length, which works in SAS. Many have forgotten this limit and been puzzled as to why their program did not read data correctly, including these authors.

database such as Microsoft Access allows flat files to be linked to one another. Bernstein and Havig have discussed the use of relational databases. Relational databases are extremely important to people who have to manage large amounts of data in order to obtain information about individual cases, but they are much less useful to data analysts who are more concerned with aggregating data as current statistical programs operate on flat files. Output from relational databases can be exported to spreadsheets. Database-generating programs, such as dBase and Fox-Pro, are also similar to spreadsheets. Their primary value is in allowing support personnel to create a database rather than in analyzing one. Spreadsheet programs can easily read data from database-generating programs.

There are several ways a database can be created and used in a spreadsheet. One advantage of using a recent version of Excel is that it offers error detection at data entry. For example, the user can designate a column (field) as numeric, in which case attempts to enter alphanumeric characters would be flagged. Limits can be placed on the range of acceptable values (e.g., only accept the numbers "1" and "2"). Earlier versions than this allow you to **filter** on a column. The filter identifies the values found in that column, which helps to identify incorrect values. These restrictions are known as **validation rules**. Error detection has been available in such programs as dBase and Access for some time. The fact that it is now available in spreadsheets means that we can concentrate on such programs to the exclusion of database creation and relational database programs in this book.

CREATING AN EXCEL FILTER

▬ 1. Select the column(s) on which you wish to generate a filter.

▬ 2. Choose **Filter** from the Data menu.

▬ 3. Because this is the simplest case, choose the Autofilter option. More advanced situations provide additional options such as limiting the range of the filtering.

▬ 4. A marker will then appear on the designated column(s). Click on it to see the various values present.

The minimum and maximum value functions are additional ways to locate out-of-range values.

Although it has its limitations, a spreadsheet is the device, par excellence, to examine a database for outliers and to perform simple analyses (which may be all a project needs). There are periodic discussions of the limits of the computational adequacy of spreadsheets in such sources as sci.stat.consult and sci.stat.edu. Spreadsheet programs are not the best way to perform complex analyses such as factor analyses even though they can be made to do so. It is far simpler to use a general-purpose statistical program such as SAS or SPSS to this end. However, it is quite instructive to perform simpler analyses in both the spreadsheet and the statistical program, whenever possible, to help get a better feel for both. In particular, the first author finds it far simpler to perform simple statistics on classroom records (test averages and the like) in Excel rather than in SAS or SPSS.

As a very simple illustration, assume that four people have answered three items and been assigned scores ranging from 0 to 9 (in entering the data, you should look for coding errors by testing for values that are negative or greater than 9). You probably will desire the following statistics (which might be provided as standard output from the software used by a bubble scanner or the like):

1. The mean for each item

2. The standard deviation for each item

3. Each person's total score (sum across items)

4. Each person's rank

5. The mean across people

6. The standard deviation across people

7. The correlation between each item and the total score

8. The estimate of the test's internal consistency as Cronbach's coefficient alpha, which can be expressed as

$$\alpha = \frac{k}{k-1} \cdot \left(1 - \frac{\sum s_i^2}{s_t^2}\right),$$

where k is the number of items, $\sum s_i^2$ is the sum of the variances of the individual items, and s_t^2 is the variance in the total scores (sum of individual item scores)

Assume that the data are as follows:

ID	Item 1	Item 2	Item 3
1	2	4	2
2	6	5	5
3	5	6	3
4	2	9	4

USING A SPREADSHEET

1. Enter the equation =SUM(B2:D2) in cell E2 to compute the total score for respondent 1. As its name indicates, this function provides the sum across the array named as its argument (cells B2, C2, and D2 in the present case).

2. Drag this equation down through the remaining respondents (cells E3–E5) to get their total scores. Dragging copies the equation into these three cells.

3. Enter the equation =RANK(E2,E$2:E$5) in cell F2 to determine the rank of this person relative to the remaining three. RANK is Excel's rank order function. It has two required arguments. The first identifies the individual cell containing the score to be ranked. The second identifies the array of cells against which the ranking is to be made. In most cases, the individual cell is addressed relatively (without the "$") so that it takes on successive values (E3, E4, and E5) as it is copied. However, the array is addressed absolutely because it remains the same (E2–E5) across all of the cells. Note that the default ranking procedure in Excel is different from that used in SAS and SPSS. For example, Excel assigns the ranks 4, 2, 2, and 1 to the numbers 10, 12, 12, and 13. Besides ranking in ascending order (which can be changed by adding a comma and any nonzero value before the closing right parenthesis), it does not follow the convention of averaging tied ranks. Both SAS and SPSS would assign ranks of 2.5 to the second and third values in the series.

4. Drag the equation through the remaining respondents' cells (F3–F5) to rank the remaining individuals.

▬ 5. Enter =AVERAGE(B2:B5) in cell B7 to compute the average for item 1. Note that a blank row has been left between the last row of data and the summary data. This is a good habit to get into. You will frequently find it necessary to sort the raw data. Inserting a blank line avoids having to eliminate the summary data from the sort manually. However, some situations, including those illustrated in the next chapter, would be complicated.

▬ 6. Drag this equation across columns C to E to get means for the remaining items and the overall mean.

▬ 7. In a like manner, enter =STDEV(B2:B5) in cell B8, =MIN(B2:B5) in cell B9, =MAX(B2:B5) in cell B10, and =CORREL(B$2:B$5, $E$2:$E$5) in cell B11. These generate the standard deviation, minimum value, maximum value for item 1, and the correlation between item 1 and the total score. Note the way in which the item 1 array is addressed relatively and the total score array is addressed absolutely in anticipation of copying this formula to the remaining items and the total score.

▬ 8. Copy the equations in columns B8 to B11 through columns C to E (note that the correlation between the total score and itself is 1.0, by definition).

▬ 9. Copy =3/2*(1 - SUMSQ(B8:D8)/E8^2) in cell B12. This is one way to implement the formula for the coefficient alpha, which is defined as

$$\alpha = \frac{k}{k-1} \cdot \left(1 - \frac{\sum s_i^2}{s_t^2}\right)$$

in summation notation, where k is the number of items, $\sum s_i^2$ is the sum of the variances of individual items, and s_t^2 is the variance in total scores (sums of item responses).

▬ 10. Enter appropriate descriptive labels and format the cells.

The completed spreadsheet appears in Figure 2.2.

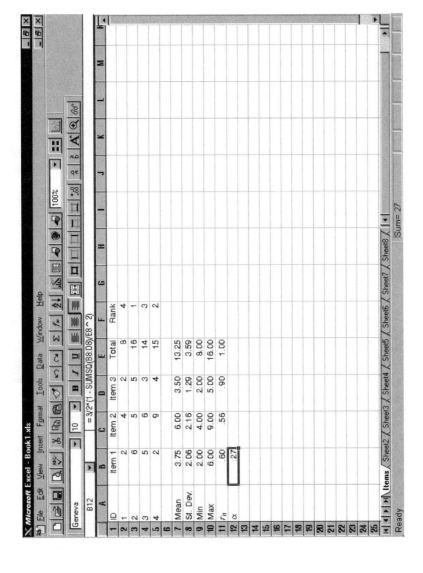

Figure 2.2. Excel spreadsheet containing data and equations for a simple test analysis.

Besides performing these basic calculations, spreadsheets can write and format reports and, what is more important in this book, create graphs. In addition, online database connectivity (ODBC) can be used to exchange data with mainframes.

Data can be input from spreadsheets to statistical programs. SPSS does so directly. You can import a spreadsheet file from Excel or similar program directly or you can copy the data to the clipboard and paste it on to SPSS' own spreadsheet for analysis. Life is a bit more complicated with SAS. SAS has an Import command that facilitates spreadsheet input, which will be discussed later in this chapter. However, the module that is needed may not be installed or may be unnecessary as there are other ways to input and output to spreadsheet programs from SAS, which will also be described later in this chapter.

Relational Databases

Spreadsheets such as Excel create two-dimensional **flat files**. A relational database such as Microsoft Access can interrelate several such flat files. For example, one file may contain student ID numbers plus other relatively permanent information about them such as their address. A second file may contain ID numbers plus the numbers of the courses they attempted, and a third file may contain course numbers and their description. Consequently, student ID can link the first two files, and course number can link the last two. The individual files would thus not be analyzed but the composite would. Though such a scenario is possible, it does not occur very often so the relational capability of these programs does not play a great practical role in statistical analysis, although such programs have important querying functions in other contexts to locate cases meeting certain criteria or tabulating their number.

Relational database programs such as dBase, FoxPro, Access, or Paradox, and, in their latest versions, spreadsheets such as Excel offer much more control. At the same time, you probably do not need to purchase a separate database program if you already have an office suite as the suite may contain one. Suppose you are using an older version of Excel and also have Access. Access can create a flat file to be exported into Excel, using its validation rules feature, which was only later incorporated into Excel.

▣ JTATIJTICAL PROGRAMJ

General Concepts

Assuming you have such standard software as a word processor and a communications package, there are two major classes of statistical software that you will need, a spreadsheet and a general-purpose statistical package. Several such programs are on the market. The two most popular are SAS and SPSS, to which major discussion will be limited. Other important packages include Statistica, Minitab, NCSS, Systat (though still an independent product, it is now a part of SPSS, Inc.), Maple, MATLAB, XLISPSTAT, and SPlus. The latter is becoming especially popular because of its incorporation of procedures for robust statistics that are useful with noisy data.

Circumstances may also dictate special-purpose software such as LIS-REL or EQS, which is used for structural equation modeling. Structural equation modeling allows specification and evaluation of models in which observed variables are described in terms of latent variables. There are a large number of structural equation models. The two most basic forms they take are **measurement** models for confirmatory factor analysis and **structural** models, a form of regression (many other models are combinations of these two). These will be considered separately in later chapters after the data they analyze are considered in the next three chapters. For now, it is noted that the interfaces for these specialized programs are unique. They were originally programs written with relatively little graphic interface. This is not necessarily undesirable from the perspective of formal modelers who tend to think in highly algebraic terms. However, as the base of users has expanded, several developers, most notably those of EQS (whose development parallels the BMDP package, both being products of UCLA), have taken a more graphic approach to broaden their base of users. At the same time, `Proc Calis`, because it is part of the widely used SAS package, needs to be given considerable attention and greater use. This procedure radically expands the range of SAS applications—it allows the large number of researchers with access to SAS the opportunity to perform latent variable modeling. Unfortunately, in our opinion, SAS has not documented this valuable product very well in the past, in that there were very few simple, straightforward examples. The online documentation that accompanies Version 8.X is an improvement.

At a bare minimum, a general-purpose statistical program creates, formats, and analyzes a database, performing the following items:

1. **Missing value handling**. You can state that certain values (e.g., "."
for numeric variables in SAS) mean that data are absent; for example, the respondent to a survey refused to answer or the question was not applicable. A missing value is different from the numerical value 0—the latter response being a quite legitimate answer to such questions as "How many children do you have?" The program also should give you options about handling missing values. For example, suppose you are correlating variables *A*, *B*, and *C*, and *A* is a missing value in a given case. You should be able to choose between deletion of the remaining two values for that case (**listwise** deletion) and allowing these values to contribute to the correlation between *B* and *C* (**pairwise** deletion).

2. **Formatting**. Formatting includes, but is not limited to, identifying the way data are to appear on input and output, for example, saying that variable *X* is to be read in or printed out with a length of four characters, one of which is a decimal point. In addition, it includes the descriptive names given to variables, so that the variable identified symbolically as Q17 can appear as "Most Recent Diagnosis" in the printout. Similarly, it also includes the descriptive names given to values that variables take, so that a response coded as "1" in the raw data would appear as "Major Depression" in the printout.

3. **File updating**. This involves such things as sorting records, creating new variables from old (e.g., `TOTAL = SUM(of X1-X10)` computes a variable called TOTAL as the sum of 10 items in SAS [similar expressions are used in other packages]), deleting unneeded variables, and merging data sets either by adding new cases or by adding new variables to old cases.

4. **Creating of permanent data sets**. By far, the bulk of time in statistical analysis is spent in organizing the database—creating the variables, ensuring they are read correctly from the data, defining missing values, detecting outliers, formatting, and so on. This typically produces a rather bulky file. In contrast, a given analysis may only require a few lines of code or equivalent menu selection. By saving the file as a permanent data set, one might simply conduct an analysis by entering code as simple as `Proc Factor DATA=SASUSER.ALPHA;`, which performs a basic factor analysis

on a previously created data set named ALPHA. Typically, though, the actual command would probably require some additional parameters such as those limiting which variables are to be included and determining which optional results should be output.

5. **Connecting analyses to one another.** You can compute a correlation matrix in one step and then factor that matrix in a second step instead of having to recompute the correlations as part of the factoring computations.

6. **Descriptive statistics.** These include means, standard deviations, and so forth on individual variables.

7. **Bivariate statistics.** These include obtaining intercorrelations among variables.

8. **Multivariate analyses for continuous variables.** These include exploratory factor analysis, regression analysis (including the analysis of variance), and discriminant analysis.

9. **Multivariate analyses for discrete variables.** These include categorical modeling, logistic regression, and log-linear modeling.

Some other desirable features include the following:

1. **Graphing.** This saves you the step of graphing the results in a program such as Excel or PowerPoint.

2. **Matrix language.** All statistical analyses can be expressed as matrix equations, though sometimes this expression is cumbersome. SAS offers its Interactive Matrix Language (`Proc IML`) as an add-on, which allows you to define your own analyses in a modified matrix notation.

3. **Other latent variable analyses.** For example, although there are separate programs that perform structural equation modeling, such as LISREL, AMOS, and EQS, it is desirable to be able to stay within a given package so you do not have to learn a new set of conventions. As noted previously, `Proc Calis` is a standard part of SAS, which performs a variety of structural equation problems, although its documentation leaves much to be desired. Of course, what is a novel analysis at one point in time later becomes a standard part of the package—all major packages now perform categorical modeling although this was not the case 15 or so years ago.

▣ BAJIC APPEARANCE OF JAJ AND JPJJ

All statistical programs typically input data in one of four ways:

1. **In line**, meaning the data are submitted along with the program statements

2. From a **separate** ASCII file

3. Via a **spreadsheet** created in Excel or equivalent program

4. Using the SAS and SPSS spreadsheet capabilities

Instructions are then given to process the data by creating a temporary or permanent file and outputting the result. The two programs of major interest to this book, SAS and SPSS, do this by different means.

Basic Elements of SAS

The most current versions of SAS are Versions 6.12 and 8.1, but 8.2 is schedule for release. The central features of the two versions, especially the printout and code (programming statements), are highly similar; the main difference is in the interface. As with any attempt to make an interface more "user-friendly," people who were familiar with the "old way of doing things" may have to make some adjustment. Moreover, some of these operations may be performed more slowly because of the greater overhead they require. Figure 2.3 contains the basic SAS windows for Version 6.12, and Figure 2.4 contains the basic SAS windows for Version 8.X. These windows may be configured differently at the user's discretion. SAS 6.12 jobs are run by entering code in a format similar to the old PL-1 language using the Program Editor window. Version 8.X has an Editor window (also called the Enhanced Editor) that formats this code to make error detection easier but also provides a traditional Program Editor window to maintain compatibility. Executed commands appear in a Log window and numeric results, if any, appear in an Output window. Version 8.X also has an Explorer window that allows the user to see the contents of the SAS environment, a Results window that identifies the output from each procedure to facilitate access, and tabs across the bottom so the user can jump to any desired window. Both versions have other

(text continues on page 39)

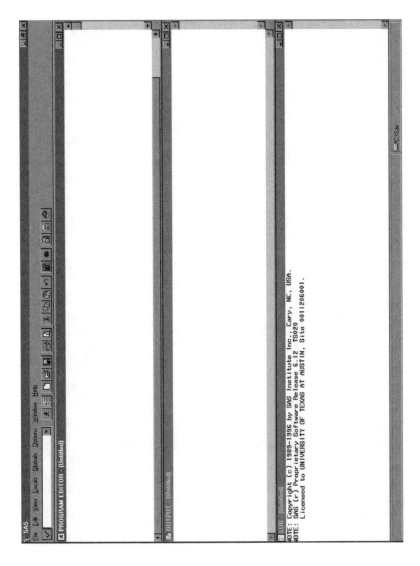

Figure 2.3. Basic SAS 6.12 for windows configuration.

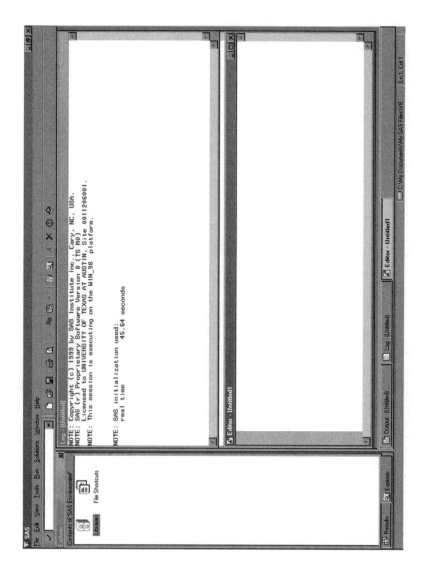

Figure 2.4. Basic SAS 8.X for windows configuration.

windows of lesser interest. Data can be printed directly, but we will provide an example in the next chapter illustrating how results may also be exported to a spreadsheet. Both versions of SAS contain a separate viewer program. This is not necessary if one simply wishes to look at input and output files, which are in plain ASCII, as any ASCII editor or word processor can read them. However, the SAS viewer also allows users to look at SAS data sets, which are not in an ASCII format.

The SAS 8.X menu options for the Editor window, when it is active, are as follows (these are also generally accessible by right-clicking within the work area). Other windows have similar, but distinct, options; for example, you cannot edit the Log or Output windows.

1. **File**. The File menu provides the typical options to open, save, preview, format, and print files found in other programs. A potentially important feature is the ability to import and export SAS data sets. However, the ability to actually use this feature depends on what other SAS options were installed. Both importing in this manner and alternative methods will be discussed later.

2. **Edit**. The Edit menu is also very similar to other applications in providing a way to select, search, replace, cut, copy, paste, and spell-check material. There are also some system options such as the number command, which automatically generates line numbers on programs.

3. **View**. The View menu provides paths to the various windows.

4. **Tools**. The Tools menu provides access to database queries and to a variety of SAS editors (table, graphics, report, image, and text). For example, the table option (Alt -T , T) opens a spreadsheet-like table for direct data entry. Once created, the data set can be saved for later access.

5. **Run**. The Run menu is one way to submit statements for execution. This involves statements in the two editor windows and clipboard. As in virtually all versions of SAS, previously submitted statements can be recalled. One can also interact with other computers using commands found here if the proper SAS modules were installed.

6. **Solutions**. The Solutions menu offers a set of programs from drop-down windows. However, these are not ones of present interest (e.g., factor analysis). Moreover, the options may take a long time to load.

7. **Windows**. The Windows menu performs such standard operations as allowing the user to tile or cascade the open windows and allows the user to change windows (e.g., to go from the Program Editor to the Log when both are displayed as full screens).

8. **Help**. The Help menu performs the standard help function. A particularly useful feature is the Keys option, which allows the user to program the function keys. First, choose this option to look at the default key assignment. If you would like to be able to clear the Output and Log windows and tile the three windows before submitting a job, enter `Output;Clear;Log;Clear;Pgm;Zoom Off;Tile;Submit;` as a key definition. Be sure to save the result.

Many of the most important SAS functions can also be handled by entering the commands as part of a program rather than by using the menus. For example, the Options command overrides system options. For example, `Options nocenter;` left-justifies output instead of having it centered.

Following is an example of SAS program created with in-line data. The `Cards` or `Datalines` statement tells SAS that what follows are data rather than program statements. Old-timers like us tend to use `Cards` in our programs, a format we will follow in this book; you may substitute `Datalines` if you wish. The term is a carryover from mainframe days when data were usually input on punched cards.

The source file is as follows, submitting the data in line:

```
Data A;
Input Id Item1 Item2 Item3;
Total = Sum(of Item1-Item3);
Cards;
1 2 4 2
2 6 5 5
3 5 6 3
4 2 9 4
;
```

```
Proc Corr Data=A;Var Item1--Total;
Proc Corr Data=A Alpha;Var Item1-Item3;
Proc Rank Data=A Out=Ranking descending;Var Total;ranks rankt;
Proc Print Data=Ranking;
Run;
```

The **Data** line identifies the succeeding steps as creating a data set in contrast to **Proc**, which identifies a procedure. This is perhaps the most basic distinction in SAS and, implicitly, in other statistical packages. This line also assigns the name "A" to the data set so that it can be distinguished from other data sets that might be created in a session. Note that a semicolon separates (delimits) statements from one another.

The **Input** line identifies the names of the variables that are being read in. Entry is freefield in the sense outlined previously—a space separates values of the four variables: ID, Item1, Item2, and Item3. These last three variables could have been equivalently denoted as Item1–Item3. In some applications, the data need be formatted so that the user would have to instruct SAS that the first variable occupies the first three columns, for example. The third line computes a new variable, Total, as the sum of Item1 to Item3, and the fourth line tells SAS that data are to follow.

SAS recognizes the line that begins with **Proc** as indicating a request for a procedure (a line beginning with **Data** would analogously signal a new data step). The first **Proc Corr** simply computes the correlations among the variables named in the **Var** part that follows or all numeric variables, by default. The **Data=** part identifies which of several possible data sets are to be analyzed. It is unnecessary here because there is only one data set, but it is good programming practice to make this explicit nonetheless. **Proc Corr** also generates univariate statistics (the mean, standard deviation, sum, minimum, and maximum values) as well as correlations and significance levels of the correlations. Had only univariate statistics been needed, **Proc Means** and **Proc Univariate** generate a variety of such results.

The second **Proc Corr** uses an option to create the coefficient alpha. It will actually generate two such values. The "raw" alpha is based on using observed response data, and the "standardized" alpha is based on converting these data to z scores. The recomputed coefficients alpha based on ignoring each of the variables in turn are also output.

Proc Rank rank-orders data and places the result in the file specified by the Out= part (Ranking). The observation with the highest value is given a rank of 1 ("descending" option). Var again states what variables are to be included in the analysis, in this case only Total, and the final portion of the line names the variable containing the rank as RANKT. Finally, Run; is a required command needed at the end of every set of program statements.

The resulting SAS log is as follows (blank lines have been deleted to conserve space):

```
65 Data A;
166 Input ID Item1 Item2 Item3;
167 Total = Sum(of Item1-Item3);
168 Cards;
NOTE: The data set WORK.A has 4 observations and 5 variables.
NOTE: The DATA statement used 0.81 seconds.
173 Proc Corr Data=A;Var Item1--Total;
NOTE: The PROCEDURE CORR used 0.05 seconds.
174 Proc Corr Data=A Alpha;Var Item1-Item3;
NOTE: The PROCEDURE CORR used 0.05 seconds.
175 Proc Rank Data=A Out=Ranking descending;Var Total;ranks rankt;
NOTE: The data set WORK.RANKING has 4 observations and 6 variables.
NOTE: The PROCEDURE RANK used 0.38 seconds.
176 Proc Print Data=Ranking;
177 Run;
NOTE: The PROCEDURE PRINT used 0.11 seconds.
```

This particular program ran without error, so it does not flag any statements. As is true most of the time, this was not the first version submitted as may be seen in the line numbers. Moreover, the results may not be as desired even if the syntax of the statements was correct. Certain programming errors that need not violate syntax may be detected by looking at the numbers of variables and observations generated in each data step. A common error is to define a variable as SUMXY in one place and SUMYX in another or the like. This will cause the number of variables in the file to differ from what you might expect. Also check to see that the number of observations is correct.

The output is as follows, again deleting blank lines:

```
Correlation Analysis
    4 'VAR' Variables:   ITEM1    ITEM2    ITEM3    TOTAL
Simple Statistics
Variable    N        Mean      Std Dev         Sum      Minimum       Maximum
ITEM1       4     3.750000    2.061553    15.000000    2.000000      6.000000
ITEM2       4     6.000000    2.160247    24.000000    4.000000      9.000000
ITEM3       4     3.500000    1.290994    14.000000    2.000000      5.000000
TOTAL       4    13.250000    3.593976    53.000000    8.000000     16.000000
Pearson Correlation Coefficients / Prob > |R| under Ho: Rho=0 / N = 4
             ITEM1         ITEM2         ITEM3         TOTAL
ITEM1      1.00000      -0.29939       0.56360       0.59611
           0.0           0.7006        0.4364        0.4039
ITEM2     -0.29939       1.00000       0.35857       0.55814
           0.7006        0.0           0.6414        0.4419
ITEM3      0.56360       0.35857       1.00000       0.89803
           0.4364        0.6414        0.0           0.1020
TOTAL      0.59611       0.55814       0.89803       1.00000
           0.4039        0.4419        0.1020        0.0
Correlation Analysis
    3 'VAR' Variables:   ITEM1    ITEM2    ITEM3
                     Simple Statistics
Variable    N        Mean      Std Dev         Sum      Minimum       Maximum
ITEM1       4     3.750000    2.061553    15.000000    2.000000      6.000000
ITEM2       4     6.000000    2.160247    24.000000    4.000000      9.000000
ITEM3       4     3.500000    1.290994    14.000000    2.000000      5.000000
Correlation Analysis
Cronbach Coefficient Alpha
for RAW variables           :   0.270968
for STANDARDIZED variables:     0.440068
                     Raw Variables              Std. Variables
Deleted       Correlation                  Correlation
Variable      with Total       Alpha       with Total         Alpha
ITEM1          0.028006      0.480000       0.160285        0.527862
ITEM2         -0.051674      0.672897       0.033463        0.720902
ITEM3          0.774597     -0.853333       0.779038       -0.854665
Pearson Correlation Coefficients / Prob > |R| under Ho: Rho=0 / N = 4
             ITEM1         ITEM2         ITEM3
ITEM1      1.00000      -0.29939       0.56360
           0.0           0.7006        0.4364
ITEM2     -0.29939       1.00000       0.35857
           0.7006        0.0           0.6414
ITEM3      0.56360       0.35857       1.00000
           0.4364        0.6414        0.0
OBS   ID   ITEM1   ITEM2   ITEM3   TOTAL   RANKT
 1     1     2       4       2       8       4
 2     2     6       5       5      16       1
 3     3     5       6       3      14       3
 4     4     2       9       4      15       2
```

SAS uses the `Infile` statement to point to an external file. This is illustrated by modifying the previous in-line input program to read an ASCII file named `Data.txt` whose path is `D:\Academic\Number Crunchin'\` as follows (the `Procs` statements remain unchanged):

```
Data A;
Infile "D:\Academic\Number Crunchin'\Data.txt";
Input ID Item1 Item2 Item3;
Total = Sum(of Item1-Item3);
```

This **Infile** statement tells SAS where to find the file. The **Cards** statement and associated information must be deleted because no data are in line. The input log contains the following information about file creation:

```
NOTE: The infile "D:\Academic\Number Crunchin'\Data.txt" is:
      FILENAME=D:\Academic\Number Crunchin'\Data.txt,
      RECFM=V,LRECL=256
NOTE: 5 records were read from the infile
       "D:\Academic\Number Crunchin'\Data.txt".
      The minimum record length was 0.
      The maximum record length was 7.
NOTE: SAS went to a new line when INPUT statement
       reached past the end of a line.
NOTE: The data set WORK.A has 4 observations and 5 variables.
NOTE: The DATA statement used 4.0 seconds.
```

Whenever you read data from a file, look carefully at the minimum and maximum record lengths to make sure the data are being read in properly.

Finally, the following statements input data from a sheet named "Items" created in Excel. The spreadsheet must be open to this page because SAS uses dynamic data exchange (DDE). A newer method known as object linking and embedding (OLE) does not require that the spreadsheet be open. The version of SAS you use may require that both the source and the destination program be open. Consider the following example. Assume that an Excel spreadsheet is open to a page with the default name of Sheet1. The data consist of an ID plus four measures named Item 1 through Item 4. They are in the second through the fifth lines (the first line might identify the names of the variables), as follows:

```
1  3  1  5  2
2  1  2  2  3
3  2  3  3  6
4  5  4  1  5
```

```
Filename Dem DDE 'Excel|Sheet1!R2C1:R5C5';
Data A;
Infile Dem;
Input ID Item1-Item4;
Total = Sum(of Item1-Item4);
Proc Means;
Run;
```

The **Filename** statement creates an alternative name (alias) for the sheet (DEM) and says that the alias is to be effected through DDE. It further says that the data begin in the second row of the first column and end in the fifth row of the fifth column. Two statements later, this alias is introduced in the data step so that the program can find the data. Had the first statement said **File** instead of **DDE**, it would have created an alias for a file named DEM and assuming the user specified a path and a file name. The program creates a variable named Total as the sum of the four items that were input, which appear in the printout associated with **Proc Means**.

Data such as the sum (Total) can be output to a spreadsheet (e.g., to a sheet named Sheet2). Consider the following modification of the previous program:

```
Filename Dem DDE 'Excel|Sheet1!R2C1:R5C5';
Filename Outm DDE 'Excel|Sheet2!R1C1:R10C10';
Data A;
Infile Dem;
Input ID Item1-Item4;
Total = Sum(of Item1-Item4);
File outm;
Put ID Total;
Output;
Run;
```

The second line adds the alias for the output file name. **File outm;** tells SAS to route the data to the spreadsheet rather than the default log, and **Put ID Total;** writes the identifier and the total to the file. The resulting contents of Sheet2 are as follows:

1   11
2   8
3   14
4   15

Note that the contents of the sheet take up less space than requested in the `Filename` program statement. This causes no problem, but requiring more space does.

SAS Import and Export Functions

Data from other sources such as spreadsheets and ASCII files can be imported to and exported from SAS in a variety of ways. For example, if you are familiar with Excel, you can simply cut and paste the data into SAS following the `Cards` statement of a data step. SAS will simply convert the column delimiter to spaces, by default, which is usually satisfactory.

As was noted previously, appropriate installation allows the user to read data from a wide variety of files, which can be either local to the computer or on another computer, such as a mainframe (if you are not planning on connecting to other computers for this end, you can delay installation of this component while installing the option to read Excel and other PC files from your own computer). Here is how you can import an Excel file with the Import options of the File menu (Alt -F, I). If you are importing data this way, the first row of the spreadsheet should contain variable names, but they must conform to SAS naming conventions. In particular, they cannot have an embedded blank: "Subject Age" is incorrect, but "Age" is not.

IMPORTING AN EXCEL FILE INTO SAS WITH THE IMPORT OPTION

▬ 1. Select **Import Data** from the File menu (Alt -F, I).

▬ 2. The Wizard will then ask you to select the type of data to be imported. In this case, the default of an Excel 97 or 2000 spreadsheet is as desired. Other options include ASCII files in which the variables are separated by commas (CSV files).

▬ 3. You will then need to supply the location and name of the file, perhaps using the Browse option.

4. The Options button allows you to name the sheet, if there is more than one in the spreadsheet, and the range within the sheet (the default is the whole sheet).

5. The next step asks for the SAS name. You may make the data set either temporary by choosing "Work" as the SAS library name, perhaps the more common choice, or permanent by pointing to the location of a permanent library.

6. You next have the option of saving the procedure you have just completed as part of Proc Import.

7. You then finish the process, which creates the data set. Note that the contents of the first row become the variable labels.

Files may be exported in an analogous manner, and those in other formats are at least as easy to import or export. A particularly useful format is the **comma-separated value** (CSV) file in which successive fields are separated by a comma. If you have Assist installed in SAS, which is accessed from the Solutions menu (Alt -S, T), it can take you through these and a host of other processes step by step.

Looking at Data Sets in SAS

Traditional ways of looking at one's data in SAS include `Proc Print` and using a `Put` statement to route values to the Log file (by default) or some other destination such as an ASCII file. The following method may prove useful:

LOOKING AT DATA SETS IN SAS

1. Go to the Explorer window. One way to do this is to select it from the View menu (Alt -V, X).

2. The right side of the window will show various Active libraries.

▬ 3. Find the location of the desired data set. Specifically, click on **Work** if the data set is a temporary one, as most will be.

▬ 4. Click on the name of your data set (unfortunately, the process of opening the data set is slow).

▬ 5. You can right-click to enter Edit mode and change the contents of specific cells.

Basic Elements of SPSS

In contrast to the traditional form of SAS (but not to its current direction of evolution), SPSS is a highly database-driven program in that its normal mode of operation is to analyze data from its own spreadsheet-like data editor (unlike a spreadsheet, computations are not automatically passed on to new cases, among other differences). Figure 2.5 contains the

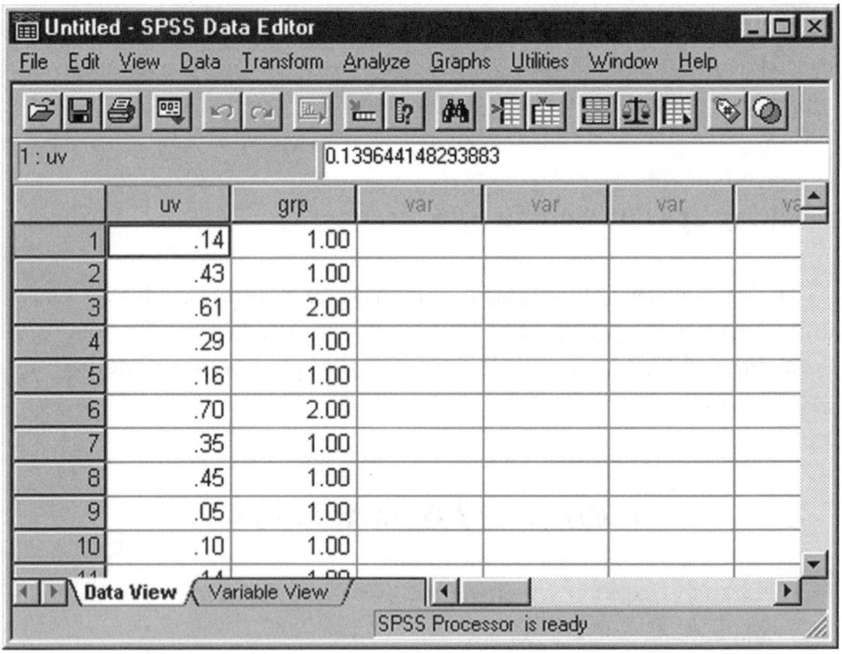

Figure 2.5. Basic SPSS for windows configuration.

opening layout of SPSS, Version 10, which we have used in this book. This data editor is one of two universally employed windows—the other containing output, but additional ones such as graphs may also be created. In particular, there is a syntax window that will be important later when program code is to be entered. This window is not needed when the user analyzes data from menu selections, which is perhaps most commonly the case. Data may be entered by hand or pasted from a spreadsheet (but not from a word processor save for doing it one value at a time). Because SPSS is spreadsheet driven, you can examine the data to make sure they are as desired. Once the basic data are entered, the SPSS programming language can be used or various menus invoked.

These menus play a greater role than their SAS counterparts and are as follows:

1. **File**. Similar to the role of the File menu in other windows-based programs, this menu is responsible for opening, closing, saving, and printing files. When the Open option is chosen ([Alt]-F, O), the suboptions are Data, Syntax, Output, Script, and Other. In addition, Open Database ([Alt]-F, B) allows a New Query, an Edit Query, and a Run Query, all of which pertain to the Structured Query Language, which we will not be concerned with. A third option is Read Text Data ([Alt]-F, R).

2. **Edit**. The Edit menu is also like other edit menus in that it is concerned with cutting, copying, and pasting data and doing searches (e.g., for particular items in a large output file).

3. **Data**. The Data menu is used to modify and insert variables and cases. It also can be used to sort a file and to merge data from multiple files. One can use the Aggregate command to create a new variable as a function of one or more existing variables so that one variable can be defined as the mean of a series of other items. In addition, one can split a file for analysis (e.g., analyze males and females separately). Furthermore, one can use a variety of rules to select some cases for further analysis instead of analyzing all cases; for example, one can select respondents to a survey who are between 40 and 50 years old for further analysis.

4. **Transform**. The Transform menu computes new variables from old (in conjunction with the aforementioned Aggregate command that is part of the Data menu). In addition to the Compute command

proper, one can also count the frequency with which variables assumed specific values (e.g., determine how many females were in
the sample). The Recode command affords another way to perform
transformations (e.g., alphanumeric values for a variable can be replaced with numeric ones).

5. **Analyze**. The Analyze menu (formerly known as the Statistics
   menu) performs calculations ranging from basic univariate and bivariate summaries to advanced ones such as factor analysis.

6. **Graphs**. Graphing is a basic part of SPSS unlike SAS, whose graphing capabilities require a separate add-on.

7. **Utilities**. Within the Utilities menu, the Variables option describes
   such things as the length of variables in the active file and whether
   they are numeric or alphanumeric. Other items in this menu are of
   minor interest.

8. **Window**. The Window menu parallels that in SAS and in most other
   windows-based programs.

9. **Help**. The Help menu performs the standard help function, including technical support numbers. It also has a very useful glossary of
   statistical terms.

Several of the more common commands are also implemented as
icons. The analysis previously performed using Excel and SAS will now be
performed in SPSS. Although the data could be entered in various ways,
they have been pasted from Excel. Actually, the data were first created
in a Microsoft Word table and then pasted into Excel—SPSS will accept
a paste from Excel but not from Microsoft Word. The variables originally appeared as var00001 to var00004. However, these names can be
changed by clicking on **Variable View** in the lower left-hand part of the
window; typing in the new name effects the change. Next, the variable
Total was obtained by going to the Transform menu. **Compute** was chosen, which opened a box. `Total` was then entered in the space marked
Target Variable and `sum (item1 to item3)` was entered in the space
marked **Numeric Expression**. The default name of SPSS spreadsheets is
Newdata, but this changes to the name under which the file is saved.

PERFORMING A BASIC ANALYSIS USING SPSS

■■ 1. Choose **Descriptive Statistics** from the Analyze menu, followed by **Descriptives** (⌊Alt⌋-S,U,D), to obtain the mean, standard deviation, and minimum and maximum values for the variables of interest (Item1–Item3 and Total). There are various options for other statistics and for the order of reporting the variables. **Variable List** was chosen for the latter. This lists variables in the order chosen by the user rather than in ascending, descending, or alphabetic order.

■■ 2. Choose **Correlate** from the Analyze menu, followed by **Bivariate** (⌊Alt⌋-S,C,B), to compute correlations among the variables of interest (again, Item1–Item3 and Total). The default options were chosen (Pearson correlation rather than rank-order statistics and pairwise rather than listwise exclusion of data). The options include means and standard deviations, but not the minimum and maximum values. None of the statistics were chosen because they were obtained in the previous step.

■■ 3. Choose **Scale** from the Analyze menu, followed by **Reliability Analysis** (⌊Alt⌋-A,R). The variables chosen were Item1 to Item3, but not Total, because that is redundant. The model chosen was **Alpha**, under Statistics. **Scale** was chosen under Descriptives and both **Correlations** and **Covariances** were chosen under Inter-Item. The latter produce what was identified as the standardized and raw alphas in SAS.

Descriptive Statistics					
	N	Minimum	Maximum	Mean	Std. Deviation
ITEM1	4	2.00	6.00	3.7500	2.0616
ITEM2	4	4.00	9.00	6.7500	2.1602
ITEM3	4	2.00	5.00	3.5000	1.2910
Valid *N* (listwise)	4				

Correlations

		ITEM1	ITEM2	ITEM3
ITEM1	Pearson correlation	1.000	−.299	.564
	Sig. (2-tailed)	.	.701	.436
	N	4	4	4
ITEM2	Pearson correlation	−.299	1.000	.359
	Sig. (2-tailed)	.701	.	.641
	N	4	4	4
ITEM3	Pearson correlation	.564	.359	1.000
	Sig. (2-tailed)	.436	.641	.
	N	4	4	4

The combined results of all three procedures are as follows:

```
****** Method 2 (covariance matrix) will be used for this analysis ******
  R E L I A B I L I T Y   A N A L Y S I S   -   S C A L E   (A L P H A)
                    Covariance Matrix
               ITEM1      ITEM2      ITEM3
ITEM1          4.2500
ITEM2         -1.3333     4.6667
ITEM3          1.5000     1.0000     1.6667
                    Correlation Matrix
               ITEM1      ITEM2      ITEM3
ITEM1          1.0000
ITEM2          -.2994     1.0000
ITEM3           .5636      .3586     1.0000
         N of Cases = 4.0
Reliability Coefficients  3 items
Alpha = .2710          Standardized item alpha = .4401
```

▣ PROBLEM*S*

2.1. Prepare a simple codebook for the project outlined in the problems for Chapter 1. Keep the number of variables to a half dozen or so. Try to incorporate the various formats (free field vs. fixed width and numeric vs. alphanumeric). Have someone else look at the book for clarity.

2.2. Enter some arbitrarily defined numbers to represent a dozen sample cases. Instead of simply making up the numbers, use a random number generator.

2.3. Assuming you are at an institution that has a bubble scanner, prepare and have some data scanned. Compare the layout of the data as you submit them to how they appear in the file.

2.4. If you have not already done so, become familiar with the essentials of your plain text editor, word processor, and spreadsheet. In particular, generate the data discussed in Problem 2.2 so that they appear in both a text file and a spreadsheet. If you are using a recent version of a spreadsheet, learn how to enter the data both directly, taking advantage of the validation rules, and by copying them from the text file.

2.5. Obtain simple descriptive statistics (means and standard deviations for quantitative variables and frequency distributions for categorical ones) on your database.

2.6. Now, open your statistical program. Import the data (both as ASCII data and as a spreadsheet, if possible) and repeat the analyses using your statistical program rather than your spreadsheet program. Explore its capacity to handle missing numbers by converting one or more valid observations to missing ones. Create a permanent data set from your input and then delete it.

Part II

3

Univariate Computer Simulations

This chapter is concerned with the empirical origin of basic statistical phenomena, such as normal distributions. These demonstrations usually require but a few programming steps, yet simulation has been minimally utilized as a didactic device. However, as noted in Chapter 1, we do not view our inductive stress on simulations to be in any way thought of as a substitute for the knowledge gained from traditional statistics textbooks.

This chapter considers the following topics:

1. A uniform random number generator, which is the "atom" of statistical simulation

2. The central limit theorem

3. The χ^2 distribution

4. The t distribution

5. The F distribution

Note: Steps are generally provided for simulations in Excel, SAS, and SPSS. However, Excel results can generally be pasted into SPSS and a variety of other packages that use spreadsheet input, so the SPSS programs are not a necessity. We have usually employed 100 observations with Excel simulations, but 1000 observations with SAS and SPSS simulations. The reason is that it requires somewhat more effort to drag equations down and obtain additional observations using Excel. However, the additional effort using SAS or SPSS is minimal, and run times for 1000 cases are but marginally longer than times for 100 cases on contemporary PCs. The reader is encouraged to expand the number used in either type of simulation, but especially the ones for Excel.

▣ UNIFORM RANDOM NUMBER GENERATION

Uniform random number generation is a process that produces alternative values over a range unpredictably and with equal probability. When generated by a computer algorithm, these numbers are not truly random. Most sources refer to them as "pseudo-random" because no deterministic process can produce a truly random outcome. However, nothing is lost and simplicity is gained by using the shorter term "random."

Standard Uniform Random Numbers

A standard uniform random number generator defines values at random over the range of 0 to 1. This is usually the basic form seen in computer programs and from which a rich variety of other types of random numbers, uniform and otherwise, are created. The specifics of the process need not be of concern save to note that it normally involves multiplying or adding two large numbers that are beyond the capability of the computer to represent, creating what is known as an **overflow**. Though this technically creates an error condition, the numbers it produces have the desired properties.

Random number generation usually allows a user to define the **seed** (starting point) of the process. This allows a sequence to be repeated and produces independent sequences with different choice of seed. To see this in SAS, consider the following data sets A, B, and C:

USING SAS TO GENERATE UNIFORM RANDOM NUMBERS

```
Data A;
Retain Y 13565;
```

```
Do I = 1 to 10;
X = Ranuni(Y);
Output;
End;
Proc Print Data=A;
Data B;
Retain Z 13565;
Do I = 1 to 10;
X = Ranuni(Z);
Output;
End;
Proc Print Data=B;
Data C;
Retain Y 13566;
Do I = 1 to 10;
X = Ranuni(Y);
Output;
End;
Proc Print Data=C;
Run;
```

Within each data step, the Retain statement assigns an initial value to a variable. The Do statement and its counterpart, End, form a loop or series of steps that is repeated 10 times in the example. Each time, a standard uniform random number is generated by the SAS function Ranuni and is placed in the data file by the Output statement. Proc Print prints out the resulting values. Note that the values in Data A and Data B are the same, even though the name of the variable used as the seed is different, but both sets differ from Data C. SAS has other standard uniform random number generators, including Uniform. The Uniform function operates in a manner similar to the Ranuni function. Call Ranuni(Seed, x) will also generate a standard uniform random number for x based on the seed value in SAS.

The following is the SPSS equivalent:

USING SPSS TO GENERATE UNIFORM RANDOM NUMBERS

Choose TRANSFORM, RANDOM SEED from the SPSS menu. Set a seed or choose Random Seed and click on OK. Then choose FILE, NEW, SYNTAX to open a syntax window. Enter the following code in the syntax window:

```
input program.
loop #i=1 to 10.
  compute uv=uniform(1).
end case.
end loop.
end file.
end input program.
execute.
```

Next, choose EDIT, SELECT ALL from the syntax menu and click on the Run button.

Because you may not have access or may only have limited access to statistical programs such as SAS or SPSS, it is useful to learn how to create uniform random numbers in a spreadsheet such as Excel. There are two basic ways to do this in Excel. One is to enter =Rand() in a cell or range of cells, which produces a uniform random number in that cell. Note what happens, however. Entering this formula in cell A1 produced a value of 0.137172. The formula was then copied to cell A2 where it produced the value 0.379054. However, the value in cell A1 changed to 0.905017! To fix the value once generated, press F9 while at the formula bar for a particular cell or select the cell or range of cells and use **Paste Special**, **Values**. Note that the Rand() function uses an internal process to generate the seed rather than a number you select so you cannot generate reproducible sequences as you can in SAS and other statistical packages. This is a significant limitation for simulation work because one usually wants to be able to reproduce random sequences. If you wish to generate uniform random numbers over the range of A to B, enter =Randbetween(A,B) to obtain an integer and the formula =Rand()*(b-a)+a to obtain a continuous variable.

Another way to obtain the integral value of a random number, in case you are using a program that does not have the equivalent of the `Randbetween` function, is to truncate a continuous variable. First, consult the program's functions to find the one that truncates a given number (random or not). Excel and several other programs call this function `Int`. For example, numbers in the first column of the following table were generated by inserting `=Rand()` in cells A1 to A5, and numbers in the second column were generated by inserting `=INT(10*A1)` in cell B1 and corresponding entries (A2, A3, ..., A5) in the remaining columns:

.3065472	3
.9794637	9
.5244738	5
.0682735	0
.7251904	7

Note that although `10*X` ranges from 0 to 10, truncated values only take on integral values from 0 to 9 because the only way to produce 10 is to get a 1.0 for the original value, but this has negligible probability. When a random sequence can only take on values of 0 or 1, the result is known as a **Bernoulli** process.

Using `Randbetween` may require you to install the Analysis ToolPak. If this package has been installed, the `Tools` menu will have an option called `Data Analysis`. If it does not, first select **Add-Ins...**, which will afford the choices available for installation. Choose Analysis ToolPak and press |Enter|.

Excel has a second means to generate random numbers that affords you more control. First, be sure that the Analysis ToolPak has been installed. Data Analysis includes the Random Number Generation option, which allows one to generate the various kinds of random numbers discussed later. The Uniform option allows you to choose the number of variables (values per columns), the number of cases (rows), the range, the seed, and determine where the values are to be placed (e.g., on the same sheet or on a different sheet).

Sampling Without Replacement

The preceding process makes selection of any given number independent of its predecessors or successors, by definition. This is **sampling**

with replacement. Think of an urn containing numbered balls. Sampling with replacement is analogous to picking a ball, noting its value, placing it back in the urn, and resampling. However, many circumstances dictate **sampling without replacement**. To continue the example, this would involve selecting a ball, noting its value, selecting a second without putting the first back, noting its value, and so on. If all balls are eventually sampled, the result is a **random permutation**. You may wish to generate these in an experiment where you have a number of conditions to which you wish to assign subjects randomly to conditions in order of appearance. For generality, assume you have four conditions denoted A, B, Γ, and Δ, and wish to assign two subjects to each group. We are using Greek rather than Roman letters in the following examples to avoid confusing the name of the column on the spreadsheet, which we continue to write in Roman, with the condition name. There is no reason why you should not use Roman letters when you run your own program. The randomization may be accomplished in SAS, SPSS, and Excel as follows:

USING EXCEL TO GENERATE RANDOM PERMUTATIONS

1. List the conditions in column A of the spreadsheet (e.g., enter A in cell A1).

2. Assign random numbers to the second column by placing =Rand() in the corresponding cells in column B.

3. Copy the cells containing the random numbers.

4. Use **Paste Special, Values** to fix the random numbers.

5. Rank-order the columns based on the values of the random numbers or, alternatively, sort the rows based on these values.

The data, ranked, formatted, and sorted, are as follows:

Condition	Random Number	Rank
B	.606363	1
A	.536480	2
A	.438970	3
Γ	.438162	4
Δ	.403337	5
Δ	.388792	6
B	.337786	7
Γ	.300502	8

Here is how the same ends could be achieved in SAS:

USING SAS TO GENERATE RANDOM PERMUTATIONS

```
Data A;
Retain seed 19982;
Input Cond $;
Rand = Uniform(seed);
Cards;
A
A
B
B
Γ
Γ
Δ
Δ
Proc Sort Data=A;by Rand;
Proc Print Data=A;
Run;
```

The dollar sign in the `Input` statement tells SAS that the variable Cond is alphanumeric. `Uniform` is like `Ranuni` in requesting a uniform random number.

Likewise, following is an SPSS program:

USING SPSS TO GENERATE RANDOM PERMUTATIONS

Proceed as before to open a syntax window. Then enter the following code:

```
input program.
loop i=1 to 8.
  compute uv=uniform(1).
end case.
end loop.
end file.
end input program.
execute.
```

Notice that the # sign has been removed from the loop index I (the # sign designates a variable as temporary in that it will not be saved). Also, the number 1000 has been changed to 8 to indicate that we want a random permutation of the integers between 1 and 8. Now proceed as before to run the program. This creates a variable I that just lists the integers 1 through 8 and a variable uv that attaches a uniform deviate to each of the eight integers. Now highlight the third column heading in the data matrix and from the SPSS Data Editor menu choose Variable View, click on the first cell of the third row, Insert name COND, type, string, Data View. Enter the variable values A A B B C C D D in variable COND. Now from the SPSS Data Editor menu choose DATA, SORT CASES, SORT BY uv, OK. This will create the desired permutation of the values in variable COND.

Nonuniform Distributions Based Directly on Uniform Processes

You may sometimes desire to have the alternatives occur with unequal probability. When a series of N trials consists of binary outcomes (0 or 1) with the probability of a 1 fixed at p for each trial, the result is called a **binomial distribution**, and numbers can be generated quite simply in Excel as one of the Random Number Generation options. It is a bit trickier when you wish to simulate a multinomial distribution where, by definition, three or more outcomes take on unequal probabilities. Assume you wish to run 100 trials with outcome *A* to occur with probability .5, outcome *B* to occur with probability .3, and outcome *C* to occur with probability .2 (the sum of the individual probabilities being 1.0).

Specifically, assume you want exactly 50 *A* outcomes, exactly 30 *B* outcomes, and exactly 20 *C* on 100 trials. In that case, use the preceding instructions for generating random permutations with that many values of each outcome, generate associated standard uniform random numbers, and sort.

However, it is probably more likely you will want a process that samples with replacement making the preceding probabilities events that are observable in the long run but perhaps not on any particular set of 100 trials.

The following steps accomplish this in Excel:

USING EXCEL TO SAMPLE WITH REPLACEMENT

1. Enter =**Rand()** the desired number of times (i.e., 100 in the present case) in column A to create a string of random numbers.

2. Enter =**IF(A1<=0.5,"A",IF(A1<= 0.8,"B","C"))** in cell B1. This equation (generated by the Excel Wizard function) makes the choice for the first trial.

3. Drag the function through the remaining rows.

To accomplish the same in SAS:

USING SAS TO SAMPLE WITH REPLACEMENT

```
Data A;
Retain err 1998;
Do I = 1 to 100;
X = Ranuni(err);
Y = "A";
If (X ge .5) then Y = "B";
If (X ge .8) then Y = "C";
```

```
Output;
End;
Run;
```

To accomplish this in SPSS:

USING SPSS TO SAMPLE WITH REPLACEMENT

Choose **TRANSFORM, RANDOM SEED** from the SPSS menu. Set a seed or choose **Random Seed** and click on **OK**. Then open a syntax window and input and run the following program:

```
input program.
string y (A1).
loop #i=1 to 100.
  compute x=uniform(1).
  compute y='A'.
  if(x ge .5) y='B'.
  if(x ge .8) y='C'.
end case.
end loop.
end file.
end input program.
execute.
```

Obtaining a Frequency Distribution From Uniform Random Numbers

Statistical theory and, perhaps, your intuition, tell you that the more observations there are in a distribution designed to simulate a mathematically defined one, the better the approximation of the relative proportions produced by the simulation to the probabilities defined by the model. Simulating distributions based on a smaller (100 in the example) and a larger number (1000) of cases using Excel can be used for illustrative purposes. The steps outlined refer to the smaller simulation, modifying it to accommodate the larger one is left as an exercise.

USING EXCEL TO PRODUCE A UNIFORM DISTRIBUTION

1. Either enter =Rand ( ) in cell A1 and drag the formula to cell A100 or use the Random Number Data Analysis option under Tools to produce the individual observations.

2. Enter .1 in cell B1.

3. Enter .2 in cell B2.

4. Select cells B1 and B2.

5. Drag through to cell B10 to produce values of .1, .2, . . . , 1.0 in column B.

6. Select **Data Analysis** from the Tools menu.

7. Select **Histogram** from the resulting menu.

8. Enter **A1 : A100** for the input range or, alternatively, click on A1 and drag through to A100.

9. Enter **B1 : B10** for the bin range (category values in which the observations are placed) or, alternatively, click on B1 and drag through B10.

10. Leave *New Output Ply:* selected.

11. Although you can select **Chart Output**, we do not find the resulting graph to be optimally informative. Graphing will be considered in the next section.

12. Press **OK**.

The resulting table will contain the bin and the frequency of observations in that bin with headers in the first row. One reason we feel the default chart output to be imperfect is that the abscissa (bin number) is the top of the bin rather than its midpoint used conventionally in the analysis of frequency distributions. This is easily overcome by inserting a column with these midpoints. Moreover, the procedure produces absolute frequencies, which are difficult to compare when distributions contain a different number of observations. Expressing the results as proportions or percentages easily rectifies this. Enter a header such as `Relative` in cell C1. The proportion of cases represented by cell C2 (bin .1) is defined as `=B2/SUM(B$2:B$11)`, and this proportion may be reformatted to a percentage using the Cells option under Format. Drag this formula down the remaining rows to produce the resulting table, which has been formatted somewhat for clarity:

Bin	Frequency	Relative
.10	12	.12
.20	9	.09
.30	4	.04
.40	11	.11
.50	5	.05
.60	16	.16
.70	9	.09
.80	7	.07
.90	20	.20
1.00	7	.07
More	0	

In other words, 12 observations fell at .1 or less, 9 between .1 and .2, and so on, and the associated relative frequencies based on the 100 observations are .12, .09, and so forth.

Now repeat the exercise by generating 1000 observations. These were placed in cells D1:D100. Repeating the previous steps gives rise to the following table:

Bin	Frequency	Relative
.10	87	.09
.20	112	.11
.30	95	.10
.40	100	.10
.50	109	.11
.60	101	.10
.70	96	.10
.80	111	.11
.90	86	.09
1.00	103	.10
More	0	

Note how much more similar the values are in this case. The relative frequencies now range from .09 to .11 instead of from .04 to .20.

The following SAS program leads to the same end. Change the `Retain` statement to vary the number of observations:

USING SAS TO PRODUCE A UNIFORM DISTRIBUTION

```
Data A;
Retain nobs 1000 seed 19842;
Do i = 1 to nobs;
X = int(10*(uniform(seed)))/10 + .1;
Output;
End;
Proc Print Data=A;
Proc Freq Data=A;Table x;
Run;
```

The rather complex statement `X = int(10*(uniform(seed)))/10 + .1;` produces numbers whose values are .1, .2, and so on, based on truncating (categorizing) the continuous values. It will begin at .1 rather than at .0 because of the addition of .1.

The Proc Freqs output is as follows:

X	Frequency	Percent	Cumulative Frequency	Cumulative Percent
0.1	93	9.3	93	9.3
0.2	98	9.8	191	19.1
0.3	110	11.0	301	30.1
0.4	105	10.5	406	40.6
0.5	80	8.0	486	48.6
0.6	95	9.5	581	58.1
0.7	114	11.4	695	69.5
0.8	101	10.1	796	79.6
0.9	97	9.7	893	89.3
1	107	10.7	1000	100.0

As can be seen, the frequency, percentage, cumulative frequency, and cumulative percentage are presented for each value of X.

USING SPSS TO GENERATE A UNIFORM DISTRIBUTION

From the SPSS menu, choose TRANSFORM, RANDOM SEED. Set your seed or choose Random Seed and click on OK. Then choose FILE, NEW, SYNTAX to open a syntax window. Enter the following code in the syntax window:

```
input program.
loop #i=1 to 1000.
  compute uv=uniform(1).
end case.
end loop.
end file.
end input program.
execute.
```

Now, from the syntax menu, choose EDIT, SELECT ALL and click on the Run button. You have now generated 1000 uniform random numbers between 0 and 1.

Graphing the Result

A graph of the distributions obtained from 100 and 1000 observations makes an instructive visualization, and this can be done quite simply in Excel. The particular type of graph to be constructed is called an "X–Y" or "scatter" graph because it plots values of a quantitative variable on the X axis or abscissa (bin values) against a second quantitative variable on the Y axis or ordinate (probabilities). The process will begin by drawing a graph of the first set of values. After this graph is completed, the second set will be added. It is quite possible to do both in the same operation, as will be noted. However, there are times when a graph needs to be constructed at one time and then modified by adding a new set of values later. Moreover, the first graph will be broken into two phases—creating the basic graph and editing its appearance.

USING EXCEL TO CREATE A BASIC GRAPH

▬ 1. Go to the Excel table containing the distribution based on 100 observations. Row 1 will contain labels. By default, the bins are in column A, the frequencies are in column B, and the relative frequencies are in column C. Use these as the column headers. Ignore or delete the row with bin label More so only numeric results appear in rows 2 to 11.

▬ 2. Although you can do all formatting later, it is useful to format at this point by assigning a custom format of #.00 or 0.00, depending on whether you want to suppress leading zeros or have them appear. This was done in creating the preceding table.

▬ 3. Change **Relative Frequency** to N = 100. This will appear in a legend in the final graph to let the reader know which set of points is being referred to.

▬ 4. Likewise, change **Bin** to **Upper Limit of X**.

▬ 5. Select the abscissa values (cells A1:A11).

▬ 6. Add C1:C11 to the selection. As the selection is noncontiguous, hold down the [Ctrl] key when you begin adding cells.

▬ 7. Using the Insert menu, choose **Chart, On New Sheet** or, alternatively, press [F11].

▬ 8. This provides step 1 of the Chart Wizard. Verify that the correct cells have been selected. If so, choose **Next**.

▬ 9. At step 2, select *X-Y (Scatter)*. From this point on, double clicking or choosing *Next* brings you to the next step. Note that you can also go back if you discover a mistake.

▬ 10. At step 3, choose option 2 to provide linear spacing of abscissa and ordinate and inclusion of data points and lines connecting them.

▬ 11. At step 4, look at the initial appearance of your graph. If it is incorrect, enter 1 for **First Column(s)** and 1 for **First Row(s)**.

▬ 12. At step 5 (the final step), select Yes for **Add a Legend**, leave **Chart Title** blank, insert the names Upper Limit of X for **Category (X)** and Relative Frequency for **Value (Y)**.

The preliminary graph appears as Figure 3.1.

Next, the data from the second distribution will be added to the graph and the result edited to a format suitable for publication:

EDITING THE BASIC GRAPH IN EXCEL

▬ 1. Prepare the spreadsheet containing the data for $N = 1000$ by titling the column of relative frequencies N = **1000** and formatting. The data to be added, which are only the ordinate values as the abscissa values are the same (.1 to 1.0), appear in cells C1 to C11.

▬ 2. Go to the sheet containing the graph and choose **New Data** from the Insert menu. A box will appear asking for the range of cells to be added.

▬ 3. Go to the spreadsheet containing the data and drag across the relevant cells (C1–C11). These cells will appear in the New Data box.

▬ 4. Press Enter.

▬ 5. A Paste Special box will appear. Press **OK**.

▬ 6. The new data will then be added to the graph.

▬ 7. Although information in the graph is complete, it is far from completed. The Excel default is to place the graph on a gray background with a border. Graphs to be used for publication are typically printed on a white background with no border. To make the change, place your mouse inside the graph. Right-click and choose **Format Plot Area**. Select None under both **Border** and **Area**.

▬ 8. Press Enter to effect the change.

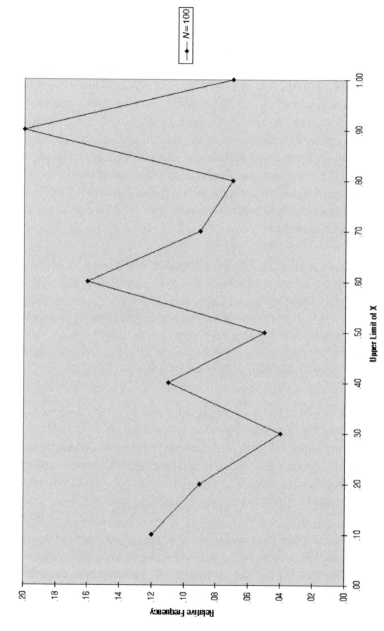

Figure 3.1. Preliminary graph of uniform frequency distributions based on 100 and 1000 observations.

▬ 9. Drag the legend inside the graph, making sure it does not overlap any part of the figure.

▬ 10. One or both sets of data points and their lines may appear in color on your screen, but they are typically printed in black and white. The symbols may also not be optimal. To change a symbol and set of points, double-click on one of the data points to open the Format Data Series box. Change the color of both line and marker to black by clicking on the box and selecting that color when alternatives drop down. In a similar manner, you can change the symbol using the alternatives under **Style** under **Marker**. One possible set of alternative markers is a filled square (black foreground and background) versus an open square (black foreground and white background). Note that you change one set of data points at a time. Press **OK** each time.

▬ 11. You may wish to change the fonts used. The American Psychological Association format requires a sans serif font such as the Arial that is supplied with Windows, which is used in the figure. The default sizes may be small (10 point) and not stand out enough. One possibility is to change all text to 12-point bold. To do so, first click near one of the corners of the figure to select the whole figure. At that point, a font change will apply to the whole figure.

▬ 12. Clicking along the abscissa provides you with various options, such as changing the range from 0 to 1.0 to .1 to 1.0, selectively changing the font along this axis, and changing the orientation of the values. Clicking along the ordinate effects analogous changes. Similarly, either axis label may be changed. However, none of these changes is necessary here.

▬ 13. The default may print a header at the top and a page number at the bottom. This should be suppressed if the figure is intended for publication in an academic journal. To do so, choose **Print Setup** from the File menu. Choose the **Header/Footer** tab and delete the entries.

▬ 14. Click on **Custom Header**.

15. Clear and clear the contents, which may be something like &[Tab], which would print the name of the sheet.

16. Do the same with **Custom Footer**.

17. You may wish to save the figure separately as a bitmap (BMP) or jpg (joint programmers group) file. To do so, first make sure the whole area is selected. Then copy the selection to the clipboard and paste it in a graphics program of your choice—Paint will do if you only require the BMP format. Because the graph is wider than it is tall (**landscape** format), either rotate it in the graphics program or print it out using the Landscape option from the appropriate menu (e.g., Page Setup in Paint or Printer Setup in LView).

The resulting graph appears in Figure 3.2.

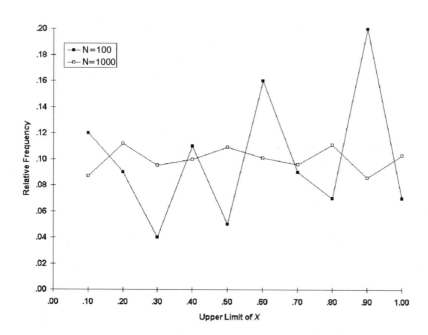

Figure 3.2. Edited version of the graph in Figure 3.1.

▣ DIƧTRIBUTIONƧ DERIVED FROM UNIFORM RANDOM NUMBER GENERATION

The **central limit theorem** is one of the most important in statistics. It states that if an observation is based on an independently derived average of k component observations, each of which has a mean of μ and a standard deviation of σ, the average will have a mean of μ and a standard deviation of $\sigma/\sqrt{k}$ and will tend toward normality as k increases. Under circumstances such as those to be described, the approximation to normality is adequate at $k = 12$.

The steps in demonstrating this theorem are as follows:

USING EXCEL TO DEMONSTRATE THE CENTRAL LIMIT THEOREM

1. Create a matrix with 1000 standard uniform random numbers in each of the first 12 columns of a new sheet.

2. It is not necessary to use uniform random numbers, but they are sufficient. As before, they can be created by entering =Rand() in cell A1 and dragging the rest of the way, but it is quicker to use the random number generation capability of the Data Analysis option under Tools.

3. If you choose the latter, enter 12 for **Number of Variables**, enter 1000 for **Number of Random Numbers**, enter Uniform for **Distribution**, leave 0 and 1 as they are under **Between**, arbitrarily choose any number for Seed, check **Output Range**, and enter A1:A1000 in the adjacent space.

4. Press Enter. Note that the process may take a bit of time.

5. Though not necessary except for clarity, enter a line from column A to column L in row 1000 to separate this set of data from what is to follow.

6. Enter =AVERAGE($A1:A1) in cell M1. This provides one observation based on one uniform random number.

7. Drag the formula in M1 through to the remaining cell X1. The contents of cell N1 will be an average of two uniform random numbers, the contents of cell O1 will be an average of three random numbers, and so on, so that successive columns will contain the average of 1, 2,... uniform random numbers.

8. Drag the formulas in cells M1:X1 through row 1000. This creates 1000 observations based on 1, 2,... standard uniform random numbers.

9. Enter =AVERAGE(M1:M1000) in cell M1001 to obtain the average of the observations in this column.

10. Enter =STDEV(M1:M1000) in cell M1002 to obtain the standard deviation.

11. Drag M1001:M1002 through column X to obtain the means and standard deviations of the remaining sets of averages.

12. Enter 1 in cell A2003.

13. Enter 2 in cell B2004.

14. Drag this equation through column X to obtain the consecutive integers from 1 to 12 in successive columns.

15. Enter =SQRT(1/(12*M1003)) in cell M1004. This is the expected standard deviation of an average based on one score drawn from a standard normal population.

16. Drag M1004 through column X to obtain the remaining expected standard deviations.

It can be shown that the expected value of the mean of any uniform distribution is $(B - A)/2$, where B is the upper limit and A is the lower limit. In the present case, this reduces to .5. Compare this value to the means in row 2001. The obtained and expected means should therefore be similar. It can likewise be shown that the variance of a standard uniform distribution is 1/12 so the standard deviation in cell M1002 should be $\sqrt{1/12}$ or .28. Likewise, the central limit theorem says that the remaining standard deviations should be $.28/\sqrt{n}$. Next, obtain frequency distributions. With 1000 observations, it is meaningful to use bin sizes of .05, that is, to make bins at .00, .05, .10. As it might be unnecessarily time consuming to calculate all 12 frequency distributions, calculate only those for 1, 6, and 12 unless you are working with others to share the labor.

Following is one possible output (after formatting):

Bin	K		
	1	6	12
.00	.00	.00	.00
.05	.06	.00	.00
.10	.04	.00	.00
.15	.05	.00	.00
.20	.05	.00	.00
.25	.05	.01	.00
.30	.06	.03	.01
.35	.06	.06	.03
.40	.05	.07	.09
.45	.04	.15	.15
.50	.04	.17	.24
.55	.05	.18	.20
.60	.07	.14	.18
.65	.05	.10	.08
.70	.04	.04	.03
.75	.04	.03	.00
.80	.05	.01	.00
.85	.05	.00	.00
.90	.05	.00	.00
.95	.05	.00	.00
1.00	.05	.00	.00

Graphing reveals the progressively more normal appearance of the frequency distribution, but this can be evaluated more formally with methods to be considered in the next section. This graph is presented in Figure 3.3.

An SAS program is as follows:

USING SAS TO DEMONSTRATE THE CENTRAL LIMIT THEOREM

```
Data A;
Retain nobs 1000 seed 19232 nitems 12;
Array Xs{*} X1-X12;
```

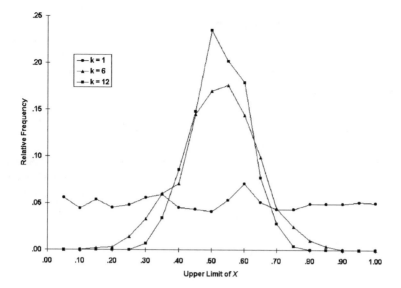

Figure 3.3. Excel graph of the central limit theorem demonstration.

```
Array Ys{*} Y1-Y12;
Do i = 1 to nobs;
Do j = 1 to nitems;
Xs(j) = uniform(seed);
Ys(j) = 0;
End;
Do j = 1 to nitems;
Do k = 1 to j;
Ys(j) = Ys(j) + Xs(k);
End;
End;
Do j = 1 to nitems;
Ys(j) = Ys(j)/j;
End;
Output;
End;
Keep y1-y12;
Proc Means Data=A;
Run;
```

Note the use of the **Array** statement. This is a way for one symbol to denote a series of related terms. SAS has two forms of array statement, explicit and implicit. We will only consider the explicit form, which is the one recommended by SAS (SAS Institute, 1990, pp. 292–306).

In general, an explicit array is declared by means of three elements: (a) the array name, (b) either the number of elements in the array or the symbol "*" enclosed in parentheses, and (c) the names of the array elements. Some arrays consist of alphanumeric elements in which case a "$" would be placed between (b) and (c). In the present case, the first array statement in the preceding programm identifies the array name as **XS** and the array elements as **X1-X12**, each of which will eventually contain a different uniform random number. Note how much simpler it is to write

```
Do j = 1 to nitems;
Xs(j) = uniform(seed);
Ys(j) = 0;
End;
```

than to write the equivalent series of 24 individual statements:

```
X1 = uniform(seed);
X2 = uniform(seed);
...
X12 = uniform(seed);
Y1 = 0;
Y2 = 0;
...
Y12 = 0;
```

Following is an SPSS equivalent:

USING SPSS TO ILLUSTRATE THE CENTRAL LIMIT THEOREM

Proceed as before to open a syntax window. Then enter and run the following program:

```
input program.
vector x(12).
vector y(12).
loop #i=1 to 1000.
```

```
loop #j=1 to 12.
  compute x(#j)=uniform(1).
  compute y(#j)=0.
end loop.
end case.
end loop.
end file.
end input program.
execute.
vector x=x1 to x12.
vector y=y1 to y12.
loop #j=1 to 12.
loop #k=1 to #j.
  compute y(#j)=y(#j)+x(#k).
end loop.
  compute y(#j)=y(#j)/#j.
end loop.
execute.
```

Next, choose **GRAPHS, HISTOGRAM, Variable Y1, Display Nor-mal Curve, OK** from the SPSS Data Editor window. Repeat this process for variables Y_2 to Y_{12}. Figure 3.4 contains the SPSS graphs with the re-sults for $N = 1$ in the top left, $N = 6$ in the bottom left, and $N = 12$ in the top right.

χ^2 Distribution

A χ^2 distribution with k degrees of freedom (df) is defined as the sum of k independent squared standard normal deviates (numbers drawn at random from a normal distribution). Algebraically, then, $\chi^2(k) = \sum_{i=1}^{k} z^2$.

The Random Number Generation tool allows this to be readily demon-strated on a spreadsheet, as follows:

USING EXCEL TO SIMULATE THE χ^2 DISTRIBUTION

▬ 1. Start a new worksheet.

▬ 2. Choose **Data Analysis** under Tools.

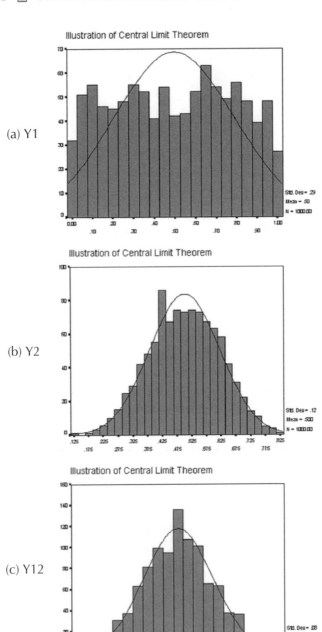

Figure 3.4. SPSS graph of the central limit theorem demonstration with the results for $N = 1$ in (a), $N = 6$ in (b), and $N = 12$ in (c).

3. Choose **Random Number Generation**.

4. Enter **4** for **Number of Variables**, enter **1000** for **Number of Random Numbers**, enter `Normal` for *Distribution*, leave the parameters **0** for **Mean** and **1** for **Standard Deviation** as they are, arbitrarily choose any number for **Seed**, check **Output Range**, and enter **A1:D1000** in the adjacent space.

5. Press Enter. Again, the process may take a bit of time.

6. Underline the bottom row for columns A:D.

7. Enter `=SUMSQ($A1:A1)` in cell E1. This will place the squared contents of cell A1 in cell E1, using the same strategy as in the central limit theorem simulation.

8. Drag this formula across to column H. This will put the sum of two squared normal deviates in cell F1, the sum of three normal deviates in cell G1, and the sum of four normal deviates in cell H1.

9. Drag E1:H1 through row 1000 to complete the four sampling distributions.

10. Enter the formulas for the mean, variance, minimum value, and maximum value in cells E1001 to E1004.

11. Drag these values across to column H.

12. Use the minimum and maximum values to create bins for the frequency distributions.

13. Obtain the resulting four frequency distributions, which are estimates of χ^2 distributions of 1 through 4 *df*.

14. Generate an *X–Y* plot of the distributions.

It can be shown that the expected value (mean) of a χ^2 distribution is 1 *df*, and the variance is 2 *df*. The following table describes the observed and expected values:

		df			
		1	2	3	4
Observed	mean	1.01	1.93	2.98	4.02
Observed	S.D.	2.03	3.35	5.67	7.99
Expected	mean	1.00	2.00	3.00	4.00
Expected	S.D.	2.00	4.00	6.00	8.00

The following is a table of these distributions:

Real Upper Limit of X	df			
	1	2	3	4
1	.67	.38	.19	.09
2	.17	.27	.23	.17
3	.07	.14	.20	.18
4	.03	.09	.13	.17
5	.03	.05	.09	.11
6	.01	.03	.07	.08
7	.01	.02	.03	.06
8	.00	.02	.03	.04
9	.00	.00	.02	.03
10	.00	.00	.01	.02
11	.00	.00	.00	.02
12	.00	.00	.00	.01
13	.00	.00	.01	.01
14	.00	.00	.00	.00
15	.00	.00	.00	.00
16	.00	.00	.00	.00
17	.00	.00	.00	.00
18	.00	.00	.00	.00

These appear graphically in Figure 3.5.

Note how the distribution goes from being J shaped at $df = 1$ to having a mode nearer the middle of the distribution at $df = 4$. The central limit theorem says that this distribution should become normally distributed as the df increases, which, in fact, it does.

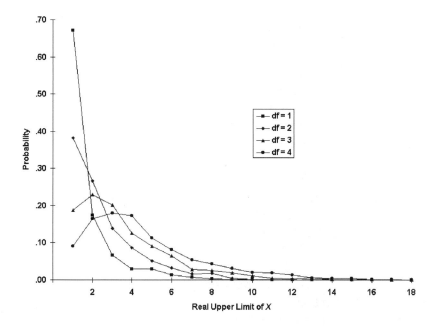

Figure 3.5. χ^2 distributions with 1, 2, 3, and 4 *df*.

An SAS equivalent is as follows:

USING SAS TO SIMULATE THE χ^2 DISTRIBUTION

```
Title 'Simulation of chi-square distributions';
Data A;
Retain nobs 1000 seed 19842 nitems 4;
Do i = 1 to nobs;
Y1 = rannor(seed)**2;
Y2 = y1+rannor(seed)**2;
Y3 = y2+rannor(seed)**2;
Y4 = y3+rannor(seed)**2;
Output;
Keep Y1-Y4;
End;
Proc Format;
```

```
Value grp    0-1='1'
             1-2 ='2'
             2-3='3'
             3-4='4'
             4-5='5'
             5-6='6'
             6-7='7'
             7-8='8'
             8-9='9'
             9-10='10'
             10-11='11'
             11-12='12'
             12-13='13'
             13-14='14'
             14-15='15'
             15-high='higher';
Proc Print Data=A;
Proc Means Data=A mean var;
Proc Freq Data=A;Table Y1-Y4;format y1-y4 grp;
Run;
```

Following is an SPSS equivalent:

USING SPSS TO SIMULATE THE χ^2 DISTRIBUTION

Open a syntax window as before. Then enter and run the following program to generate 1000 χ^2 random deviates with 4 df:

```
input program.
loop #i=1 to 1000.
  compute c2v=rv.chisq(4).
end case.
end loop.
end file.
end input program.
execute.
```

To get the mean and variance of these, select ANALYZE, DESCRIP-TIVE STATISTICS, DESCRIPTIVES, Variables C2V, OPTIONS, mean, variance, continue, OK from the SPSS Data Editor menu.

---

t Distribution

Assume one has a sample mean $(\overline{X})$ of size N and wishes to test the null hypothesis that it was drawn from a population with mean μ. If the population standard deviation could be assumed to be equal to σ, this null hypothesis could be tested with the test statistic

$$z = \frac{(\overline{X} - \mu) \cdot \sqrt{N}}{\sigma}.$$

However, it is usually the case that the sample standard deviation is estimated from the data. In this case, the proper test statistic is Student's t:

$$t = \frac{(\overline{X} - \mu) \cdot \sqrt{N}}{s}$$

(Hays, 1988, pp. 288–296). Student's t produces a family of distributions, each defined by $N - 1$, the number of degrees of freedom (df). For simplicity, assume a standard normal distribution in which $\mu = 0$ and $\sigma = 1$, symbolized as $N(0, 1)$ in the literature. In this case, the two equations reduce to $\overline{X}\sqrt{N}$ and $(\overline{X}\sqrt{N})/s$, respectively. Student's t has a $\mu = 0$ and a $\sigma = \sqrt{(N-1)/(N-3)}$. It is a relatively simple matter to compare the two distributions, which will be done here with $N = 4$ observations $(df = 3)$. To make the standard deviations of the two distributions comparable, we will define Z as $\sqrt{(N-1)/(N-3)}\overline{X}$. The steps are as follows:

USING EXCEL TO SIMULATE THE t DISTRIBUTION

▬ 1. Using previously discussed methods, generate four columns of 1000 random normal deviates in cells A1:D1000.

▬ 2. Enter the formula =SQRT(3)*AVERAGE(A1:D1) in cell F1.

▬ 3. Enter the formula =SQRT(4)*AVERAGE(A1:D1)/STDEV(A1:D1) in cell G1.

▬ 4. Drag the two down through cells F1000 and G1000.

▬ 5. Enter the formulas for the mean, standard deviations, and minimum and maximum values in row 1002.

▬ 6. Use these data to generate an X–Y graph.

A representative frequency distribution is as follows:

Apparent Upper Limit of X	z	t
−6	.00	.00
−5	.00	.00
−4	.00	.01
−3	.00	.01
−2	.02	.03
−1	.11	.10
0	.39	.36
1	.36	.32
2	.11	.11
3	.01	.03
4	.00	.01
5	.00	.01
6	.00	.00

Likewise, the graph is presented in Figure 3.6. The point to note is the longer tails of the t distribution. This is known as greater leptokurtosis (more positive kurtosis).

The following SAS statements will generate the basic data to compare z and t:

USING SAS TO SIMULATE THE t DISTRIBUTION

```
Filename Dem DDE 'Excel|Sheet6!R1C1:R1000C2';
Data A;
```

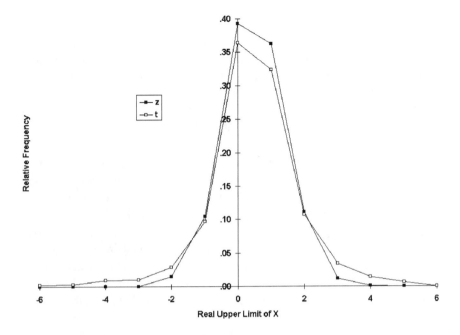

Figure 3.6. *z* vs. *t* distributions.

```
Retain nobs 1000 Nt 4 err 19992;
Do I = 1 to nobs;
Do J = 1 to Nt;
X = rannor(err);
Output;
End;
End;
Keep I X;
Proc Means Noprint Data=A Mean T;
Output Out=B Mean=z T=T;by I;
Data C;
Set B;
File Dem;
NT = 4;
Z = SQRT((NT-1)/(NT-3))*Z;
Put z T;
Run;
```

As before, data are going to be routed to an Excel spreadsheet using the Filename command. Data set A will generate a number of trials defined by nobs (1000) with Nt observations per trial (4) and random number seed 19992. There are two loops. The outer one proceeds nobs times and the inner one proceeds Nt times. Each pass of the inner loop outputs (a) the value of the outer loop (I) and (b) a random normal deviate so data set A will contain 1000 * 4 observations on two variables ordered by the value of nobs (I).

Proc Means computes two statistics, the mean and the value of t. The Noprint option suppresses printing, though you may wish to look at the individual values in the Output file. Instead, the output statement creates a new file (B) containing the mean (labeled z) and the t statistic (labelled t). Data set B will contain 1000 observations on five variables: I, t, z, and two additional variables created by Proc Means called _Type_ and _Freq_. _Type_ is not of interest here because it always equals 0, though it may be of interest if there are several types of output. _Freq_ is always 4 because both t and z are based on four observations. The statement By I causes this to be done separately for each of the 1000 values of I in A. Data set C then simply reads the data output from Proc Means, and the combination of File and Put routes two columns of 1000 observations to Excel where it may be further analyzed.

Following is an SPSS equivalent:

USING SPSS TO SIMULATE A t DISTRIBUTION AND A NORMAL DISTRIBUTION WITH THE SAME MEAN AND VARIANCE

Open a syntax window as before to run the following program:

```
input program.
loop #i=1 to 1000.
  compute tv=rv.t(3).
  compute zv=normal(sqrt(3)).
end case.
end loop.
end file.
end input program.
execute.
```

This produces the variable tv from a t distribution with 3 df and a variable zv from a normal distribution with mean 0 and standard deviation $\sqrt{3}$. Proceed as before to get the mean and variance of the variables.

F Distribution

The F distribution with $N_1 - 1$ and $N_2 - 1$ df begins with one sample of N_1 observations and a second sample of N_2 observations. It is defined as the ratio of the variance of the first set of observations to the variance of the second set of observations. This is such a straightforward extension of the previous discussion that we will only present the outline, leaving the details as an exercise. Let $N_1 = 4$, $N_2 = 2$, and assume 1000 observations.

USING EXCEL TO SIMULATE AN F DISTRIBUTION ON A SPREADSHEET (OUTLINE)

▬ 1. Enter random normal deviates in cells A1 to F1000.

▬ 2. Enter =VAR(A1:D1) in cell G1.

▬ 3. Enter =VAR(E1:F1) in cell H1.

▬ 4. Enter =G1/H1 in cell I1.

▬ 5. Drag cells G1:I1 down to row 100.

▬ 6. Obtain the resulting frequency distribution of observations in cells G1:I1000.

This may be accomplished in several ways in SAS. One outline, based on the preceding, is as follows:

USING SAS TO SIMULATE AN F DISTRIBUTION (OUTLINE)

▬ 1. Two data sets corresponding to A will be created, to be denoted A1 and A2.

▬ 2. Change **err** to any other arbitrary number and **NT** from 4 to 2 in data set A2, leaving A1 as is.

▬ 3. **Proc Means** will be run twice, once on A1 and once on A2. Each time, you will only be interested in the variance and not the mean or z. Consequently, change **Mean T** in each of the two new **Proc Means** statements proper to **Var** and change **Out=B Mean=z T=T** to **Out=B1 Var=Var1** for data set A1 and to **Out=B2 Var=Var2** for data set A2.

▬ 4. Change **Set B** to **Merge B1 B2; by I;**.

▬ 5. Following this line, add **F = Var1/Var2**.

▬ 6. Change **Put z T to Put F**. For diagnostic purposes, also output Var1 and Var2 using a separate **Proc Print Data=C;** statement. These values should both vary about 1.0.

USING SPSS TO SIMULATE AN *F* DISTRIBUTION

Open a syntax window and run the following program:

```
input program.
loop #i=1 to 1000.
  compute f=rv.F(4,2).
end case.
end loop.
end file.
end input program.
execute.
```

This generates random numbers from an *F* distribution with 4 *df* in the numerator and 2 *df* in the denominator.

▣ PROBLEM*/*

3.1. (*Note*: As far as time permits, do as many of the following problems both on your spreadsheet and on your statistical program. In addition, repeat all demonstrations presented in the text with a different random number seed.) Generate a series of 1000 uniform random integers from 1 to 10 and plot their frequency distribution.

3.2. Take the digits from 1 to 10 and create 100 random permutations.

3.3. Create one or more graphs from the data in your simulated database.

3.4. Create a distribution of 1000 random normal deviates.

3.5. Repeat the demonstration of the central limit theorem using sums based on 1, 4, 7, 10, and 13 observations.

3.6. Modify the data in Problem 3-5 to create a χ^2 distribution with that number of *df*. Plot the distributions on a common set of axes. Compare your results to those presented in the text.

3.7. Do the same to create a *t* and an *F* distribution.

3.8. Finally, generate a normal distribution.

4

Basic Correlational Simulations

This chapter extends the types of simulations presented in the previous chapter to the bivariate and multivariate cases. The topics to be considered are as follows:

1. Simulating a correlation from both **regression** and **structural** (latent variable) models

2. Simulating a set of variables that fits a unifactor model

3. Simulating sets of variables that fit multifactor models (two simulations involve one case in which variables forming a cluster are uncorrelated with variables forming a second cluster; the second involves a case in which the variables forming the two clusters are correlated)

4. Simulating profiles

▣ JIMULATING A CORRELATION WITHIN JAMPLING ERROR

Regression Simulation

Equation 4.1 provides one of the many formulas for the sample Pearson product–moment correlation (r, which will simply be called the correlation hereafter):

$$r = \frac{s_{xy}}{s_x s_y},$$ (4.1)

where s_{xy} is the covariance of X and Y and s_x and s_y are their respective standard deviations.

Equation 4.2 defines one way to obtain a correlation and covariance of magnitude b between variables X and Y (subscripts are often included to denote the subject number, but this is not necessary here):

$$Y = b \cdot X + \sqrt{1 - b^2}\, e.$$ (4.2)

In this equation,

1. Assume (without loss of generality) that X and e are standardized. Also, think of X as something measured without error ("truth") and e as error in the sense of being unrelated to X. Because Y then becomes something affected by both truth (X) and error (e) but X is pure "truth," this is known as a **regression** model.

2. Assume that e does not affect X and vice versa; that is, the two values are generated independently.

3. Because of this independence, the two parts of Equation 4.2 contribute additively to Y.

4. It can be shown that $s_{xy} = b$.

5. It can further be shown that the variance of $b \cdot X$ is b^2.

6. It can be shown that the variance of $\sqrt{1 - b^2} \cdot e$ is $1 - b^2$.

7. Because of the independence of X and e, the variances of the two components add, but $b^2 + 1 - b^2$ is simply 1 so the variance of Y is 1.0. From point 2, it can also be shown that $\overline{Y}$ is .0 so Y is also standardized.

8. This means that $r = b$ within sampling error. Following standard practice, the population correlation, which equals b exactly, will be symbolized as ρ.

To simulate a correlation on a spreadsheet such as Excel, do the following:

USING EXCEL TO PRODUCE A CORRELATION FROM A REGRESSION MODEL

▬ 1. Enter the expected (desired) value of $b(\rho)$ in cell A1.

▬ 2. Generate two columns of 100 random normal deviates, each starting in cells A2:B101. These represent 100 values each of X and e.

▬ 3. Enter =A$1*A2 + sqrt(1 - A$1^2)*B2 in cell C2. This represents the first value of Y as defined by Equation 4.2.

▬ 4. Drag cell C2 through cell C102. This defines the remaining values of Y.

▬ 5. Enter =Correl(A2:A101,C2:C101) in cell B1. This is the observed correlation, and its value should fall within the sampling error of A1.

▬ 6. The theoretical values of the slope and intercept of the regression line predicting Y from X are b and 0. The obtained values are s_{xy}/s_x^2 and $\overline{Y} - b\overline{X}$. In Excel, these become =COVAR(A2:A101,C2:C101)/VAR(A2:A101) and =AVERAGE(C2:C101) - C1*AVERAGE(A2:A101). Place these in cells C1 and D1, respectively.

▬ 7. Enter =C$1*A2 - D$1 in cell D2. This provides the predicted value ($\hat{Y}$) for the first observation.

▬ 8. Drag cell D2 down through row 101 to obtain the remaining predicted values.

▬ 9. Enter =C2-D2 in cell E2. This provides the residual (Y^*) for the first observation. Note that $Y^* = Y - \hat{Y}$.

▬ 10. Drag cell E2 through row 101 to obtain the remaining predicted values.

▬ 11. Enter =AVERAGE(A2:A101) in cell A102 to obtain $\overline{X}$.

▬ 12. Drag cell A102 through column E to produce the remaining means. These should all be 0 within sampling error.

▬ 13. Enter =STDEV(A2:A101) in cell A102 to obtain s_x.

▬ 14. Drag cell A102 through column E to produce the remaining standard deviations. The expected values for these standard deviations are given in various sources on correlational analysis; see, for example, Nunnally and Bernstein (1994, p. 146).

▬ 15. Although you already have the value of r_{xy}, which is most central to the simulation, obtain correlations among all the variables. One way to do so is to use the Correlation option under Data Analysis. The expected values for these correlations are given in the same sources.

▬ 16. As an additional exercise, draw a scatterplot containing A2:A101 and C2:C101 as points. This involves formatting each paired value with a suitable marker (e.g., a circle) and choosing **None** for the line type.

▬ 17. Place the regression line as a solid line in this graph by inserting points A2:A101 and D2:D101 as a new series. This time, choose **None** for the marker and use the default line type.

▬ 18. Note how the data change as different values are inserted in field A1.

Data in which a criterion value, Y, depends on a single underlying quantity, X plus error are called **unifactor** data.

An equivalent SAS program is as follows:

USING SAS TO PRODUCE A CORRELATION
FROM A REGRESSION MODEL

```
Data A;
Retain err 1996 n 500 b .8;
```

```
Bres = sqrt(1 - b**2);
Do I = 1 to N;X = rannor(err);
Y = b*X + bres*rannor(err);
YPred = b*X;
YRes = Y - YPred;
Output;
End;
Keep X Y YPred YRes;
Proc Corr Data=A;
Run;
```

In this program, b and bres are the weights respectively applied to X and e in Equation 4.2. Rannor(err) is the random normal deviate function. For simplicity, $\hat{Y}$, and therefore Y^*, were computed from the expected rather than obtained correlation. The following values were obtained:

	Variable			
	X	Y	$\hat{Y}$	Y^*
Mean	−.02	.04	−.02	.05
S.D.	1.00	1.01	.80	.61
r with X	1.00	.80	1.00	.01
r with Y	.80	1.00	.80	.61
r with $\hat{Y}$	1.00	.80	1.00	.01
r with Y^*	.01	.61	.01	1.00

Following is the SPSS equivalent:

USING SPSS TO PRODUCE A CORRELATION
FROM A REGRESSION MODEL

Open a syntax window and enter and run the following program:

```
input program.
loop #i=1 to 500.
  compute #b=.8.
```

```
compute #bres=sqrt(1-#b**2).
compute x=normal(1).
compute y=#b*x+#bres*normal(1).
compute ypred=#b*x.
compute yres=y-ypred.
end case.
end loop.
end file.
end input program.
execute.
```

The actual correlations may be obtained by choosing ANALYZE, COR-RELATE, BIVARIATE from the SPSS Data Editor menu.

Structural Models

In contrast to having Y depend on X and error, suppose that the following hold:

$$X = b_x t + \sqrt{1 - b_x^2} \cdot e_x, \qquad (4.3a)$$

$$Y = b_y t + \sqrt{1 - b_y^2} \cdot e_y. \qquad (4.3b)$$

Thus,

1. Both X and Y depend on a common element, t ("truth"), which is in standard normal form.

2. Both X and Y are also affected by error, but the error affecting $X(e_x)$ is independent of the error affecting $Y(e_y)$. Both error terms are in standard normal form.

3. This dependency of the two on "truth" is expressed by weights of b_x and b_y (note that the regression model is a special case in which $b_x = 1$ and $b_y = b$). In particular, assume that $b_x = b_y = \sqrt{b}$.

4. Let $\sqrt{1-b}$ be the weight assigned to each of the two error terms so that $X = \sqrt{b} \cdot t + \sqrt{1-b} \cdot e_x$ and $Y = \sqrt{b} \cdot t + \sqrt{1-b} \cdot e_y$.

5. As in the regression model, X and Y will be standardized in the population.

"Truth" (t) is a **latent** variable causing X and Y, and Equations 4.3a and 4.3b form a **structural** model. Modifications of the spreadsheet and SAS program are left as an exercise. One hint for the spreadsheet is that you will need to start with three rather than two columns of random normal deviates to represent t, e_x, and e_y.

Simulating a Sampling Distribution

Either of the preceding simulations may be expanded upon to create a sampling distribution of r. The regression model was chosen arbitrarily. The idea is to generate samples of a given size repeatedly with independent error. This is cumbersome on a spreadsheet, but very simple to do in SAS:

USING SAS TO CREATE A SAMPLING DISTRIBUTION OF r

```
Filename Dem DDE 'Excel|Sheet1!R1C1:R1000C1';
Data A;
Retain err 1996 n 50 b .8 nrep 1000;
Bres = sqrt(1 - b**2);
Do j = 1 to nrep;
Do I = 1 to N;
X = rannor(err);
Y = b*X + bres*rannor(err);
Output;
End;
End;
Keep J X Y;
Proc Sort Data=A;by J;
Proc Corr Data=A Noprint outp=B;by J;Var X Y;
Data C;
File Dem;
Set B;
If (_Type_ = "CORR") and (_Name_ = "Y") then Put X;
Run;
```

The reason for the `If (_Type_ = ···)` statement is that the data set B contains two variables, X and Y, and generates five observations per value of J. The respective values of `_TYPE_` and `_NAME_` are as follows:

```
MEAN
STD
N
CORR  X
CORR  Y
```

The values of MEAN, STD, and N are simply the mean, standard deviation, and number of observations for X and Y. These are not of present interest. The value of CORR for `_NAME_` = X is 1.0 for variable X, by definition, as is the value of CORR for `_NAME_` = Y and variable Y. Consequently, only the value of X is needed for `_NAME_` = Y (or the converse).

Figure 4.1 shows the distribution of values of r obtained from the 1000 trials. Two main points to note are that the average of these observations is nearly identical to the population parameter that generated them (.80) and that the sampling distribution is negatively skewed. The negative skew comes about because values can only extend .2 units above the

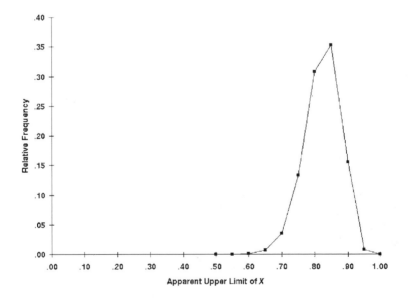

Figure 4.1. Sampling distribution of r.

parameter in this case, but can fall 1.8 units below it, albeit with increasingly small probability.

One way to eliminate the negative skew in the distribution is to convert r to Fisher's Z':

$$Z' = \ln \frac{1}{2} \cdot \frac{1+r}{1-r}.$$

This variable should be normally distributed. The computation, for example, =LN(0.5*(1 + A1)/(1 - A1)), in Excel is straightforward, and the resulting graph of these Z' values appears in Figure 4.2.

Following is an SPSS equivalent:

USING SPSS TO CREATE A SAMPLING DISTRIBUTION OF r

Open a syntax window and enter and run the following program:

```
input program.
loop #j=1 to 1000.
loop #i=1 to 50.
  compute j=#j.
  compute #b=.8.
  compute #bres=sqrt(1-#b**2).
  compute x=normal(1).
  compute y=#b*x+#bres*normal(1).
end case.
end loop.
end loop.
end file.
end input program.
execute.
split file by j.
correlations variables=x y /matrix=out(*).
select if (rowtype_ eq 'CORR') and (varname_ eq 'Y').
execute.
```

Delete variable j and obtain descriptive statistics on variable x as previously demonstrated.

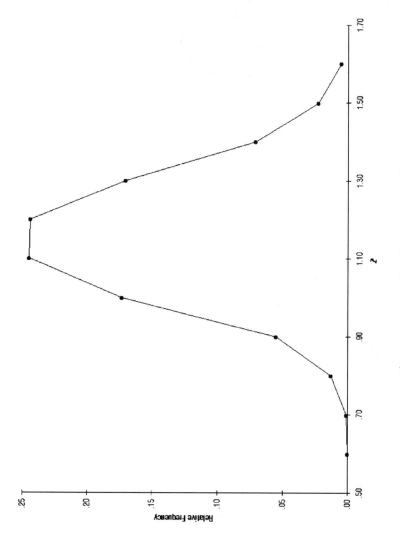

Figure 4.2. Sampling distribution of Z'.

Range Restriction

One of the basic phenomena affecting correlations is that the more restricted the range of values on either variable is, the lower the correlation. The following SAS program demonstrates this phenomenon (the Excel version is cumbersome):

USING SAS TO SIMULATE RANGE RESTRICTION

```
Data A;
Retain err 1996 n 500 b .8;
Bres = sqrt(1 - b**2);
Do I = 1 to N;
T = rannor(err);
X = b*T + bres*rannor(err);
If (X GT -2.5) then X1 = X;else X1 = .;
If (X GT -2.0) then X2 = X;else X2 = .;
If (X GT -1.5) then X3 = X;else X3 = .;
If (X GT -1.0) then X4 = X;else X4 = .;
If (X GT -0.5) then X5 = X;else X5 = .;
If (X GT -0.0) then X6 = X;else X6 = .;
If (X GT 0.5) then X7 = X;else X7 = .;
If (X GT 1.0) then X8 = X;else X8 = .;
If (X GT 1.5) then X9 = X;else X9 = .;
If (X GT 2.0) then X10 = X;else X10 = .;
If (X GT 2.5) then X11 = X;else X11 = .;
Output;
End;
Keep T X X1-X11;
Proc Corr Data=A;Var T;With X X1-X11;
Run;
```

Thus, T is the "true" component that should correlate $b = .8$ with X, whose range is not restricted. The variables X_1 to X_{11} are progressively more restricted.

The results are as follows:

Variable	r	N
X	.80	500
X_1	.79	496
X_2	.77	486
X_3	.75	466
X_4	.70	424
X_5	.64	359
X_6	.56	257
X_7	.39	160
X_8	.37	85
X_9	.29	38
X_{10}	.53	9
X_{11}	.56	3

Note the progressive decline in r with progressive restriction of range. This is a consequence of the built-in linear relationship between the two variables, as there can be both increases and decreases in r with range restriction if the relation is nonlinear. The apparent reversal for X_{10} and X_{11} is due to the instability of the small sample size.

Following is an SPSS equivalent:

USING SPSS TO SIMULATE RANGE RESTRICTION

Open a syntax window and enter and run the following program:

```
input program.
loop #i=1 to 500.
  compute #b=.8.
  compute #bres=sqrt(1-#b**2).
  compute t=normal(1).
  compute x=#b*t+#bres*normal(1).
if (x gt -2.5) x1 = x.
if (x gt -2.0) x2 = x.
if (x gt -1.5) x3 = x.
```

```
if (x gt -1.0) x4 = x.
if (x gt -0.5) x5 = x.
if (x gt -0.0) x6 = x.
if (x gt 0.5) x7 = x.
if (x gt 1.0) x8 = x.
if (x gt 1.5) x9 = x.
if (x gt 2.0) x10 = x.
if (x gt 2.5) x11 = x.
end case.
end loop.
end file.
end input program.
execute.
```

To see the correlations, choose ANALYZE, CORRELATE, BIVARIATE, (choose all variables), OK from the SPSS Data Editor menu.

Regression Toward the Mean

Another basic phenomenon is regression toward the mean—observations initially below the mean will increase and observations above the mean will decrease relative to the group mean, on average, in a retest. The following SAS program illustrates this purely statistical effect. Among other things, it accounts for part of the improvement in any therapy because patients being treated are usually, by definition, below the mean on a relevant health criterion so should improve regardless of what is done to them (assuming, as is reasonable, that the mean on the measure over all individuals in the population, which includes those who are treated and those who are not, remains the same). Again, the Excel version is cumbersome.

USING SAS TO SIMULATE REGRESSION TOWARD THE MEAN

```
Data A;
Retain err 1996 nobs 500 b .7;
Bres = sqrt(1 - b**2);
Do i = 1 to nobs;
```

```
Pre = rannor(err);
Post = b*pre + bres*rannor(err);
If (pre > 0) then Group = 1;Else Group = 0;
Output;
End;
Keep pre post group;
Proc Sort Data=A;by Group;
Proc Means Data=A;by Group;
Run;
```

Thus, subjects in group 0 fell below the mean on the initial test, and subjects in group 1 fell above the mean on the initial test ("pre," as opposed to "post"). The means are presented in Figure 4.3, where it can be seen that the mean score for group 0 increased and the mean score for group 1 decreased in the postmeasure relative to the premeasure.

Following is an SPSS equivalent:

USING SPSS TO SIMULATE REGRESSION TOWARD THE MEAN

Open a syntax window and enter and run the following program:

```
input program.
loop #i=1 to 500.
  compute #b=.7.
  compute #bres=sqrt(1-#b**2).
  compute pre=normal(1).
  compute post=#b*pre+#bres*normal(1).
  do if (pre gt 0).
    compute group=1.
  else.
    compute group=0.
  end if.
end case.
end loop.
end file.
end input program.
execute.
```

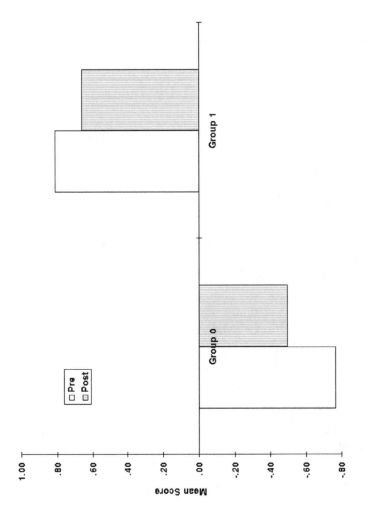

Figure 4.3. Increases in a posttest for a "group" defined by performance that is below chance on a pretest (group 0) and decreases in the same posttest for a "group" defined by performance that is above chance on a pretest (group 1) illustrating regression toward the mean.

The pre- and postmeans for the two groups may be obtained by choosing DATA, SPLIT FILE, ORGANIZE OUTPUT BY GROUPS, GROUPS BASED ON Group, SORT FILE BY GROUPING VARIABLE, OK from the SPSS Data Editor menu. Then choose ANALYZE, DESCRIPTIVE STATISTICS, DESCRIPTIVES, Variables Pre and Post, OK from this menu.

Reliability Coefficient and Reliability Index

When both X and Y of Equation 4.3 are thought of as independent, fallible estimates of a true score component, t, their correlation is known as a **reliability coefficient**, variously symbolized as r_{xx}, r_{oo}, r_{tt}, or r_{ii}. For example, one might generate two tests of 20 items each by selecting pairs of five digits and asking grammar school students to complete each of the two tests. The process of generating the two sets of items randomly from a prescribed population of items is known as **domain sampling**. Conceptually, at least, it is the process assumed in most test construction. In contrast, the correlation between X or Y and t is the correlation between a fallible estimate of a true score and the true score itself. It is known as a **reliability index** and commonly symbolized as r_{xt} or r_{ot}.

It can be shown that, under a wide variety of circumstances,

$$r_{xx} = r_{xt}^2, \qquad (4.4a)$$
$$r_{xt} = \sqrt{r_{xx}}. \qquad (4.4b)$$

To demonstrate this, the basic correlation simulation will be modified slightly, as follows:

1. Notationally, we will change X (the true component) to T.

2. Similarly, we will change Y (the fallible component) to X_1. Call its error component E_1.

3. A second fallible variable, X_2, will be added. It will be defined in a parallel manner to X_1 and have the same true score component. However, its error component, E_2, will be generated independently.

USING EXCEL TO DEMONSTRATE THE RELATIONSHIP BETWEEN THE RELIABILITY INDEX AND THE RELIABILITY COEFFICIENT

1. Enter the desired value of b in cell A1.

2. Generate three columns of random normal deviates starting in cell A2 (i.e., enter 300 numbers in cells A2:C101). These represent 100 values of T, E_1, and E_2, respectively.

3. Enter `=$A$1*$A2 + sqrt(1 - $A$1^2)*B2` in cell D2. This represents the first value of X_1.

4. Drag the equation in cell D2 to cell E2 to produce the first value of X_2.

5. Drag the equations in cells D2 and E2 through row 101 to produce the remaining values of X_1 and X_2.

6. Enter `=CORREL($A2:$A101,D2:D101)` in cell B1. This produces the correlation between T and X_1. It should fall (within sampling error) to the value entered in A1 illustrating how $r_{xt} = b$.

7. Drag the equation in cell B1 to cell C1. This represents the correlation between T and X_2. It should be similar to those in cells A1 and A2, thus illustrating how $b = r_{xt}$, as well as illustrating how $r_{xt} = b$.

8. Enter `=CORREL(D2:D101,E2:E101)` in cell D1. It should approximately equal the square of the values in A1 to C1, illustrating how $r_{xx}(r_{xt}^2) = b^2$.

The corresponding SAS program is as follows:

USING SAS TO DEMONSTRATE THE RELATIONSHIP BETWEEN THE RELIABILITY INDEX AND THE RELIABILITY COEFFICIENT

```
Data A;
Retain err 1996 n 500 b .8;
```

```
Bres = sqrt(1 - b**2);
Do I = 1 to N;
T = rannor(err);
X1 = b*T + bres*rannor(err);
X2 = b*T + bres*rannor(err);
Output;
End;
Keep T X1 X2;
Proc Corr Data=A;
Run;
```

In the real world, though, t is unknowable. However, it can be estimated through successive approximation. Define a series of numbers,
$x_1, x_2, \ldots, x_n$, as before. Now, define x'_2 as $x_1 + x_2$, x'_3 as $x_1 + x_2 + x_3$,
and so on. Correlate x_1 with x'_2 to x'_n in turn (for purposes of the simulation, a value of $n = 12$ will suffice). This generates a series of fallible
scores in which the error component averages out. Note how successive
values increase from b^2 (r_{xx}) to b (r_{xt}).

The corresponding SPSS program is as follows:

USING SPSS TO DEMONSTRATE THE RELATIONSHIP
BETWEEN THE RELIABILITY INDEX AND THE
RELIABILITY COEFFICIENT

Open a syntax window and enter and run the following code:

```
input program.
loop #i=1 to 500.
  compute #b=.8.
  compute #bres=sqrt(1-#b**2).
  compute t=normal(1).
  compute x1=#b*t+#bres*normal(1).
  compute x2=#b*t+#bres*normal(1).
end case.
end loop.
end file.
end input program.
execute.
```

The correlations among T, X_1, and X_2 may be found as previously shown.

USING EXCEL TO DEMONSTRATE THE EFFECTS OF NUMBER OF OBSERVATIONS ON RELIABILITY

▬ 1. Enter the desired population correlation in cell A1 (e.g., .7).

▬ 2. Generate 13 columns of 100 random numbers each in cells A2 to M101. Think of column A as defining true scores and columns B through M as defining unique error for a series of observations.

▬ 3. Enter =$A$1*$A2 + SQRT(1 - $A$1^2)*B2 in cell N2. This makes the first observation a linear combination of the true score in cell A2 and the error in cell B2, weighted by the population correlation in cell A1.

▬ 4. Drag N2 through column Y to make these columns linear combinations of column A (true scores) and the error in C2:M2, respectively. Consequently, these form a series of 12 parallel measures.

▬ 5. Drag N2:Y2 down to row 101 to provide the remaining observations.

▬ 6. Enter 1 in cell Y1.

▬ 7. Enter 2 in cell Z1.

▬ 8. Drag 1 and 2 jointly through column AJ. These numbers will be used later as abscissa values for a graph.

▬ 9. Enter =SUM($N2:O2) in cell Z2. This provides an observation based on the sum of two parallel measures.

▬ 10. Drag Z2 through column AJ. This provides observations based on 3, 4, ..., 12 parallel measures.

▬ 11. Drag Z2:AJ2 through row 101 to complete the data.

▬ 12. Enter =AVERAGE(A2:A101) in cell A102 to provide the mean of the true scores.

▬ 13. Enter =STDEV(A2:A101) in cell A103 to provide the standard deviation of the true scores. This should be 1.0 within sampling error.

▬ 14. Drag A102:A103 through column AJ to provide the remaining means and standard deviations. The averages of all the corresponding averages will all be close to .0 (if, however, the mean of the true scores is positive, by chance, the means in cells Y102:AJ102 will appear to increase and vice versa if the mean of the true scores is negative). The more interesting thing to observe is the increase in the standard deviation. Starting with column Y, the values will increase as a square root function of the number of measures being summed. Paraphrasing results given in Nunnally and Bernstein (1994, p. 219), this can be expressed as

$$s_k = \sqrt{k + k^2 b}, \tag{4.5}$$

where s_k is the standard deviation of a sum of k measures and b is the true score weight, that is, the correlation (cell A1).

▬ 15. Enter =CORREL($A2:$A101,B2:B101) in cell B104. This is the correlation between a true score and an error score so it should be .0 within sampling error.

▬ 16. Drag the formula through row AJ. The values in cells B104:M104 should be 0 within sampling error for the same reason. However, the values in N104:Y104, which are correlations between a true score and fallible scores, should equal the value in cell A1 for the reasons noted in the original correlation simulation at the beginning of this chapter These correlations are reliability **indexes**, symbolized as values of r_{xt} (among other symbols). The remaining values in cells Z104:AJ104 are also correlations between a true score and fallible scores and thus reliability indexes, but the fallible scores are based on sums of 2, 3, ..., 12 observations, each containing independent error, so the correlations should increase in magnitude to approach 1.0.

▬▬ 17. Enter =**CORREL($X2:$X101,Y2:Y101)** in cell Y105. These represent correlations between two fallible scores. The second of these scores is a progressively increasing sum containing independent error. Consequently, they, too, should increase in magnitude. Being correlations between fallible scores, they are called reliability **coefficients** and symbolized as r_{xx} (again, among various other symbols). However, note that they start at $\sqrt{r_{xt}}$ (within sampling error) and asymptotically reach r_{xt}.

Figure 4.4 presents values of the reliability index and reliability coefficient as a function of the number of observations on which they are based.

An analogous SAS program is as follows:

USING SAS TO DEMONSTRATE THE EFFECTS OF NUMBER
OF OBSERVATIONS ON RELIABILITY

```
Data A;
b = .7;bres = sqrt(1 - b**2);
Retain err 1998 nobs 1000;
Array xs{*} x1-x12;
Do i = 1 to nobs;
t = rannor(err);
xo = b*t + bres*rannor(err);
Do j = 1 to 12;xs{j} = 0.;end;
End;
Do j = 1 to 12;do k = 1 to 12;
If (k gt j) then goto endlp;
xs{j} = xs(j) + b*t + bres*rannor(err);
Endlp: end;
End;
Output;
End;
Keep t xo x1-x12;
Proc Corr Data=a;var t xo x1-x12;
Run;
```

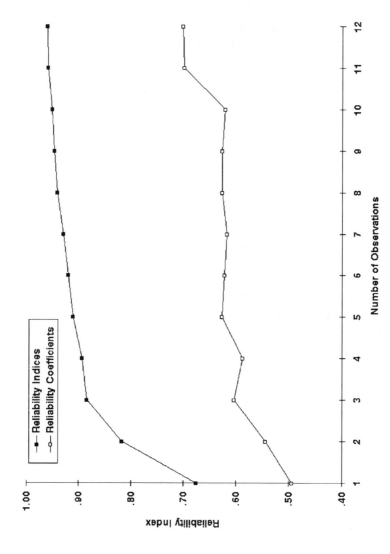

Figure 4.4. Reliability indexes and reliability coefficients as a function of the number of observations on which a fallible score is based.

These data are almost like test items, but test items are discrete (categorical) rather than continuous. We will now consider the effects of the categorization inherent in simulating item distributions starting with a simple correlation between two values.

Following is the SPSS equivalent:

USING SPSS TO DEMONSTRATE THE EFFECTS OF NUMBER OF OBSERVATIONS ON RELIABILITY

Run the following program from a syntax window:

```
input program.
vector x(12).
loop #i=1 to 1000.
  compute #b=.7.
  compute #bres=sqrt(1-#b**2).
  compute t=normal(1).
  compute x0=#b*t+#bres*normal(1).
loop #j=1 to 12.
  compute x(#j)=0.
end loop.
loop #j=1 to 12.
loop #k=1 to #j.
  compute x(#j)=x(#j)+#b*t+#bres*normal(1).
end loop.
end loop.
end case.
end loop.
end file.
end input program.
execute.
```

Obtain correlations among t and x_0 to x_{12}.

Effects of Categorization

To appreciate this process, let x be a normally distributed random variable. Now, let x_2 be a variable derived from x by dichotomization, x_3 be one derived from x by dividing it into three parts, and so on. For simplicity, assume that, in each case, the division cuts the distribution in equal parts. This means that x_2 is 0 if x is negative and 1 if x is greater than or equal to 0. Likewise, x_3 will equal 0 if x is $-.44$ or less, 1 if it falls between $-.44$ and $+.44$, and 2 if it falls at .44 or above. Finally, x_4 will equal 0 if x is less than $-.67$, 1 if it falls between $-.67$ and $.0$, 2 if it falls between .0 and $+.67$, and 3 if it falls at .67 or above.

The following is one way to accomplish this in Excel:

USING EXCEL TO DEMONSTRATE THE EFFECTS OF CATEGORIZATION UPON r

▬ 1. Place random normal deviates in cells A1:A100 using the methods discussed previously.

▬ 2. Enter `=IF(A1 < 0,0,1)` in cell B1. This will dichotomize observation A1.

▬ 3. Enter `=IF(A1 < -0.44,0,IF(A1 < 0.44,1,2))` in cell C1. This will place observation A1 into one of three categories.

▬ 4. Enter `=IF(A1 < -0.67,0,IF(A1 < 0,1,IF(A1 < 0.67,2,3)))` in cell D1. This will place observation A1 into one of four categories.

▬ 5. Drag B1:D1 down to row 100.

▬ 6. Enter `=CORREL($A1:$A100,B1:B100)` in cell A101.

▬ 7. Drag cell A101 through column D.

The chosen random number seed produced values of .796, .889, and .918, so the difference between dichotomized data and data divided into three categories was greater than the difference between data divided into three categories and data divided into four categories. However, it is useful to explore this issue when data are not divided into equal categories. This can be accomplished by changing the cutoffs in the "If" statements. Data can be placed into greater numbers of categories to simulate Likert-type and related item responses, but the "If" statements become increasingly cumbersome. This is less of a problem in SAS.

The following is an SAS equivalent:

USING SAS TO DEMONSTRATE THE EFFECTS OF CATEGORIZATION UPON *r*

```
Data A;
Retain err 1994 nobs 1000;
Do I = 1 to nobs;
X = rannor(err);
X2 = 0;If X > 0 then X2 = 1;
X3 = 0;If X > -.44 then X3 = 1;
If X > .44 then X3 = 2;
X4 = 0;If X > -.67 then x4 = 1;
If X > 0 then X4 = 2;
If X > .67 then X4 = 3;
Output;
Keep X X2-X4;
End;
Proc Corr Data=A;
Proc freq;tables X2-X4;
Run;
```

Following is the SPSS equivalent:

USING SPSS TO DEMONSTRATE THE EFFECTS OF CATEGORIZATION UPON *r*

Run the following program from a syntax window:

```
input program.
loop #i=1 to 1000.
```

```
compute x=normal(1).
compute x2=0.
if (x gt 0) x2=1.
compute x3=0.
if (x gt -.44) x3=1.
if (x gt .44) x3=2.
compute x4=0.
if (x gt -.67) x4=1.
if (x gt 0) x4=2.
if (x gt .67) x4=3.
end case.
end loop.
end file.
end input program.
execute.
```

Obtain correlations by choosing ANALYZE, CORRELATE, BIVARIATE, Variables x, x2, x3, x4, OK from the SPSS Data Editor menu. Obtain frequencies by choosing ANALYZE, DESCRIPTIVE STATISTICS, FRE-QUENCIES, Variables x2, x3, x4, OK.

Related material on categorization will be employed in Chapters 6 and 11 where simulated responses to test items will be considered.

▣ CORRELATION MATRICEƒ

It is a simple step from creating a single correlation between two variables to the creation of an entire matrix. Such matrices are basic to the under-standing of factor analysis and related models. The first demonstrations involve a matrix of correlations that are homogeneous within sampling error, which are produced by a series of measures that are all defined by a single factor, have equal loading on that factor, and have identical uni-variate distributions (i.e., are univariate parallel). Such data will consist of a series of variables, Y_i, each of which satisfies Equation 4.2. For his-toric reasons, one might label the common portion of each variable as f (for factor score), t (for true score), or g (for general intelligence), but the symbol X will be retained. Note that the weight applied to it (b) does not need subscripting because it is the same for each Y_i. The examples in this section will use $k = 6$ variables.

The methods we have chosen here are used because they are relatively simple to follow. There is another approach based on what is known as Cholesky decomposition that can also be used, but we will not present it here.

Homogeneous Matrix

The following will create a matrix of correlations that are homogeneous within sampling error:

USING EXCEL TO CREATE A HOMOGENEOUS CORRELATION MATRIX

▬ 1. Enter the desired population correlation value in cell A1.

▬ 2. Enter =SQRT(1 - A1^2) in cell B1.

▬ 3. Enter a series of random normal deviates in cells A2 to G101.

▬ 4. Leave row 102 blank.

▬ 5. Enter =$A$1*$A2 + $B$1*B2 in cell B103. This defines the score for the first observation on the first measure in accord with Equation 4.2 and related discussion.

▬ 6. Drag B103 through column G to create the remaining measure on the first observation.

▬ 7. Drag B103:G103 down through row 202 to create the remaining observations.

▬ 8. Leave row 203 blank.

▬ 9. Compute means in B204:G204 as before. These should be approximately .0.

▬ 10. Compute standard deviations in B205:G205 as before. These should be approximately 1.

▬ 11. Although the **CORREL** function can be invoked repeatedly to generate the various correlations, it is more convenient to employ the Data Analysis option under Tools (Alt -T , D).

▬ 12. Choose **Correlation** from the resulting menu.

▬ 13. Enter **$B$103:$I$202** for the input range.

▬ 14. Leave **Columns** selected under **Grouped By**.

It is not critical which Output option you choose as long as you do not overwrite lines containing data.

With $b = .7$, the population correlations are all $.49(.7^2)$. The sample correlations ranged from .38 to .56.

The following is a corresponding SAS program:

USING SAS TO CREATE A HOMOGENEOUS CORRELATION MATRIX

```
Data One;
Title "Single factor homogeneous correlations";
Retain b .7 nobs 1000 nitems 6 err 1998;
Bres = sqrt(1 - b**2);
Array Y(*) s1-s6;
Do I = 1 to nobs;
F = rannor(err);
Do j = 1 to 6;
Y(j) = b*F + bres*rannor(err);
End;
Keep s1-s6;
Output;
End;
Proc Corr Data=One;
Run;
```

Following is the SPSS equivalent:

USING SPSS TO CREATE A HOMOGENEOUS CORRELATION MATRIX

Run the following program from a syntax window and then obtain correlations as before:

```
input program.
vector y(6).
loop #i=1 to 1000.
  compute #b=.7.
  compute #bres=sqrt(1-#b**2).
  compute f=normal(1).
loop #j=1 to 6.
  compute y(#j)=#b*f+#bres*normal(1).
end loop.
end case.
end loop.
end file.
end input program.
execute.
```

Single-Common-Factor Model (Spearman's g)

It is but a simple step from a homogeneous matrix to one that fits the single-common-factor model. Such a matrix has long been known to meet the **vanishing tetrad** criterion (Spearman, 1904). Let i and j denote any two rows and k and l denote any two columns of a correlation matrix **R**. This combination of rows and columns will contain four possible correlations: r_{ik}, r_{il}, r_{jk}, and r_{jl}. The criterion is that $r_{ik} \cdot r_{jl} - r_{il} \cdot r_{jk} = 0$ for all possible tetrads. This model is more general than the previous one. The difference is that, in the former case, the weights b and $\sqrt{1-b^2}$ were the same for each variable and now they are allowed to vary. In the previous case, the common variance (variance shared with other variables) was the same; now it is free to vary, subject to the restriction that each variable still can only be influenced by a single common source. Subscripts, which previously were omitted in the formal equation, are now needed to specify which variable is being considered. Consequently, Equation 4.2 will be rewritten as

$$Y_i = b_i \cdot X_i + \sqrt{1 - b_i^2} \cdot e_i. \tag{4.6}$$

The following steps can create a matrix conforming to this model. Begin by copying the spreadsheet that created the homogeneous matrix. This can be done as follows in Excel:

USING EXCEL TO SIMULATE SPEARMAN'S g

1. Right-click on the sheet name.

2. Choose **Move or Copy**.

3. Check **Create a Copy**.

4. Take the copy and modify row 1 by inserting the desired values of b_i starting in column B, right above the first set of what served as error terms previously.

5. Leave the data in rows 2 to 102 as they were.

6. Change the entry in B103 to =B$1*$A2 + SQRT(1 - B$1^2)*B2. This provides the first score for the first variable.

7. Drag this equation through column G.

8. Drag columns B to G down to row 202.

9. Check the values of the means and standard deviations in rows 204 and 205 to make sure they are approximately .0 and 1., respectively.

10. Use the **CORREL** function as before to intercorrelate the data in $B$103:$I$202. The correlation between variables in columns i and j should approximately equal $b_i \cdot b_j$.

The corresponding SAS program is as follows:

USING SAS TO SIMULATE SPEARMAN'S *g*

```
Data One;
Retain b1-b6 (.8 .8 .8 .4 .4 .4);
Retain nobs 1000 nitems 6 err 1998;
Array Y(*) s1-s6;
Array Bs(*) b1-b6;
Array Bress(*) bres1-bres6;
Do I = 1 to nitems;bress(I) = sqrt(1 - bs(I)**2);
End;
Do I = 1 to nobs;
F = rannor(err);
Do j = 1 to nitems;
Y(J) = Bs(J)*F + Bress(J)*rannor(err);
End;
Keep s1-s6;
Output;
End;
Proc Corr Data=One;
Run;
```

Following is an SPSS equivalent:

USING SPSS TO SIMULATE SPEARMAN'S *g*

Run the following program from a syntax window and obtain correlations as before:

```
input program.
vector s(6).
vector #b(6).
vector #bres(6).
```

```
loop #i=1 to 1000.
  compute #b1=.8.
  compute #b2=.8.
  compute #b3=.8.
  compute #b4=.4.
  compute #b5=.4.
  compute #b6=.4.
  compute f=normal(1).
loop #j=1 to 6.
  compute #bres(#j)=sqrt(1-#b(#j)**2).
  compute s(#j)=#b(#j)*f+#bres(#j)*normal(1).
end loop.
end case.
end loop.
end file.
end input program.
execute.
```

Note that the correlations in these single-factor data will vary because the measures vary in reliability. Those involving S_1 to S_3 will be higher than those involving a member of this group and one from S_4 to S_6. In turn, these correlations will be higher than those involving S_4 to S_6. You can verify that these data fulfill Spearman's famous tetrad criterion—take the values in any two columns and rows other than those falling along the diagonal (e.g., columns 1 and 3 and rows 2 and 4). You should find these approximately fulfill Spearman's vanishing tetrad criterion stated previously. The reason that this is an approximation rather than an exact result is that error has been added to the measures. It would be exact in the absence of error.

Data Forming Multiple Clusters (Multifactor Data)

A more general description of data than afforded by a single-factor model is as follows:

$$Y_i = b_{Ij}X_I + b_{IIj}X_{II} + b_{IIIj}X_{III} + \cdots + \sqrt{1 - \sum_j b_k^2}\, e_i. \qquad (4.7)$$

The main difference is that each observed variable, Y_i, is a function of several latent variables, X_{ij}. Following standard practice, these are subscripted with roman numerals (F is often used rather than X to denote that these may be regarded as **factor scores**). Each factor score, X_i, has a corresponding b_i, known as a pattern weight. The term $\sqrt{1 - \sum_j b_k^2}$ standardizes the Y_i. In an **orthogonal** solution, the latent variables are uncorrelated; in an **oblique** solution, they are allowed to correlate. However, completely different sets of latent variables may be generated to produce the same observed variables. The process relating these sets of latent variables to one another is known as **rotation** from its geometric representation; the matrix of X_j, **X**, is not unique. However, certain structures lend themselves to an orthogonal solution, but either an orthogonal or an oblique solution can represent others. In particular, suppose one set of variables depends on X_1 and no other latent variables, a second set depends on X_2, and so on, and the X_i themselves are orthogonal. It can be shown that the correlations among the observed variables will form clusters in which variables dependent on the same latent variable (X_i) will correlate highly with one another but the variables dependent on different latent variables will be uncorrelated within sampling error. The collective results may be thought of as orthogonal clusters. In contrast, if observed variables depend on more than one latent variable, the resulting clusters will be oblique: If two variables each depend most highly on the same Y_i, they will correlate highly, but if they depend on different Y_i, their correlation will be lower. A basic equation in factor analysis states that the correlation between observed variables X_i and X_j, r_{ij}, is

$$r_{ij} = b_{Ii} \cdot b_{Ij} + b_{IIi} \cdot b_{IIj} + b_{IIIi} \cdot b_{IIIj} + \cdots. \tag{4.8}$$

The following spreadsheet will be capable of demonstrating both basic outcomes using six observed variables that depend on two latent variables:

USING EXCEL TO DEMONSTRATE DATA WITH TWO FACTORS

▬ 1. Enter .8 in cells C1:E1.

▬ 2. Enter .0 in cells F1:H1.

▬ 3. Enter .0 in cells C2:E2.

■ 4. Enter .8 in cells F1:H1. The net result is that observed variables Y_1–Y_3 will depend on X_1 alone and observed variables Y_4–Y_6 will depend on X_2 alone.

■ 5. Using previously discussed methods, place random normal deviates in cells A3:H102. Columns A and B contain the X variables and columns C to H contain the error terms.

■ 6. Leave row 103 blank.

■ 7. Obtain means in row 104 and standard deviations in row 105.

■ 8. Leave row 106 blank.

■ 9. Enter =$A3*C$1 + B3*C$2 + SQRT(1 - C$1^2 - C$2^2)*C3 in cell C107. This creates the first observation on Y_1.

■ 10. Drag this formula through to column H. This creates the first observations on Y_2 to Y_6.

■ 11. Drag the row through to row 206.

■ 12. Compute means and correlations as before.

The entries in the first two rows will cause the correlations among variables Y_1–Y_3 and among variables Y_4–Y_6 to equal $.8^2 = .64$. However, correlations between variables taken from the two sets of variables should be close to 0. Now, change the .0 entries in rows 1 and 2 to .5. Note that this increases the correlations both within and between the two groups.
The following is a SAS program:

USING SAS TO DEMONSTRATE DATA WITH TWO FACTORS

```
Data Two;
Retain bI1-bI6 (.8 .8 .8 .0 .0 .0);
Retain bII1-bII6 (.0 .0 .0 .8 .8 .8);
```

```
Retain nobs 1000 nitems 6 err 1998;
Array Ys(*) y1-y6;
Array BIs(*) bI1-bI6;
Array BIIs(*) bII1-bII6;
Array Bress(*) bres1-bres6;
Do I = 1 to nitems;
Bress(I) = sqrt(1 - bIs(I)**2 - bIIs(I)**2);
End;
Do I = 1 to nobs;
X1 = rannor(err);
X2 = rannor(err);
Do j = 1 to nitems;
Ys(J) = BIs(J)*X1 + BIIs(J)*X2 + Bress(J)*rannor(err);
End;
Output;
Keep y1-y6;
End;
Proc Corr Data=Two;
Run;
```

Following is an SPSS equivalent:

USING SPSS TO DEMONSTRATE DATA WITH TWO FACTORS

Run the following program from a syntax window and obtain correlations as before:

```
input program.
vector y(6).
vector #bi(6).
vector #bii(6).
vector #bres(6).
loop #i=1 to 1000.
  compute #bi1=.8.
  compute #bi2=.8.
  compute #bi3=.8.
  compute #bi4=.0.
```

```
compute #bi5=.0.
compute #bi6=.0.
compute #bii1=.0.
compute #bii2=.0.
compute #bii3=.0.
compute #bii4=.8.
compute #bii5=.8.
compute #bii6=.8.
compute #x1=normal(1).
compute #x2=normal(1).
loop #j=1 to 6.
  compute #bres(#j)=sqrt(1-#bi(#j)**2-#bii(#j)**2).
  compute y(#j)=#bi(#j)*#x1+#bii(#j)*#x2+#bres(#j)*
    normal(1).
end loop.
end case.
end loop.
end file.
end input program.
execute.
```

▣ PROFILEſ

A profile is a set of descriptive values, for example, the scores one obtains on scales of a personality test. As such, it is totally equivalent to a vector in mathematics. Multivariate texts consider such issues as describing the similarity between vectors. In this section, our concern will be limited to creating a series of profiles that have properties of interest.

Creating a Matrix of Profiles

The following spreadsheet consists of profiles, which were generated by the following model:

$$X_{ij} = \mu_{ji} + \sigma_{ij}\left(b_{ji}F_{ij} + \sqrt{1 - b_{ij}^2}e_{ij}\right), \tag{4.9}$$

where X_{ij} is the ith profile element in the jth group (there is a third sub-script denoting replication number, which we will ignore for simplicity), μ_{ij} is the population mean of the element (prototype), σ_{ij} is the population standard deviation of the element, b_j is the factor weight for that element, F_{ij} is the factor score (true component), and e_{ij} is the error component.

USING EXCEL TO CREATE A MATRIX OF PROFILES

The following Excel statement can be used to generate the data. In the example, there are 20 replications of three prototypical profiles, each of which has three elements. In the following example, the three prototype profiles are (15, 20, 25), (18, 23, 28), and (20, 25, 30). All values of s_{ij} are 3, and all values of b_j are .8. The random numbers are in standard normal form.

1. Enter the three values of μ_{i1} in cells E1:G1.

2. Enter the three values of σ_{i1} in cells E2:G2.

3. Enter the three values of b_{i1} in cells E3:G3.

4. Enter random normal deviates in cells A4:D23. The first column represents F_{i1} the remaining columns represent e_{ij}, $j = 1$–3.

5. Enter =E$1 + E$2*(E$3*$A4 + SQRT(1 - E$3^2)*B4) in cell E4 to obtain the first replication of X_{11}.

6. Drag this formula through column G to obtain the first replication of X_{21} and X_{31}.

7. Drag cells E4:G4 through row 23 to obtain the remaining replications of the profiles in group 1.

8. Repeat steps 1 to 4 to obtain the parameters and random normal deviates for group 2. The parameters will appear in cells L1:N3 and the random normal deviates will appear in cells H4:K23.

9. Enter =L$1 + L$2*(L$3*$H4 + SQRT(1 - L$3^2)*I4) in cell L1 to obtain the first replication of X_{12}.

▬ 10. Repeat steps 6 and 7 first using columns M and N, again copying through row 23 to obtain the remaining values of X_{i2}.

▬ 11. Repeat steps 1 to 4 to obtain the parameters and random normal deviates for group 3. The parameters will appear in cells S1:U3 and the random normal deviates will appear in cells O4:R23.

▬ 12. Enter =S$1 + S$2*(S$3*$O4 + SQRT(1 - S$3^2)*P4) in cell O1 to obtain the first replication of X_{13}.

▬ 13. Repeat steps 6 and 7 first using columns T and U, again copying through row 23 to obtain the remaining values of X_{i3}.

The following is a SAS equivalent:

USING SAS TO CREATE A MATRIX OF PROFILES

```
Data Clusters;
Retain err 1998;
Retain Mu11 Mu21 Mu31 Mu12 Mu22 Mu32 Mu13 Mu23 Mu33
  (15 20 25 18 23 28 20 25 30);
Retain S11 S21 S31 S12 S22 S32 S13 S23 S33
  (3 3 3 3 3 3 3 3 3);
Retain b11 b12 b13 b21 b22 b23 b31 b32 b33
  (.8 .8 .8 .8 .8 .8 .8 .8 .8);
Do I = 1 to 60;
f = rannor(err);
If I le 20 then A = Mu11 + S11*(b11*f + sqrt(1 - b11)*
  rannor(err));
If I le 20 then B = Mu21 + S21*(b21*f + sqrt(1 - b21)*
  rannor(err));
If I le 20 then C = Mu31 + S31*(b31*f + sqrt(1 - b31)*
  rannor(err));
If I le 20 then Group = 1;
If I gt 20 and I le 40 then A = Mu12 + S12*(b12*f
  + sqrt(1 - b12)*rannor(err));
If I gt 20 and I le 40 then B = Mu22 + S22*(b22*f
  + sqrt(1 - b22)*rannor(err));
```

```
If I gt 20 and I le 40 then C = Mu32 + S32*(b32*f
  + sqrt(1 - b32)*rannor(err));
If I gt 20 and I le 40 then Group = 2;
If I gt 40 then A = Mu13 + S13*(b13*f + sqrt(1 - b13)*
  rannor(err));
If I gt 40 then B = Mu23 + S23*(b23*f + sqrt(1 - b23)*
  rannor(err));
If I gt 40 then C = Mu33 + S33*(b33*f + sqrt(1 - b33)*
  rannor(err));
If I gt 40 then Group = 3;
Output;
End;
```

Following is an SPSS equivalent:

USING SPSS TO CREATE A MATRIX OF PROFILES

```
input program.
loop #i=1 to 60.
  compute mu11=15.
  compute mu21=20.
  compute mu31=25.
  compute mu12=18.
  compute mu22=23.
  compute mu32=28.
  compute mu13=20.
  compute mu23=25.
  compute mu33=30.
  compute s11=3.
  compute s21=3.
  compute s31=3.
  compute s12=3.
  compute s22=3.
  compute s32=3.
  compute s13=3.
  compute s23=3.
```

```
compute s33=3.
compute b11=.8.
compute b12=.8.
compute b13=.8.
compute b21=.8.
compute b22=.8.
compute b23=.8.
compute b31=.8.
compute b32=.8.
compute b33=.8.
compute f=normal(1).
if(#i le 20) a=mu11+s11*(b11*f+sqrt(1-b11)*normal(1)).
if(#i le 20) b=mu21+s21*(b21*f+sqrt(1-b21)*normal(1)).
if(#i le 20) c=mu31+s31*(b31*f+sqrt(1-b31)*normal(1)).
if(#i le 20) group=1.
if((#i gt 20) and (#i le 40)) a=mu12+s12*(b12*f
  +sqrt(1-b12)*normal(1)).
if((#i gt 20) and (#i le 40)) b=mu22+s22*(b22*f
  +sqrt(1-b22)*normal(1)).
if((#i gt 20) and (#i le 40)) c=mu32+s32*(b32*f
  +sqrt(1-b32)*normal(1)).
if((#i gt 20) and (#i le 40)) group=2.
if(#i gt 40) a=mu13+s13*(b13*f+sqrt(1-b13)*normal(1)).
if(#i gt 40) b=mu23+s23*(b23*f+sqrt(1-b23)*normal(1)).
if(#i gt 40) c=mu33+s33*(b33*f+sqrt(1-b33)*normal(1)).
if(#i gt 40) group=3.
end case.
end loop.
end file.
end input program.
execute.
```

▣ PROBLEMſ

4.1. Simulate correlations of .5 from both a regression model and a latent variable model using a sample size of 1000. Be sure to specify your random number seed. Now change the value of the seed and repeat the exercise.

4.2. Create sampling distributions of r using a population correlation (ρ) and a sample size of (a) .0 and 50, (b) .0 and 100, (c) .0 and 200, (d) .0 and 1000, (e) .4 and 50, (f) .4 and 100, (g) .4 and 200, (h) .4 and 1000, (i) .8 and 50, (j) .8 and 100, (k) .8 and 200, (l) .8 and 10, and (m) −.8 and 50. Plot the distributions and compute each of the standard deviations of the correlations (standard errors).

4.3. Repeat the range restriction demonstration using a population correlation (ρ) of (a) .6, (b) .4, (c) .2, and (d) 0.

4.4. Repeat the regression toward the mean demonstration using a population correlation (ρ) of (a) .6, (b) .4, (c) .2, and (d) 0.

4.5. Repeat the demonstration of the relationship between the reliability coefficient and the reliability index using a population correlation of (a) .6, (b) .4, (c) .2, and (d) 0.

4.6. Repeat the demonstration of the relationship between the reliability coefficient and the number of observations using a population correlation of (a) .6, (b) .4, (c) .2, and (d) 0.

4.7. Repeat the demonstration of the effects of categorization. This time, divide the categories unequally. Specifically, divide the dichotomy so that the probabilities are .8 and .2. Then, divide the three-categories so that the probabilities are .5, .25, and .25. Finally, divide the four categories so that the probabilities are .5, .125, .125, and .125.

4.8. Generate homogeneous correlation matrices with correlations of .4 and .9.

4.9. Repeat the simulation of Spearman's g using the following sets of population correlations: (.5, .5, .5, .5, .5, .5) and (.8, .8, .8, .4, .4, .0). Examine the resultant raw data.

4.10. Use the two-factor simulation to create correlation matrices based on the following **B** (pattern) matrices:

$$
\begin{bmatrix}
.7 & .4 \\
.7 & .4 \\
.7 & .4 \\
.4 & .7 \\
.4 & .7 \\
.4 & .7
\end{bmatrix}'
\begin{bmatrix}
.5 & .0 \\
.5 & .0 \\
.5 & .0 \\
.0 & .5 \\
.0 & .5 \\
.0 & .5
\end{bmatrix}
\quad \text{and} \quad
\begin{bmatrix}
.8 & .0 \\
.8 & .0 \\
.8 & .0 \\
.0 & .5 \\
.0 & .5 \\
.0 & .5
\end{bmatrix}.
$$

4.11. Take the program used to create profiles at the end of the chapter and modify it so that (a) the correlations are reduced to .4, (b) the separation between comparable elements in the three groups is reduced from 5 to 3, and (c) the standard deviations are made larger for the three elements of group 1 than those of group 2 or 3. What effects do these manipulations have on the resulting profiles?

5

Multivariate Correlational Simulations

This chapter further extends data simulation to the following cases:

1. Multiple regression

2. Analysis of variance (ANOVA) and analysis of covariance (ANCOVA)

3. Within, between, and total correlation matrices

4. Canonical correlation analysis, which is also used as an introduction to structural equation modeling

▣ MULTIPLE REGRESSION

In multiple regression, a criterion (Y) is predicted from two or more predictors. The prediction equation is of the form:

$$Y = \beta_1 X_1 + \beta_2 X_2 + \cdots + \beta_k X_k + e. \tag{5.1a}$$

Thus, Y is a linear combination of the predictors, $X_1, X_2, \ldots, X_i, X_k$. The **beta weights**, $\beta_1, \beta_2, \ldots, \beta_i, \ldots, \beta_k$ determine the relative contribution of each predictor, and e is error in the model. Another way to write this is to define the predicted outcome as

$$\hat{Y} = \beta_1 X_1 + \beta_2 X_2 + \cdots + \beta_k X_k. \tag{5.1b}$$

The correlation between the actual outcome Y and the predicted outcome $\hat{Y}$ then defines the **multiple correlation** R. This chapter will create a data set with three mutually orthogonal (independent) predictors.

The steps in producing data with three orthogonal predictors on a spreadsheet are as follows:

USING EXCEL TO SIMULATE MULTIPLE REGRESSION

▬ 1. Enter the population values of β_1 to β_3 in cells A1 to C3.

▬ 2. Enter =SUMSQ(A1:C1) in cell D3. This function returns the sum of squares of β_1, β_2, and β_3, $\sum b_i^2$, which equals the population value of R^2. This value must be less than or equal to 1.0.

▬ 3. Enter =SQRT(D1) in cell D4. This is the population value of R.

▬ 4. Enter random normal deviates in cells A2:D101. These respectively define the three predictors (X_1–X_3) and the model's error (e) for the first observation.

▬ 5. Enter =A$1*A2 + B$1*B2 + C$1*C2 in cell E2. This is the true component of the criterion for the first observation ($\hat{Y}$).

▬ 6. Enter =E2+SQRT(1-D$1^2)*D2 in cell F2. This is the observed criterion (Y) value for the first observation (i.e., the true component plus random error).

▬ 7. Drag cells A2:F2 down through row 101 to generate the remaining observations.

▬ 8. Compute means, standard deviations, and correlations among the variables, starting in row 102. The means among all variables should be close to 1.0. The standard deviations of the three predictors, the measurement error, and the true component of the criterion should be close to 1.0. However, the standard deviation for e should be close to $1 - R^2$. The correlations among the three predictors and the error should be close to .0. The correlations among the three predictors and $\hat{Y}$ should be the respective values of B, and the correlation between $\hat{Y}$ and Y should be R. Note that there is a substantial correlation between Y and the measurement error ($\sqrt{1 - R^2}$). This holds true for a simple reason. Some values of Y are very positive because they reflect high scores on X_1 to X_3, but others are large because of positive measurement error and vice versa for values of Y that are very negative. In other words, some people get where they are because of skill; others do so because of luck.

The corresponding SAS program is as follows:

USING SAS TO SIMULATE MULTIPLE REGRESSION

```
Data A;
Retain nobs 1000 b1-b3 (.4 .5 .6) err 3213;
Bres = sqrt(1 - b1**2 - b2**2 - b3**2);
Array Xs(*) X1-X3;
Do I = 1 to nobs;
Do J = 1 to 3;
Xs(J) = rannor(err);
End;
Yp = b1*x1 + b2*x2 + b3*x3;
Y = Yp + bres*rannor(err);
Output;
End;
Run;
```

When the data are subjected to multiple regression analysis (see Chapter 9), these uncorrelated predictors will be used in their present form, but will also be recombined to form new, correlated predictors.

The following is an SPSS equivalent:

USING SPSS TO SIMULATE MULTIPLE REGRESSION

```
input program.
vector x(3).
loop #i=1 to 1000.
  compute b1=.4.
  compute b2=.5.
```

```
compute b3=.6.
compute bres=sqrt(1-b1**2-b2**2-b3**2).
loop #j=1 to 3.
  compute x(#j)=normal(1).
end loop.
compute yp=b1*x1+b2*x2+b3*x3.
compute y=yp+bres*normal(1).
end case.
end loop.
end file.
end input program.
execute.
```

▣ ANALYSIS OF VARIANCE AND ANALYSIS OF COVARIANCE

Simple or One-Way Analysis of Variance

The **analysis of variance** (ANOVA) is a special form of multiple regression using predictors that can only assume a limited number of discrete values. Observations in the various groups may be independent of one another or related in some way; for example, they may represent scores for the same participant on each of several trials.

Simulations relevant to the ANOVA are rather easy, especially when there is but a single predictor, the **simple** or **one-way** ANOVA. Note that the predictor is not quantitative as it consists of the three categories, A, B, and C. Proc ANOVA, which performs this analysis in SAS, also converts these categories to an appropriate form and will be discussed in Chapter 9.

From a mathematical standpoint, the ANOVA assumes that effects are **additive**. In the case of the simple ANOVA, the following three effects are assumed:

1. The **universal** effect (U) is a constant that applies to every observation. It is rarely of interest, but, in exceptional circumstances, one may test that it is .0.

2. The **treatment** effect (t_i) is the same for members of a given group but may vary across groups. With a loss of generality that is almost

always minor, it is assumed that these add to 0 over groups, that is, $\sum t_i = 0$. The null hypothesis is that all of these treatment effects are .0 (i.e., $t_i = .0$ for all i). The groups are usually chosen to represent what is known as a **fixed effect** in which the analysis is concerned with specific levels of that variable and only those levels. For example, in a drug study, the investigator might employ three different drugs, including a control (placebo). The conclusions apply only to those specific drugs. Another way of looking at a fixed effect is that the investigator is interested in estimating the **means** of the various conditions. A fixed effect will be contrasted with a **random** effect later in this chapter.

3. The error effect (e_{ij}) is specific to each observation and accounts for the fact that observations vary about their own group mean. The mean error is normally assumed to be 0, but the variance can be formulated in different ways to accommodate the scaling of the data.

Equation 5.2 is perhaps the simplest way to describe the model:

$$Y_{ij} = U + t_i + e_{ij}. \tag{5.2}$$

The following spreadsheet provides simulated data conforming to the model using three groups and five observations per group. It represents data that are assumed to come from a series of independent observations.

USING EXCEL TO SIMULATE DATA
FOR THE SIMPLE ANOVA

■ 1. Enter U in cell A1.

■ 2. Enter t_1 and t_2 in cells A2 and B2.

■ 3. Because the three treatment effects must add to 0, enter =-SUM(A2:B2) in cell C2.

■ 4. Enter random normal deviates in cells A3:C7. These are the error effects.

■ 5. Enter =$A$1 + A$2 + A3 in cell D3. This defines the first observation in the first group.

6. Drag the equation in D3 through columns E and F to define the first observation in the second and third groups.

7. Drag D3:F3 down to row 7 to complete the simulated observations.

8. The population standard deviations for each group will be 1.0. The observations may be all multiplied by a constant to change this unit.

The format for the data depends on how they are to be analyzed. Some programs, such as Excel's Data Analysis for a Single Factor ANOVA, will accept the data in the form in which they presently exist, that is, as

A	B	C
4.74	4.45	4.87
5.68	5.86	3.24
4.87	5.21	3.15
5.57	5.64	5.37
4.60	4.97	4.53

On the other hand, programs such as **Proc Reg** in SAS look for the data arrayed in two columns—one denoting the predictor (group) and the other the criterion—as follows:

A	4.74
A	5.68
A	4.87
A	5.57
A	4.60
B	4.45
B	5.86
B	5.21
B	5.64
B	4.97
C	4.87
C	3.24
C	3.15
C	5.37
C	4.53

The same thing may be done using SAS, as follows:

USING SAS TO SIMULATE DATA FOR THE SIMPLE ANOVA

```
Data Simple;
Retain U 5 err 1223 t1 .3 t2 .6;
t3 = -sum(of t1-t2);
Put t3;
Array ts (*) t1-t3;
Do I = 1 to 3;
Do J = 1 to 5;
Group = I;
e = rannor(err);
Y = U + ts(I) + e;
Output;
Keep Group Y;
End;
End;
Run;
```

Following is an SPSS equivalent:

USING SPSS TO SIMULATE DATA FOR THE SIMPLE ANOVA

```
input program.
vector #t(3).
loop #i=1 to 3.
  compute #u=5.
  compute #t1=.3.
  compute #t2=.6.
  compute #t3=-sum(#t1,#t2).
loop #j=1 to 5.
  compute group=#i.
  compute #e=normal(1).
  compute y=#u+#t(#i)+#e.
```

```
end case.
end loop.
end loop.
end file.
end input program.
execute.
```

Two-Factor (Two-Way Factorial With Independent Groups) Analysis of Variance

As the name indicates, observations in the two-way ANOVA can be influenced by two separate sources. It is also possible that the effect of combining the two may be different from their ordinary sum; that is, the two may **interact**. The treatment effects for the two separate (main) effects, denoted A and B, will be denoted ta_i and tb_j and the treatment effects for their interaction will be denoted $ta_i b_j$. As in the case of the simple analysis of variance, the parameters sum to 0: $\sum ta_i = \sum tb_j = \sum ta_i b_j = 0$. The last of these constraints means that values of $ta_i b_j$ add to 0 over both A and B so that

$$a_1 b_1 + a_2 b_1 + \cdots + a_{n(a)} b_1 = a_1 b_1 + a_1 b_2 + \cdots + a_1 b_{n(b)} = 0,$$

where $n(a)$ and $n(b)$ are the respective number of levels of A and B. As before, the null hypotheses for the three effects are that the individual treatment effects for that effect are all .0; that is, all $ta_i = .0$ is the null hypothesis for the A effect. As before, U is the universal effect, and error (the disparity between an observation and its group mean) will be denoted e_{ijk}.

Equation 5.3 describes the model:

$$Y_{ijk} = U + ta_i + tb_j + ta_i b_j + e_{ijk}. \tag{5.3}$$

This form of the two-way ANOVA assumes that treatments are **crossed** in that the meaning of a given level of one variable is the same across levels of the second variable. For example, one might combine three dosage levels of one drug (none, low, and high) with three dosage levels of a second drug (again, none, low, and high). In this design, it is assumed that the actual low dosage of the first drug, for example, is the same regardless of which dosage of the second level is used. Suppose, however, that there are three main types of drugs used to treat a given disease, but, within each main type, there are several brands. Brand 1 of the first drug type has no logical relation to brand 1 of the second drug type. In this case, the brand of drug is **nested** within drug type. This book will only consider crossed designs.

The present example employs three levels of each of the two treatments so there are nine groups. There are five participants per group. Data may be simulated in Excel as follows:

USING EXCEL TO SIMULATE A TWO-WAY ANOVA

1. Enter U in cell A1 as before.

2. Enter values for ta_1 and ta_2 in cells A2 and B2.

3. Enter $=-SUM(A2:B2)$ in cell C2 to compute the value of ta_3.

4. Enter values for tb_1 and tb_2 in cells A3 and B3.

5. Enter $=-SUM(A3:B3)$ in cell C2 to compute the value of tb_3.

6. Enter values for ta_1b_1 and ta_1b_2 in cells A4:B4.

7. Enter $=-SUM(A4:B4)$ in cell C4 to provide the value of ta_1b_3.

8. Enter values for ta_2b_1 and ta_2b_2 in cells A5:B5.

9. Enter $=-SUM(A5:B5)$ in cell C5 to provide the value of ta_2b_3.

10. Enter values for ta_2b_1 and ta_2b_2 in cells A5:B5.

11. Enter $=-SUM(A5:B5)$ in cell C5 to provide the value of ta_2b_3.

12. Enter $=-SUM(A4:A5)$ in cell A6 to provide the value of ta_3b_1.

13. Drag this equation through columns B and C to provide the values of ta_3b_2 and ta_3b_3.

14. Enter random normal deviates in cells A7:I11 to provide the error terms.

▬ 15. Enter =$A$1 + $A$2 + $A$3 + $A$4 + A7 in cell A12 to create X_{111}.

▬ 16. Enter =$A$1 + $A$2 + $B$3 + $B$4 + B7 in cell B12 to create X_{121}.

▬ 17. Enter =$A$1 + $A$2 + $C$3 + $C$4 + C7 in cell C12 to create X_{131}.

▬ 18. Enter =$A$1 + $B$2 + $A$3 + $A$5 + D7 in cell D13 to create X_{211}.

▬ 19. Enter =$A$1 + $B$2 + $B$3 + $B$5 + E7 in cell E13 to create X_{221}.

▬ 20. Enter =$A$1 + $B$2 + $C$3 + $C$5 + F7 in cell F12 to create X_{231}.

▬ 21. Enter =$A$1 + $C$2 + $A$3 + $A$5 + G7 in cell G13 to create X_{311}.

▬ 22. Enter =$A$1 + $C$2 + $B$3 + $B$6 + H7 in cell H13 to create X_{321}.

▬ 23. Enter =$A$1 + $C$2 + $C$3 + $C$6 + I7 in cell I13 to create X_{331}.

▬ 24. Drag A13:I13 through 16 to create the remaining observations.

Using .20, .40, and −.60 for ta_i, .25, −.70, and .45 for tb_j, and 0 for all $ta_i b_j$, the data, rounded to one decimal point, were as follows:

A1B1	A1B2	A1B3	A2B1	A2B2	A2B3	A3B1	A3B2	A3B3
11.3	10.0	11.5	9.9	8.8	11.5	9.5	9.3	8.4
8.7	8.7	10.7	9.9	8.6	10.2	9.1	7.5	9.0
10.5	10.2	11.1	12.1	10.3	10.3	8.9	7.7	10.1
10.0	8.9	9.2	10.1	8.2	11.0	9.3	10.2	10.1
9.9	10.2	10.6	9.7	10.9	10.0	10.8	8.0	8.0

The following creates data using SAS:

USING SAS TO SIMULATE A TWO-WAY ANOVA

```
Data Twoway;
Retain U 5 err 1224 ta1 .20 ta2 .40 tb1 .25 tb2 -.70;
Retain tab11 tab12 tab21 tab22 (0. 0. 0. 0.);
Ta3 = -sum(of ta1-ta2);
Tb3 = -sum(of tb1-tb3);
Tab13 = -sum(tab11,tab12);
Tab23 = -sum(tab21,tab22);
Tab31 = -sum(tab11,tab21);
Tab32 = -sum(tab12,tab22);
Tab33 = -sum(tab13,tab23);
Array TAs(*) ta1-ta3;
Array TBs(*) tb1-tb3;
Array TABs(*) tab11 tab12 tab13 tab21 tab22 tab23
  tab31 tab32 tab33;
Do Conda = 1 to 3;
Do Condb = 1 to 3;
Do Subj = 1 to 5;
Int = 3*(Conda - 1) + Condb;
e = rannor(err);
Treata = Tas(Conda);
Treatb = Tbs(Condb);
Treatab = Tabs(Int);
Y = U + Treata + Treatb + Treatab + e;
Output;
Keep Conda Condb Int Y;
End;
End;
End;
Run;
```

Following is an SPSS equivalent:

USING SPSS TO SIMULATE A TWO-WAY ANOVA

```
input program.
vector #ta(3).
vector #tb(3).
vector #tab(9).
vector #ab(9).
loop #i=1 to 3.
loop #j=1 to 3.
loop #k=1 to 5.
  compute i=#i.
  compute j=#j.
  compute #ta1=.20.
  compute #ta2=.40.
  compute #ta3=-sum(#ta1,#ta2).
  compute #tb1=.25.
  compute #tb2=-.70.
  compute #tb3=-sum(#tb1,#tb2).
  compute #tab1=0.
  compute #tab2=0.
  compute #tab3=-sum(#tab1,#tab2).
  compute #tab4=0.
  compute #tab5=0.
  compute #tab6=-sum(#tab4,#tab5).
  compute #tab7=-sum(#tab1,#tab4).
  compute #tab8=-sum(#tab2,#tab5).
  compute #tab9=-sum(#tab3,#tab6).
  compute #ab1=11.
  compute #ab2=12.
  compute #ab3=13.
  compute #ab4=21.
  compute #ab5=22.
  compute #ab6=23.
  compute #ab7=31.
  compute #ab8=32.
  compute #ab9=33.
```

```
compute ij=3*(i-1)+j.
compute index2d=#ab(ij).
compute #treata=#ta(i).
compute #treatb=#tb(j).
compute #treatab=#tab(ij).
compute #u=5.
compute y=#u+#treata+#treatb+#treatab+normal(1).
end case.
end loop.
end loop.
end loop.
end file.
end input program.
execute.
```

To get the cell means, choose DATA, SPLIT FILE. Organize output by groups based on index2d, file is already sorted, OK. Then, choose ANALYZE, DESCRIPTIVE STATISTICS, DESCRIPTIVES, variable y, OK.

Repeated-Measure Analysis of Variance

In the previous designs, different participants were present in the different groups. We now consider the case in which the same participants are observed in all conditions, as when scores are obtained on the first, second, ..., kth trial of a learning experiment.

The role that participants play is different from the role that the treatment variable normally plays. As was noted, treatment variables are normally **fixed** effects—levels that are specifically chosen because of the investigator's interest in them—and the intent of the study is to estimate the means of these specific treatment levels. In contrast, investigators are usually only interested in participants in so far as they represent a larger population of participants from which they are sampled—they are neither interested in them as specific observations nor interested in their mean. Rather, investigators are interested in their **variance**. Participants are an example of a **random** effect. Indeed, they are the most commonly encountered such example. However, it is possible for other variables to be random effects. For example, suppose that an investigator was interested in how similar different brands of aspirin are and that the investigator had a list of all possible brands that were on the market on a given date.

The study would probably not permit all to be included in the design. One option would be to sample a small number of brands at random. Assuming random sampling, it would be appropriate for the investigator to generalize the resulting estimate of the variance among brands to the larger population. Likewise, it would be proper to generalize the variance in participants' scores to the population from which they are sampled.

Assuming that a participant is tested once, the model again consists of a universal effect (U); a fixed treatment effect of the treatment A_i, denoted ta_i as before; the random effect of participants $(S$, still in common use on historical grounds despite the fact that "participant" has replaced "subject" in some journals), denoted ts_j; and their interaction, denoted $ta_i s_j$, which denotes individual differences in the treatment. As before, $\sum ta_i = \sum ts_j = \sum ta_i s_j = .0$. The null hypothesis for A is also again that the individual $ta_i = 0$. There is usually little interest in the participant effect in laboratory experiments, but there may be if the participants are raters. In that case, one may wish to examine interrater reliabilities.

The model may be stated as:

$$Y_{ijk} = U + ta_i + ts_j + ta_i s_j + e_{ijk}. \tag{5.4}$$

The term e_{ijk} describes individual differences in response to combinations of A and S and can only be estimated directly when multiple observations are made on each participant per level of A. An important feature of this design is that there is normally a correlation between observations made on the same individual over conditions, which we will call r_w. For example, if the conditions were trials, we would normally expect participants who performed well on the first trial to perform well on subsequent trials, even though the mean over all participants might increase through learning.

An Excel spreadsheet for three treatment levels and five participants may be created as follows:

USING EXCEL TO SIMULATE A REPEATED-MEASURE ANOVA

▬ 1. Enter a value for U in cell A1.

▬ 2. Enter a value for r_w in cell B1.

▬ 3. Enter values for ta_1 to ta_2 in cells A2:B2.

4. Enter =-SUM(A2:B2) in cell C3 to compute ta_3.

5. Enter random normal deviates in cells A3:D7. The first column of data represents the true component of participants' individual differences and the remaining three columns are the unique components over conditions. This is yet another application of the principles used to generate correlations of a given magnitude.

6. Enter =$A$1 + A$2 + $B$1*$A3 + SQRT(1 - $B$1^2)*B3 in cell A8. This defines X_{11}. The first term represents U, the second represents ta_1, and the final two combine to form $ta_i s_j$.

7. Drag A8 through column C to define X_{21} to X_{31}.

8. Drag A8:C8 through row 12 to obtain the remaining observations.

9. Though not necessary, it is useful to compute the treatment means in cells A13:C13 and the subject means in cells D8:D12 to verify that their values are reasonable.

Using a value of $U = 15.0$, values of $A = -.50$, $.10$, and $-.40$, and $r_w = .80$, the observations to one decimal point were

14.2	14.5	15.2
15.6	16.9	17.2
14.2	16.6	16.9
12.6	12.2	13.5
15.3	16.0	15.7

The corresponding SAS program is as follows:

USING SAS TO SIMULATE A REPEATED-MEASURE ANOVA

```
Data Repeated;
Retain U 15. ta1-ta2 (-.5 .1) npart 5 ncond 3 err 2135
   rw .8;
Array tas (*) ta1-ta3;
```

```
Tas(ncond) = 0;
Do Cond = 1 to ncond - 1;
Tas(ncond) = tas(ncond) - tas(Cond);
End;
Do Subj = 1 to npart;
S = rannor(err);
Tast = rannor(err);
Do Cond = 1 to ncond;
TAP = rw*tast + sqrt(1 - rw**2)*rannor(err);
Y = U + tas(Cond) + S + TAP;
Output;
End;
End;
Keep Y Subj Cond U err;
Run;
```

Following is an SPSS equivalent:

USING SPSS TO SIMULATE A REPEATED-MEASURE ANOVA

```
input program.
vector #ta(3).
loop #i=1 to 5.
  compute #ncond=3.
  compute #rw=0.8.
  compute #ta1=-0.5.
  compute #ta2=0.1.
  compute #ta(#ncond)=0.
loop #k=1 to #ncond-1.
  compute #ta(#ncond)=#ta(#ncond)-#ta(#k).
end loop.
  compute #s=normal(1).
  compute #tat=normal(1).
loop #j=1 to #ncond.
  compute i=#i.
  compute j=#j.
```

```
compute u=15.
compute #tap=#rw*#tat+sqrt(1-#rw**2)*normal(1).
compute y=u+#ta(#j)+#s+#tap.
end case.
end loop.
end loop.
end file.
end input program.
execute.
```

Higher Order Design (One Between-Group Measure and One Within-Group Measure)

The ANOVA is a topic of its own as attested to by references such as Winer, Brown, and Michels (1991). One illustration of a more complex design arises when there is a single group factor and a single repeated measure. For example, one might have one or more experimental groups and a control group studied over trials. This is equivalent to what is known in agricultural statistics as a **split-plot** design. Data of this form may be examined in two ways. One is through what is basically a conventional ANOVA—this approach will be considered in Chapter 9. The alternative is through the multivariate analysis of variance (MANOVA), which will be considered in Chapter 10. For now, we will consider how the data are generated. The basic relationships are as follows, which is a slight modification of that presented in Winer et al. (1991, p. 510):

$$Y_{ijk} = U + ta_i + tb_j + ta_ib_j + t\pi_{k(i)} + tb\pi_{k(i)} + e_{m(ijk)}. \tag{5.5}$$

Participants are randomly assigned to one level of A but appear at all levels of B. In this equation, $t\pi_{k(i)}$, $tb\pi_{k(i)}$, and $e_{m(ijk)}$ all represent what is known as a nested effect. This is because a given participant appears in only one of the groups so that the first subject in group 1 is a different person from the first subject in group 2, and so on. The first term, $t\pi_{k(i)}$, describes the individual differences in A for a given group and is pooled over groups. The second term, $tb\pi_{k(i)}$, describes how these individual differences differ across combinations of A and B. Finally, $e_{m(ijk)}$ is random error that would be present if a respondent could be tested more than once at each combination. It cannot be estimated directly as this is not the case. As before, $\sum ta_i = \sum t\pi_{k(i)} = \sum tb_j = \cdots = \sum e_{m(ijk)} = 0$.

The following simulates data for a design with two groups, five levels for the repeated variable and five subjects/group:

USING EXCEL TO SIMULATE DATA FOR A HIGHER ORDER DESIGN

1. Enter a value for U in cell A1.

2. Enter a value for r_w in cell B1.

3. Enter a value for ta_1 in cell A2.

4. Enter =-A2 in cell B2 to compute ta_2.

5. Enter values for tb_1 to tb_4 in cells A3:D3.

6. Enter values for $ta_1b_1:ta_1b_4$ in cells A4:D4.

7. Enter =-SUM(A4:D4) in cell E4 to obtain ta_1b_5.

8. Enter =-A4 in cell A5 to obtain ta_2b_1.

9. Drag A5 through cell E5 to obtain $ta_2b_2:ta_2b_5$.

10. Enter random normal deviates in cells A6:A15 to obtain values of $t\pi_{k(i)}$.

11. Enter random deviates in cells A16:F25. The first column of data represents the true component of participants' individual differences and the remaining three columns are the unique components over conditions.

12. Enter =$A$1 + $A$2 + A$3 + A$4 + $A6 + $B$1*$A16 + SQRT(1 - $B$1^2)*B16 in cell A26. This defines X_{111}. The first term represents U, the second represents ta_1, and the final two combine to form ta_1s_1.

13. Drag A26 through column E to define X_{121} to X_{151}.

14. Drag A26:E26 through row 30 to obtain the remaining observations for level 1 of *A*.

15. Enter =$A$1 + $B$2 + A$3 + A$5 + $A11 + $B$1*$A21 + SQRT(1 - $B$1^2)*B21 in cell A31. This defines X_{211}.

16. Drag A31 through column E to define X_{221} to X_{251}.

17. Drag A31:E31 through row 35 to obtain the remaining observations for level 2 of *A*.

18. Again, it is useful, but not necessary to compute means for various effects as before.

For purposes of simulation, a value of 10 was entered for U, .20 for a_1 (and therefore –.20 for a_2), –2.40, –1.00, .10, and .04 for b_1 to b_4, –2.40, –1.00, .10, and .04 for a_1b_1 to a_1b_4, and all zeros for a_2b_1 to a_2b_4. This simulates a learning task in which group 1 improved over trials, but group 2 (perhaps a control group) did not.

The following is the SAS equivalent:

USING SAS TO SIMULATE DATA FOR A HIGHER ORDER DESIGN

```
Data Ho;
Retain err 12102 nbet 2 nwithin 5 nwg 5 rw .8;
Retain U 10 a1 1. b1-b4 (-.7 -.5 .1 .04 .0);
Retain a1b1-a1b5 (-.8 -.3 .1 .3 .0) a2b1-a2b5
  (.0 .0 .0 .0 .0) a2b1-a2b5;
rwres = sqrt(1 - rw**2);
Array As (*) a1-a2;
Array Bs (*) b1-b5;
Array A1Bs (*) A1B1-A1B5;
Array A2Bs (*) A2B1-A2B5;
b5 = 0;
A2 = -A1;
Do I = 1 to nwithin - 1;
```

```
b5 = b5 - Bs(I);
End;
Do I = 1 to nwithin - 1;
A2Bs(i) = -A1Bs(i);
End;
Do i = 1 to nwithin - 1;
a1b5 = a1b5 - A1Bs(i);
a2b5 = a2b5 - A2Bs(i);
End;
Do swg = 1 to nwg;
Do B = 1 to nbet;
Berr = rannor(err);
Werr = rannor(err);
Do W = 1 to nwithin;
Werri = rw*werr + rwres*rannor(err);
If B = 1 then AB = A1Bs(W);else AB = A2Bs(W);
Y = U + As(B) + Bs(W) + AB + werri + Berr;
keep B W swg Y;
Output;
End;
End;
End;
```

Following is an SPSS equivalent:

USING SPSS TO SIMULATE DATA FOR A HIGHER ORDER DESIGN

```
input program.
vector #a(2).
vector #b(5).
vector #a1b(5).
vector #a2b(5).
  compute #nwg=5.
loop #swg=1 to #nwg.
  compute #nbet=2.
  compute #nwithin=5.
```

```
  compute #rw=0.8.
  compute #rwres=sqrt(1-#rw**2).
  compute #u=10.
  compute #a1=1.
  compute #a2=-#a1.
  compute #b1=-0.7.
  compute #b2=-0.5.
  compute #b3=0.1.
  compute #b4=0.04.
  compute #b5=0.
loop #i=1 to #nwithin-1.
  compute #b5=#b5-#b(#i).
end loop.
  compute #a1b1=-0.8.
  compute #a1b2=-0.3.
  compute #a1b3=0.1.
  compute #a1b4=0.3.
  compute #a1b5=0.
  compute #a2b5=0.
loop #i=1 to #nwithin-1.
  compute #a2b(#i)=-#a1b(#i).
  compute #a1b5=#a1b5-#a1b(#i).
  compute #a2b5=#a2b5-#a2b(#i).
end loop.
loop #bt=1 to #nbet.
  compute #berr=normal(1).
  compute #werr=normal(1).
  loop w=1 to #nwithin.
    compute swg=#swg.
    compute bt=#bt.
    compute #werri=#rw*#werr+#rwres*normal(1).
    do if (#bt eq 1) .
      compute #ab=#a1b(w).
    else.
      compute #ab=#a2b(w).
    end if.
    compute y=#u+#a(#bt)+#b(w)+#ab+#werri+#berr.
end case.
end loop.
end loop.
```

```
end loop.
end file.
end input program.
execute.
```

Analysis of Covariance

The **analysis of covariance** (ANCOVA) is an extension in which additional predictor(s), usually continuous, are present and their covariance with the criterion measure is extracted first. The simplest case involves one covariate and one factor:

$$Y_{ij} = U + t_i + \beta_{y \cdot x} \cdot x_{ij} + e_j. \qquad (5.6)$$

Four terms, Y_{ij}, U, t_i, and e_j, maintain their previous meanings as criterion measure, universal effect, treatment effect, and error. The $\beta_{y \cdot x}$ is the slope relating the criterion to the covariate (x_{ij}, which is assumed to be in standardized form in this particular equation, unlike the criterion).

To create the spreadsheet in the simplest manner, make a copy of the spreadsheet previously used for the simple ANOVA and proceed as follows:

USING EXCEL TO SIMULATE DATA FOR AN ANCOVA

■ 1. Place the population correlation between criterion and covariate in cell G1.

■ 2. Enter values that represent the error in the covariate (with respect to the criterion measures) in cells H3:J7.

■ 3. Enter =$G$1*A3 + SQRT(1 - $G$1^2)*H3 in cell K3. This represents the covariate score for the first observation in the first group.

■ 4. Drag the equation in cell K3 to cells L3:M3 to create the covariate scores for the first observations in the remaining groups.

▬ 5. Drag K3:M3 down to row 7 to produce the remaining covariate
scores.

▬ 6. As before, copy these *x* scores alongside the vector of *Y* scores
to facilitate creating the data step in programs that read data in this
form.

The SAS equivalent is as follows:

USING SAS TO SIMULATE DATA FOR AN ANCOVA

```
Data Ancov;
Retain U 5 err 1223 t1 .3 t2 .6;
Retain b .8;
Bres = sqrt(1 - b**2);
t3 = -sum(of t1-t2);
Array ts(*) t1-t3;
Do I = 1 to 3;
Do J = 1 to 5;
Group = I;
e = rannor(err);
Y = U + ts(I) + e;
X = b*e + bres*rannor(err);
Keep Group Y X;
Output;
End;
End;
Run;
```

Following is an SPSS equivalent:

USING SPSS TO SIMULATE DATA FOR AN ANCOVA

```
input program.
vector #t(3).
```

```
compute #u=5.
compute #t1=0.3.
compute #t2=0.6.
compute #t3=-sum(#t1,#t2).
compute #b=0.8.
compute #bres=sqrt(1-#b**2).
loop #i=1 to 3.
  loop #j=1 to 5.
  compute group=#i.
  compute #e=normal(1).
  compute y=#u+#t(#i)+#e.
  compute x=#b*#e+#bres*normal(1).
end case.
end loop.
end loop.
end file.
end input program.
execute.
```

▣ CREATING WITHIN, BETWEEN, AND TOTAL CORRELATION MATRICES

Many circumstances involve two or more variables studied in two or more groups. For example, one might obtain verbal and quantitative scores at a university with a college of engineering and a college of liberal arts. Suppose one were to correlate the two scores, **ignoring** the college from which they were obtained. This is known as a **total** correlation, and it confounds two influences:

1. The relationship between the two measures **within** colleges. In this case, one would expect a positive relationship because measures related to intellectual functioning usually are positively related. This is the **within-group** correlation. Within-group correlations are obtained from scores expressed as deviations about their group means.

2. The relationship **between** the college means themselves. One possibility is that it might be negative—the mean quantitative score

in engineering might exceed the mean verbal score, but the reverse might be true in liberal arts. This is the **between-group** correlation. Between-group correlations are obtained by correlating means over groups.

The differences among total, between-group, and within-group correlations are important to the understanding of many apparent statistical anomalies. For example, suppose that differences in the rate of hiring members of two groups, X and Y, are at issue in employment discrimination litigation and that people are hired in one of several departments. It is quite possible that a higher overall proportion of group X members is hired than group Y members, while, at the same time, a higher proportion of group Y is hired relative to group X in each and every department! Despite the paradox, this simply implies a positive total correlation (the phi coefficient between membership in one group or the other and being hired vs. not being hired) and a negative within-group correlation or vice versa. It arises because the between-group correlation is both strong and in the opposite direction from the within-group correlation. The effect is known as Simpson's paradox (Simpson, 1951). For various demonstrations, see Bickel, Hammel, and O'Connell (1975), Hintzman (1980), Nunnally and Bernstein (1994), and Paik (1985). Bickel et al. examine a historically very important lawsuit affecting the University of California system.

Computationally, these correlations are obtained as an extension of the computations that provide the sum of squares within groups, the sum of squares between groups, and the total sum of squares in the ANOVA. The extension is to consider sums of products (terms of the form $\sum x_I x_J$) as well as the perhaps more familiar sums of squares $(\sum x_i^2)$. These sums of products and sums of squares are then divided by the appropriate df to produce corresponding covariances and variances. The df for the within terms is $\sum (n_i - 1)$, where n_i is the number of participants in group i. The df for the between terms is $n_g - 1$, where n_g is the number of groups. Finally, the df for the total terms is $N - 1$, where N is the total number of participants. Note that $N - 1 = \sum (n_i - 1) + n_g - 1$, as in the simple ANOVA. Finally, correlations are obtained by dividing covariances by the square root of the product of the respective variances. Discriminant analysis and the multivariate analysis of variance are based on the relationships among the three types of correlation matrices.

A spreadsheet can demonstrate the dynamics of the relationships among the three types of correlations as it can allow a change in parameters to change the relationships among the three types of correlations and their associated scatterplots. The particular demonstration is an extension of the one that was used to demonstrate correlations in Chapter 4. It is necessarily complex, but the concepts it illustrates are vital. Both the SAS and SPSS equivalents lack this interactive feature.

USING EXCEL TO SIMULATE BETWEEN, WITHIN, AND TOTAL CORRELATION MATRICES

1. Enter a value for ρ in cell A1. This is the common population correlation for each of the two groups.

2. Enter values in cells D1 and F1 that will determine the amount by which the X and Y values in the second group are shifted in z-score units. It is useful to make these values relatively large (e.g., 4.0) to have the mean differences be readily apparent.

3. Enter 400 random normal deviates in cells A2:D101.

4. Insert a blank column between columns B and C. Columns A and B will then contain the X and error scores for group 1. Columns D and E will contain data that form the basis of the scores for group 2 but will not be these scores themselves.

5. Enter =$A$1*A2 + SQRT(1 - $A$1^2)*B2 in cell C2. This contains the first Y value for group 1.

6. Drag C2 down to row 101 to produce the remaining Y values for group 1.

7. Copy C2 to cell F2.

8. Drag F2 to cell F101.

9. Enter =$D$1 + D2 in cell G2. This is the first X value for group 2.

10. Drag G2 down to row 101 to produce the remaining X values for group 2.

▬▬ 11. Copy G2 to I2. This is the first Y value for group 2 (leaving column H blank will facilitate later copy operations).

▬▬ 12. Drag I2 to row 101 to produce the remaining Y values.

▬▬ 13. Enter =AVERAGE(A2:A101) in cell A103. This mean should be close to 0.

▬▬ 14. Enter =STDEV(A2:A101) in cell A104. This standard deviation should be close to 1.0.

▬▬ 15. Drag the contents of cells A103 and A104 through column I. You will get an error message in column H, because there are no data to input to the two functions, so simply delete the equations in these two cells. The averages should be close to .0 for all but columns G and I, where the averages should be close to the contents of G1 and I1, respectively. The standard deviations should all be close to 1.0.

▬▬ 16. Enter =CORREL(A2:A101,C2:C101) in cells A105, D105, and G105. These will provide the correlations between X and Y in group 1 for the interim data and between X and Y in group 2. All should be similar to the entry in cell A1.

▬▬ 17. Enter =COUNT(A2:A101) in cell A106. This provides the number of observations (100 in the example). This and the next several steps provide interim results.

▬▬ 18. Enter =SUM(A2:A101) in cell A107. This provides the sums of the variables.

▬▬ 19. Enter =SUMSQ(A2:A101) in cell A108. This provides the sums of squares.

▬▬ 20. Drag the contents of cells A106:A108 through column I, again deleting the equations in column H.

▬▬ 21. Enter =SUMPRODUCT(A2:A101,C2:C101) in cell A109.

▬▬ 22. Copy A109 to G109. These equations provide uncorrected sums of squares and cross products (quantities of the form $\sum X_i^2$ and

$\sum X_I X_J)$, which are used to compute corrected sums of squares and cross products (quantities of the form $\sum x_i^2$ and $\sum x_I x_J$). The latter, in turn, are used to compute variances, covariances, and correlations, as will be noted in the succeeding steps.

■ 23. Enter =SUM(A108,G108) - SUM(A107,G107)ˆ2/ SUM(A106,G106) in cell A111.

■ 24. Copy the equation in cell A111 to cell C111.

■ 25. Enter =SUM(A109,G109) - (SUM(A107,G107)* SUM(C107,I107))/SUM(A106,C106) in cell A112. In conjunction with the previous two steps, this yields the **total corrected sums of squares and cross products**.

■ 26. Enter =A108 - A107ˆ2/A106 in cell A114.

■ 27. Copy the equation in cell A114 to cells C114, G114, and I114.

■ 28. Enter =A109 - (A107*C107)/A106 in cell A115.

■ 29. Copy the equation in cell A115 to cell G115. In conjunction with the previous three steps, this yields the **corrected sums of squares and cross products within the two groups**.

■ 30. Enter =SUM(A114,G114) in cell A117.

■ 31. Copy cell A117 to cells C117 and A118. These two steps yield the **pooled within-group corrected sums of squares and cross products**.

■ 32. Enter =A111-A117 in cell A120.

■ 33. Copy cell A120 to cells C120 and A121. These two steps yield the **between-group corrected sums of squares and cross products**.

■ 34. Enter =A111/SUM(A$106,G$106) to cell A123.

■ 35. Copy cell A123 to cells C123 and A124. These two steps yield the **total variances and covariances** (**total covariance matrix**).

▬ 36. Enter =A117/SUM((A$106 - 1),(G$106 - 1)) to cell A126.

▬ 37. Copy cell A126 to cells C126 and A127. These two steps yield the **pooled within-group variances and covariances (within-group covariance matrix)**.

▬ 38. Enter =A120 in cell A129.

▬ 39. Copy cell A129 to cells C129 and A130. These two steps yield the **between-group variances and covariances (between-group covariance matrix)**.

▬ 40. Enter =A124/SQRT(A123*C123) in cell 132. This yields the **total correlation** (it forms a matrix with more than two predictors).

▬ 41. Enter =A127/SQRT(A126*C126) in cell 135. This yields the **pooled within-group correlation** (it likewise forms a matrix with more than two predictors) and should fall at intermediate values relative to the two individual within-group correlation matrices presented in cells A105 and G105. It should also be similar to the population value entered in cell A1.

▬ 42. Enter =A130/SQRT(A129*C129) in cell 138. This yields the **between-group correlation**. With only two groups, it will take on the value of +1 if the entries in columns D1 and F1 are of the same sign and −1 if they are of different sign.

▬ 43. Plot the observations in cells A2:A101 against the observations in cells C2:C101. First, choose the *X–Y* plot option and then choose option 1 if the line connecting the points is absent (scatterplot). When you have generated the graph, modify the markers to black solid circles.

▬ 44. Use Insert, New Data ($\boxed{\text{Alt}}$-I, N) to enter the observations from cells G2:G101 versus I2:I101. Excel may enter these as two separate sets of data, with A2:A101 as the corresponding *X* values. If this happens, delete one of the two sets. Then, click on one of the points from the remaining set. The equation =SERIES(,'Between, Within, Total'!$A$2:$A$101, 'Between, Within, Total'!$G$2:$G$101,2) or =SERIES(,'Between,

Within, Total'!$A$2:$A$101, 'Between, Within, Total'!$I$2:$I$101,2) will then appear (depending on which set of points you deleted). Edit this to read =SERIES(,'Between, Within, Total'!$G$2:$G$101, 'Between, Within, Total'!$I$2:$I$101,2). Format the markers for the points as an open circle.

▬ 45. Explore positive and negative values of various magnitudes in cells A1, D1, and F1 and observe how the scatterplots change.

For example, Figure 5.1 shows the results of using the previous parameters (within-group correlation=.5, separation of +4 on X, and separation of –4 on Y). Note that the correlation within each of the two groups is clearly positive, but the two groups fall at the top left and bottom right so that a line connecting the joint X and Y means of the groups (their bivariate centroids) would have negative slope. Conversely, Figure 5.2 shows the result of setting the within-group correlation to .0 and using separations of +4 on both X and Y. The observations within each of the two groups become uncorrelated and a line connecting the two bivariate centroids of the two groups would have positive slope.

The following program creates analogous data in SAS save for the scatterplot:

USING SAS TO SIMULATE BETWEEN, WITHIN, AND TOTAL CORRELATION MATRICES

```
Data A;
Retain nobs 1000 b .5 dx .5 dy -.5 err 13334;
bres = sqrt(1 - b**2);
Do I = 1 to 1000;
If (I gt nobs/2) then group = 2;else group = 1;
X = rannor(err);
Y = b*X + bres*rannor(err);
If group = 2 then X = X + dx;
If group = 2 then Y = Y + dy;
Output;
```

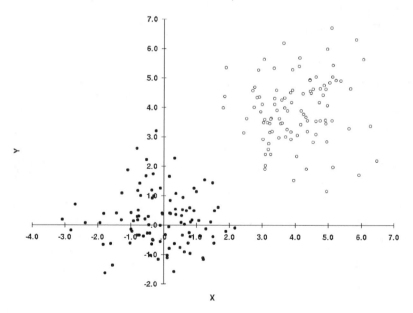

Figure 5.1. Scatterplot representing two groups containing a within-group correlation = .5, separation of +4 on X, and separation of −4 on Y.

Figure 5.2. Scatterplot representing two groups containing a within-group correlation = .0 and separation of +4 on both X and Y.

```
End;
Proc Means Data=A;var x y;by group;
Proc Candisc Data=A pcorr wcorr bcorr tcorr;
Class group;
Var x y;
Run;
```

The quantity b determines the common within-group correlation, and the values of dx and dy determine the between-group correlation by shifting the group 2 centroid by these amounts.

Following is an SPSS equivalent:

USING SPSS TO SIMULATE BETWEEN, WITHIN, AND TOTAL CORRELATION MATRICES

```
input program.
compute #nobs=1000.
loop #i=1 to 1000.
  compute b=0.5.
  compute dx=0.5.
  compute dy=-0.5.
  compute bres=sqrt(1-b**2).
  compute group=1.
  if(#i gt #nobs/2) group=2.
  compute x=normal(1).
  compute y=b*x+bres*normal(1).
  if(group eq 2) x=x+dx.
 if(group eq 2) y=y+dy.
end case.
end loop.
end file.
end input program.
execute.
```

To get the total correlations between x and y, from the SPSS Data Editor menu choose ANALYZE, CORRELATE, BIVARIATE, variables x y,

OK. To get the pooled within-group correlation matrix, from the SPSS Data Editor menu choose ANALYZE, CLASSIFY, DISCRIMINANT, grouping variable group, define range, min 1 max 2, CONTINUE, independents x y, ANALYZE, within-groups correlation, CONTINUE, OK. To get the separate within-group correlations, from the SPSS Data Editor menu choose DATA, SPLIT FILES, organize output by groups, groups based on group, OK. Then choose ANALYZE, CORRELATE, BIVARIATE, variables x y, OK. To get the between-group correlations, first get the means of each variable for each group by choosing ANALYZE, DESCRIPTIVE STATISTICS, SUMMARIZE, DESCRIPTIVES, variables x y, OK. Next, enter the means in a new data window. Choose FILE, NEW, DATA. You will now be asked if you want to save the current data matrix. If you do, answer "Yes" and supply the necessary file information. If not, answer "No." Enter the group means for variable x in a column and the corresponding group means for y in another column. Get correlations for these variables as before. These will be the between-group correlations.

Concentrated Versus Diffuse Structures

Discriminant analysis is concerned with finding linear combinations of variables that maximally discriminate among groups. These optimally discriminating linear combinations are the eigenvectors of $\mathbf{BW}^{-1}$, where $\mathbf{B}$ is the between-group covariance matrix and $\mathbf{W}^{-1}$ is the inverse of the within-group covariance matrix. If there are k groups and p variables, there will be anywhere from 0 to the minimum of $k - 1$ and p nonzero eigenvalues in the population. If there is 0, the multivariate group means on the various measures (group centroids) are identical. If there is 1, the group differences are said to be **concentrated** along a single dimension. If there are several, the group differences are said to be **diffuse**. Because of sampling error, the concept of 0 is replaced with "small" in referring to sample data, and appropriate test statistics exist to define "small." Put in less technical language, if the reason that group A differs from group B is the same as the reason that group B differs from group C, then the group differences are concentrated along a common dimension. Conversely, if the reasons for the two group differences are different, then the structure is diffuse.

Another way to look at this is to define group membership in terms of $k - 1$ predictors. There are an infinite number of ways of doing this, but one simple procedure is **dummy coding**. If there are $k = 3$ groups, there will be two dummy codes. A participant who belongs to the first group is assigned scores of 1 and 0 on the resulting two dummy codes. Likewise, a participant who belongs to the second group is assigned scores of 0 and 1, and a participant who belongs to the third group is assigned scores of 0 and 0. With additional groups, the rule is to assign a score of 1 to the group to which the participant belongs unless it is the last (kth) group and a 0 to all other groups. Participants who belong to the last group are assigned scores of 0 on all groups. With this coding, it is possible to express any comparison between groups as a linear combination of these dummy codes. Call the scores X_1 and X_2 for the three-group situation. Then, $X_1 - X_2$ denotes the difference between the first two groups, and X_1 denotes the difference between the first and third group. The problem of discriminant analysis then becomes that of relating optimum linear combinations of the dummy codes to optimum linear combinations of the criterion variables. In a concentrated structure, only one such pair of optimal linear combinations exists; in a diffuse structure, more than one such combination exists.

One way to create a concentrated structure is to have the criteria reflect a single factor using the simulation procedure of the previous chapter. Groups are then made to vary along the latent dimension. Note that a structure can be concentrated although the criteria are multidimensional as long as these additional dimensions are unrelated to group differences. Conversely, the structure may not be concentrated although the criteria vary along a single dimension because the groups may not differ on this latent variable. However, we will create the structure using unidimensional latent variables with three groups and three variables, as follows:

USING EXCEL TO SIMULATE A CONCENTRATED STRUCTURE

▬ 1. Enter the population group means in cells A1:C1. Values of .0, .5, and 1.0 were used in this example.

▬ 2. Enter the loadings (pattern elements, b) for the within-group correlations in cells A2:C2. This example assumes that the structure is the same for each group, but allows the individual criteria to have different weights so values of .8, .7, and .85 were used.

3. Enter =SQRT(1 - A2^2) in cell A3 as in previous situations involving the generation of weights applied to error terms for standardized variables having a desired correlation magnitude.

4. Drag the preceding equation to cells B3 and C3 to obtain the other two error term weights.

5. Generate standard normal deviates in cells A5:L104. These can be thought of as forming three groups of four variables each. The first variable in each group is the factor score, and the remaining three are the error observations.

6. Enter =A$2*($A5 + $A$1) + A$3*B5 in cell B106. This computes X_1 for the first observation in the first group.

7. Drag this equation across the next two columns to obtain X_2 and X_3 for the first observation in the first group.

8. Enter =A$2*($F5 + $B$1) + A$3*F5 in cell F106. This computes X_1 for the first observation in the second group.

9. Drag this equation across the next two columns to obtain X_2 and X_3 for the first observation in the second group.

10. Enter =A$2*($I5 + $C$1) + A$3*J5 in cell J106. This computes X_1 for the first observation in the third group.

11. Drag this equation across the next two columns to obtain X_2 and X_3 for the first observation in the third group.

12. Drag row 106 through row 205 to create a total of 100 observations/group.

13. Enter the formulas for the means of each of the nine criterion variables in row 207, leaving row 206 blank. The means should equal the product of the group means and the corresponding loadings (row 1 times row 2).

▬ 14. Enter the formulas for the standard deviations of each of the nine criterion variables in row 208. These should all approximately equal 1.0.

▬ 15. Enter =CORREL($B106:$B205,C106:C205) in cell B209. This provides the correlation between X_1 and X_2 for group 1.

▬ 16. Copy this equation to cell B210 to obtain the correlation between X_1 and X_3 in group 1.

▬ 17. Enter =CORREL(C106:C205,D106:D205) in cell B211 to obtain the correlation between X_2 and X_3 in group 1.

▬ 18. Copy B209:C210 to F209:G210 to obtain the correlations within group 2.

▬ 19. Copy B209:C210 to J209:K210 to obtain the correlations within group 3.

The SAS counterpart is as follows:

USING SAS TO SIMULATE A CONCENTRATED STRUCTURE

```
Data Conc;
Retain err 1998 b1 .8 b2 .7 b3 .85 nobs 100 d1 .5 d2 1.;
Array bs (*) b1-b3;
Array bress (*) bres1-bres3;
Do i = 1 to 3;
Bress(i) = sqrt(1 - bs(i)**2);
End;
Do I = 1 to nobs;
Grp = mod(i,3) + 1;
F = rannor(err);
If (Grp = 2) then F = F + d1;
If (Grp = 3) then F = F + d2;
X1 = b1*F + bres1*rannor(err);
X2 = b2*F + bres2*rannor(err);
```

```
X3 = b3*F + bres3*rannor(err);
Output;
Keep x1-x3 grp;
End;
Proc Sort Data=Conc;by grp;
Proc Corr Data=Conc;by grp;
Run;
```

The function `mod(i,3)` returns the remainder after dividing *i* by 3; for example, it takes on the value of 0 if *i* is exactly divisible by 3. The final two statements are a check that the means and correlations are as they should be.

Following is an SPSS equivalent:

USING SPSS TO SIMULATE A CONCENTRATED STRUCTURE

```
input program.
vector #b(3).
vector #bres(3).
loop #i=1 to 100.
  compute #b1=.8.
  compute #b2=.7.
  compute #b3=.85.
  compute #d1=.5.
  compute #d2=1.
  loop #j=1 to 3.
    compute #bres(#j)=sqrt(1-#b(#j)**2).
  end loop.
  compute grp=mod(#i,3)+1.
  compute #f=normal(1).
  if (grp eq 2) #f=#f+#d1.
  if (grp eq 3) #f=#f+#d2.
  compute x1=#b1*#f+#bres1*normal(1).
  compute x2=#b2*#f+#bres2*normal(1).
  compute x3=#b3*#f+#bres3*normal(1).
end case.
```

```
end loop.
end file.
end input program.
execute.
```

To get the correlations by groups, you must first "split" the data. To do this, from the SPSS Data Editor menu choose DATA, SPLIT FILE, organize output by groups, groups based on grp, sort the file by grouping variable, OK. Get correlations among x_1, x_2, and x_3 as before.

To obtain a diffuse structure, make the criterion variables have at least two factors within groups and associate different group differences with different factors. Note that if the within-group correlations are unifactor, you cannot have a diffuse structure. In the present case, the three criterion measures will be mutually orthogonal in the population. The groups will not differ on X_1, group 2 will have a high mean score on X_2 relative to groups 1 and 3, and group 3 will have a high mean score on X_3 relative to groups 1 and 2.

USING EXCEL TO SIMULATE A DIFFUSE STRUCTURE

1. Enter the population means in cells A1:I1. To obtain the mean differences indicated previously, enter a 0 in all columns save E and I, and enter .75 in those two cases.

2. Enter the pattern elements for the three criterion variables in cells A2:C2 as in the previous case.

3. Enter =SQRT(1 - A2^2) in cell A3 as before to generate the error term weight for X_1.

4. Drag the preceding equation to cells B3 and C3 to obtain the other two error term weights.

5. Generate standard normal deviates in cells A5:R104. These can be thought of as forming nine pairs of variables. The first variable in each pair is the factor score, and the second is the error observation.

The nine pairs represent what will ultimately be X_1 to X_3 for groups 1, 2, and 3, respectively.

6. Enter =$A$2*(A5 + $A$1) + $A$3*B5 in cell B105 to generate the first observation for X_1 in group 1.

7. Enter =$B$2*(C5 + $B$1) + $B$3*D5 in cell D105 to generate the first observation for X_2 in group 1.

8. Enter =$C$2*(E5 + $C$1) + $C$3*F5 in cell F105 to generate the first observation for X_3 in group 1.

9. Enter =$A$2*(G5 + $D$1) + $A$3*H5 in cell H105 to generate the first observation for X_1 in group 2.

10. Enter =$B$2*(I5 + $E$1) + $B$3*J5 in cell J105 to generate the first observation for X_2 in group 2.

11. Enter =$C$2*(K5 + $F$1) + $C$3*L5 in cell L105 to generate the first observation for X_3 in group 2.

12. Enter =$A$2*(M5 + $G$1) + $A$3*N5 in cell N105 to generate the first observation for X_1 in group 3.

13. Enter =$B$2*(O5 + $H$1) + $B$3*P5 in cell P105 to generate the first observation for X_2 in group 3.

14. Enter =$C$2*(Q5 + $I$1) + $C$3*R5 in cell R105 to generate the first observation for X_3 in group 3.

15. Drag row 105 through row 204 to create the remaining observations.

16. Enter formulas for the mean, standard deviation, and correlation as before. The means should be .0 within sampling error, the standard deviations should be 1.0, and the correlations should be .0.

The corresponding SAS program is as follows:

USING SAS TO SIMULATE A DIFFUSE STRUCTURE

```
Data Diff;
Retain err 1998 b1 .9 b2 .8 b3 .7 nobs 100;
Array bs (*) b1-b3;
Array bress (*) bres1-bres3;
Array xs (*) x1-x3;
Do i = 1 to 3;
Bress(i) = sqrt(1 - bs(i)**2);
End;
Do I = 1 to nobs;
Grp = mod(i,3) + 1;
F = rannor(err);
Do j = 1 to 3;
xs(j) = bs(j)*F + bress(j)*rannor(err);
End;
If (Grp = 2) then X2 = X2 + .85;
If (Grp = 3) then X3 = X3 + .75;
Output;
Keep x1-x3 grp;
End;
Proc Sort Data=Diff;by grp;
Proc Corr Data=Diff;by grp;
Run;
```

Following is an SPSS equivalent:

USING SPSS TO SIMULATE A DIFFUSE STRUCTURE

```
input program.
vector #b(3).
vector #bres(3).
vector x(3).
loop #i=1 to 100.
```

```
compute #b1=.9.
compute #b2=.8.
compute #b3=.7.
loop #j=1 to 3.
  compute #bres(#j)=sqrt(1-#b(#j)**2).
end loop.
compute grp=mod(#i,3)+1.
compute #f=normal(1).
loop #j=1 to 3.
  compute x(#j)=#b(#j)*#f+#bres(#j)*normal(1).
end loop.
if (grp eq 2) x2=x2+.85.
if (grp eq 3) x3=x3+.75.
end case.
end loop.
end file.
end input program.
xecute.
```

Split the file and proceed as in the concentrated structure. After you have gotten the within-group correlations, do the following to run a discriminant analysis. From the SPSS Data Editor menu, choose DATA, SPLIT FILE, highlight grp in groups based on box and remove it from the box, pick analyze all cases, do not create groups, OK. Choose ANALYZE, CLASSIFY, DISCRIMINANT, grouping variable grp, define range min 1 max 3, independents x1 x2 x3, CLASSIFY, use covariance matrix within-groups, CONTINUE, STATISTICS, means, within-group correlations, CONTINUE, OK.

▣ CANONICAL CORRELATION

Canonical correlation involves two sets of variables, which we will designate the "*X* set" and the "*Y* set" (SAS refers to these as "VAR" and "WITH" variables). It is concerned with constructing pairs of linear combinations, one from each set, that maximally explain the available variance.

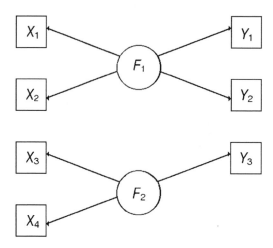

Figure 5.3. Path diagram (in nonstandard form to emphasize the relationship of canonical analysis to classical multivariate statistics). This depicts the relationships among seven observables (X_1–X_4 and Y_1–Y_3) and two latent variables (F_1 and F_2). Note that X_1, X_2, Y_1, and Y_2 relate to F_1 and not to F_2, whereas X_3, X_4, and Y_3 relate to F_2 and not to F_1.

The X set will have four members, X_1, X_2, X_3, and X_4, and the Y set will have three members, Y_1, Y_2, and Y_3. All seven variables will be functions of two latent variables, F_1 and F_2. As a result, there will be pairs of two linear combinations in the population that explain true variance. The first pair will be denoted I_{x1} and I_{y1}, and the second pair will be denoted I_{x2} and I_{y2}. The I_{x1} and I_{x2} will each be linear combinations of the X set, and the I_{y1} and I_{y2} will each be linear combinations of the Y set. Moreover, I_{x1} and I_{x2} will be uncorrelated, as will I_{y1} and I_{y2}. For purposes of the example, we have defined the parameters such that X_1, X_2, Y_1, and Y_2 relate to F_1 and not to F_2, whereas X_3, X_4, and Y_3 relate to F_2 and not to F_1. This is depicted in Figure 5.3. This is a form of **structural model**, though the notation is somewhat different to emphasize links to classical multivariate analysis problems. Moreover, the model is not terribly interesting. There are only two latent variables and these are unrelated to one another. This is unlike more general models containing a potentially rich network of interrelationships. At the same time, our simple model will make visualization of the canonical analysis easier.

All structural models are constructed to examine the correlations, covariances, corrected sums of products (terms of the form $\sum xy$ and $\sum x^2$), or raw sums of products (terms of the form $\sum XY$ and $\sum X^2$). Unlike classical canonical correlation analysis, which is exploratory, structural modeling is highly confirmatory. As is consistent with the plan of this book, we will concentrate on constructing the data here but delay its analysis to a later point (Chapter 9). A more interesting structural model will be introduced at that point.

USING EXCEL TO SIMULATE CANONICAL CORRELATION DATA

1. Enter the population values of β_{x11} to β_{x13} in cells C1:F1.

2. Enter the population values of β_{y11} to β_{y13} in cells G1:I1.

3. Enter the population values of β_{x21} to β_{x24} in cells C2:F2.

4. Enter the population values of β_{y21} to β_{y23} in cells G1:I1.

5. Enter =SQRT(1 - C1^2 - C2^2) in cell C3. This is the unique error weight for X_1.

6. Drag the contents of cell C3 through column I to produce the remaining unique error weights.

7. Enter random normal deviates in cells A4:I103. Columns A and B are the true score (F_1 and F_2) measures, columns C through F are the unique errors for the X observations, and columns G through I are the unique errors for the Y observations.

8. Enter =C$1*$A4 + C$2*$B4 + C$3*C4 in cell J4 to produce the first X_1 observation.

9. Drag J4 through cell P4 to produce the first observations for the remaining variables.

10. Drag J4:P4 through row 103 to produce the remaining observations.

11. Generate means in row 104. These should be approximately .0.

12. Generate standard deviations in row 105. These should be approximately 1.

▬ 13. Enter =CORREL($A4:$A103,A4:A103) in cell A106. This produces the correlation between the F_1 measures and themselves, which, of course, will be 1.

▬ 14. Drag A106 through column P. The values in cells B106:I106 will be close to 0 as they are correlations between pairs of sets of random normal deviates. The remaining values should be similar to cells C1:I1.

▬ 15. Enter =CORREL($B4:$B103,A4:A103) in cell A107. This produces the correlation between F_1 and F_2, which should be close to 0.

▬ 16. Drag A106 through column P. The value in cell B106 should be 1 because it represents the correlation between the F_2 measures and themselves. The values in cells B106:I106 will be close to 0 as they are correlations between pairs of sets of random normal deviates. The remaining values should be similar to cells C2:I1.

▬ 17. It will be useful for you to compute the multiple correlations between X_1–X_4 and F_1–F_2 and between Y_1–Y_4 and F_1–F_2. These data should also include the zero-order correlations between the observables and the two latent variables (which should be similar to the weights specified in rows 1 and 2). The importance of these regression analyses will be made explicit in Chapter 9.

The corresponding SAS program is as follows:

USING SAS TO SIMULATE CANONICAL CORRELATION DATA

```
Data Canon;
Retain err 1999 nobs 1000 bx11-bx14 (.4 .3 .0 .0)
  bx21-bx24 (.0 .0 .5 .7) by11-by13 (.5 .7 .0)
  by21-by23 (.0 .0 .8);
bx1r = sqrt(1 - bx11**2 - bx21**2);
bx2r = sqrt(1 - bx12**2 - bx22**2);
bx3r = sqrt(1 - bx13**2 - bx23**2);
```

```
bx4r = sqrt(1 - bx14**2 - bx24**2);
by1r = sqrt(1 - by11**2 - by21**2);
by2r = sqrt(1 - by12**2 - by22**2);
by3r = sqrt(1 - by13**2 - by23**2);
Do I = 1 to nobs;
F1 = rannor(err);
F2 = rannor(err);
X1 = bx11*f1 + bx21*f2 + bx1r*rannor(err);
X2 = bx12*f1 + bx22*f2 + bx2r*rannor(err);
X3 = bx13*f1 + bx23*f2 + bx3r*rannor(err);
X4 = bx14*f1 + bx24*f2 + bx4r*rannor(err);
Y1 = by11*f1 + by21*f2 + by1r*rannor(err);
Y2 = by12*f1 + by22*f2 + by2r*rannor(err);
Y3 = by13*f1 + by23*f2 + by3r*rannor(err);
Output;
End;
Keep f1 f2 x1-x4 y1-y3;
Proc Reg Data=Canon corr;
Model F1-F2=X1-X4;
Model F1-F2=Y1-Y3;
Model Y1-Y3=X1-X4;
Model X1-X4=Y1-Y3;
Run;
```

Following is an SPSS equivalent:

USING SPSS TO SIMULATE CANONICAL CORRELATION DATA

```
input program.
compute #nobs=1000.
compute #bx11=.4.
compute #bx12=.3.
compute #bx13=.0.
compute #bx14=.0.
compute #bx21=.0.
compute #bx22=.0.
```

```
compute #bx23=.5.
compute #bx24=.7.
compute #by11=.5.
compute #by12=.7.
compute #by13=.0.
compute #by21=.0.
compute #by22=.0.
compute #by23=.8.
compute #bx1r=sqrt(1-#bx11**2-#bx21**2).
compute #bx2r=sqrt(1-#bx12**2-#bx22**2).
compute #bx3r=sqrt(1-#bx13**2-#bx23**2).
compute #bx4r=sqrt(1-#bx14**2-#bx24**2).
compute #by1r=sqrt(1-#by11**2-#by21**2).
compute #by2r=sqrt(1-#by12**2-#by22**2).
compute #by3r=sqrt(1-#by13**2-#by23**2).
loop #i=1 to #nobs.
  compute f1=normal(1).
  compute f2=normal(1).
  compute x1=#bx11*f1+#bx21*f2+#bx1r*normal(1).
  compute x2=#bx12*f1+#bx22*f2+#bx2r*normal(1).
  compute x3=#bx13*f1+#bx23*f2+#bx3r*normal(1).
  compute x4=#bx14*f1+#bx24*f2+#bx4r*normal(1).
  compute y1=#by11*f1+#by21*f2+#by1r*normal(1).
  compute y2=#by12*f1+#by22*f2+#by2r*normal(1).
  compute y3=#by13*f1+#by23*f2+#by3r*normal(1).
end case.
end loop.
end file.
end input program.
execute.
```

After you run the program from the SPSS syntax window, to get correlations for all variables, from the SPSS Data Editor menu choose ANALYZE, CORRELATE, BIVARIATE. To do regressions, choose ANALYZE, REGRESSION, LINEAR.

It is important that you see the relationships between canonical correlation and discriminant analysis as well as with structural equation modeling. Consider the fact that you can represent k groups as a series of $k-1$

codes (e.g., dummy codes). This means that each observation will consist of these codes and the predictors of group membership, which are simply predictors of the codes. These constitute the two sets of variables used in discriminant analysis.

▣ PROBLEM*ſ*

5.1.   Using the appropriate program to generate multiple regression data, intercorrelate the predictors and the criterion (do not do a multiple regression analysis yet, as this will be discussed in Chapter 9).

5.2.   Using the appropriate program to perform a simple ANOVA, obtain group means (do not do an ANOVA yet, as this will be discussed in Chapter 9). Then increase the difference among the population means by .5 z-score units and repeat the exercise.

5.3.   Repeat this (including the change in population means) with the data for the two-way ANOVA.

5.4.   Repeat this (including the change in population means) with the data for the repeated-measure ANOVA.

5.5.   Repeat this (including the change in population means) with the data for higher order design.

5.6.   Repeat this (including the change in population means) with the data for the ANCOVA.

5.7.   Simulate between, within, and total correlation matrices using a within-group correlation of $-.7$ and group mean differences of $+.5$ on both variables. Generate the various correlation matrices and a scatterplot.

5.8.   Repeat this for a within-group correlation of 0, using the same group mean differences.

5.10. Generate concentrated and diffuse structures with appropriate parameters that are different from those used in the text.

6

Simulations With Categorical Data

CHAPTER OBJECTIVES

This chapter considers the following simulations of categorical data:

1. Logistic functions

2. Psychophysical and psychometric applications

3. Thurstone's "law" of comparative judgment, signal detection theory, and other categorical data

▣ LOGISTIC FUNCTIONS

A logistic function relates a continuous independent variable, X, to the probability of a designated response, p, as follows:

$$p = \frac{e^{da(X-b)}}{1 + e^{da(X-b)}}. \tag{6.1}$$

In this equation, e is the constant $2.7182818\ldots$ and d is a scaling constant that is usually set either at 1.0 or at 1.7. When set at 1.7, the resulting function will closely match the cumulative normal distribution. The parameters a and b respectively define the function's **slope** and **location**. The slope describes how **discriminating** the stimulus (physical event or item) is, in the sense of how rapidly it rises from near 0 to near 1 (0 and 1 are asymptotic values, which the function never reaches as long as X is finite). The location determines the **threshold**, or value of X for which p is .5, and is also known as the **difficulty** in psychometric theory. See Hambleton and Swaminathan (1985) and Hambleton, Swaminathan, and Rogers (1991) for further discussion.

The logistic function is sigmoid (S-shaped or ogival), a commonly encountered mathematical form. Sigmoidal shapes used to be modeled with the cumulative normal distribution. This was referred to as the "phi-gamma hypothesis" in some contexts and as "probit analysis" in others. However, logistic functions are numerically similar but mathematically easier to work with (see Chapter 11). Consequently, they have supplanted the cumulative normal distribution in most formal theories, particularly in psychometric applications. When data fit a logistic function, $\text{logit}(p) = \ln[p/(1-p)]$ and X are linearly related, ln is the natural-log function using the base of $e = 2.71828+$. Logits are also known as **log-odds ratios** (LOR) because they involve taking the natural logarithm of the odds favoring an event. The solid line in Figure 6.1 describes data from a logistic function.

The following can be used to simulate a logistic function:

USING EXCEL TO SIMULATE A LOGISTIC FUNCTION

1. Enter the threshold value in cell G1. A value of .0 is good to start with because it should make $p(0) = p(1) = .5$.

2. Enter =IF(F2 > G$1,1,0) in cell G2. This dichotomizes the first of the continuous measures, cell F2.

3. Drag G2 through the remaining data cells (i.e., to row 101).

4. Drag the formula for the mean in cell F103 to cell G103. The value should be close to the probability expected from the term entered in G1. In the present case, the value should be close to .5. Changing G1 to 1.0 should reduce this value to approximately .16.

The SAS program used earlier with a continuous criterion may likewise be modified simply as follows:

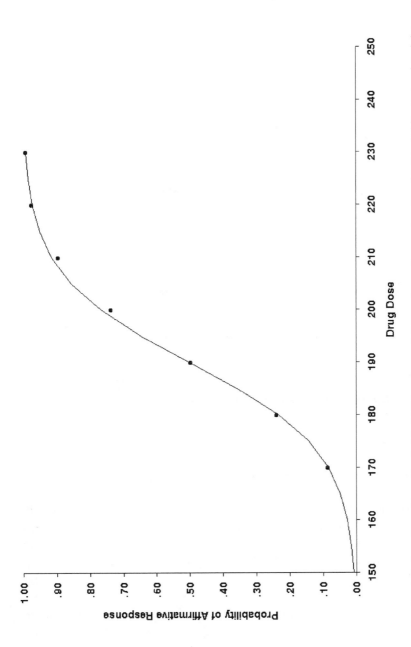

Figure 6.1. Logistic function with a threshold of 190, a slope of .07, and a scaling parameter of 1.7 (solid line); simulated data points (filled circles).

USING SAS TO SIMULATE A LOGISTIC FUNCTION

▬ 1. Enter `Retain Thresh .0;` immediately after the previous `Retain` statement.

▬ 2. Enter `If (Yp > thresh) then Yc = 1;else Yc = 0;` immediately before the `Output` statement.

Following is an SPSS equivalent:

USING SPSS TO SIMULATE A LOGISTIC FUNCTION

Enter the following two lines after the `compute y =` line in the multiple regression simulation program:

```
compute yc=0.
if (yp gt 0) yc=1.
```

▣ P/YCHOMETRIC FUNCTION/, ITEM OPERATING CHARACTERI/TIC/ (TRACE LINE/), AND DO/E-RE/PON/E CURVE/

Psychometric functions relate stimulus magnitudes to judgments about the stimuli and are important in the study of both perception and psychometric theory. Indeed, they serve as a link between the two through the concept of an **item operating characteristic** (IOC, also known as a **trace line**). A psychometric function for an absolute threshold plots the probability of reporting that a stimulus was present against its physical magnitude. A psychometric function for a difference threshold plots the probability of reporting that a comparison stimulus was of a stimulus or the difference in magnitude between a comparison stimulus and a standard. An IOC plots the probability of answering a test item in a keyed direction against the magnitude of the underlying trait or skill. In classical psychometrics, the **observed** test score defines trait magnitude. In modern psychometrics (item response theory, IRT, Hambleton & Swaminathan, 1985; Hambleton, Swaminathan, & Rogers, 1991), trait magnitude is a latent variable usually symbolized as θ (rather than p as in Equation 6.1). Both psychometric functions and IOCs thus plot the probability of a discrete event against a continuous variable. IOCs are discussed more fully later in this chapter.

Yet another situation in which a logistic function is found is in evaluating therapeutic effects, particularly those involving drugs. In this case, X is the dose of a particular drug and p is the probability of a particular response. In particular, assume that doses in magnitude of 170 to 230 (in units that need not be of concern) are each presented 1000 times and some response that can be coded as "1" (present) or "0" (absent) is obtained. The function in this case is known as a **dose–response curve**. When the response is the death of the organism, the threshold is known as the LD-50 to denote that the probability of a lethal dose is .5. The filled circles in Figure 6.1 describe simulated data from a dose–response study.

The Excel simulation is as follows:

USING EXCEL TO SIMULATE A PSYCHOMETRIC FUNCTION/DOSE–RESPONSE CURVE

▬ 1. Enter a value representing the population mean of the function in cell A1. The present example uses 190.

▬ 2. Enter a value representing the population standard deviation in cell B1.

▬ 3. Enter values representing magnitudes in cells A2:G2. These should fall above and below the value in A1.

▬ 4. Enter 100 random normal deviates in cells A3:G102.

▬ 5. Enter =IF((A$2>$A$1 + $B$1*A3),1,0) in cell A103. This generates a binary outcome that is partially random.

▬ 6. Drag the equation through column G. This generates outcomes for the remaining magnitudes.

▬ 7. Drag row 103 down to row 202. This generates the remaining outcomes.

▬ 8. Enter =AVERAGE(A103:A202) in cell A203. This computes the probability of a favorable outcome for the lowest magnitude.

▬ 9. Drag this equation through to column G. This produces the remaining probability. The values should increase monotonically.

▬ 10. Generate an X–Y plot of row 2 versus row 203.

The equivalent SAS program is as follows:

USING SAS TO SIMULATE A PSYCHOMETRIC FUNCTION/DOSE–RESPONSE CURVE

```
Data Pmetric;
Retain nobs 1000 Gmn 190 GSd 15 err 2234 x1-x7
  (170 180 190 200 210 220 230);
Array xs(*) x1-x7;
Array ys(*) y1-y7;
Do I = 1 to nobs;
Do j = 1 to 7;
If (xs(j) gt gmn + gsd*rannor(err)) then ys(j) = 1;else
ys(j) = 0.;
Keep x1-x7 y1-y7;
End;
Output;
End;
Run;
```

These data will be analyzed in Chapter 11 using SAS **Proc Catmod**. Instead of having an observation consist of the entire set of X and the entire set of Y values, one typically needs it to contain only one X and its associated Y values. This may be done as follows:

MODIFYING SAS OUTOUT FOR INPUT TO PROC CATMOD

```
Data A1;set Pmetric;x = x1; y = y1; keep x y;output;
Data A2;set Pmetric;x = x2; y = y2; keep x y;output;
Data A3;set Pmetric;x = x3; y = y3; keep x y;output;
Data A4;set Pmetric;x = x4; y = y4; keep x y;output;
Data A5;set Pmetric;x = x5; y = y5; keep x y;output;
Data A6;set Pmetric;x = x6; y = y6; keep x y;output;
Data A7;set Pmetric;x = x7; y = y7; keep x y;output;
Data Pmetrica;Set A1 A2 A3 A4 A5 A6 A7;
Proc Means Data= Pmetrica;by X;var Y;
Run;
```

The original data set, Pmetric, had 1000 observations, because the parameter nobs = 1000. It had 14 pairs of variables (X1 and Y1, X2 and Y2, ..., X7 and Y7). Each data set derived from this (A1, A2, ..., A7) also contained 1000 observations. However, it was limited to only one of these pairs, and the generic names X and Y were used instead of specific ones like X_1 and Y_1. Finally, data set Pmetrica aggregated all seven data sets and therefore contained 7000 observations, but with only one pair of variables. The Proc Means was used to verify that Y increases from a value near .0 to a value near 1.0 as X increases. These values are as follows:

X	$f(X)$
170	.087
180	.240
190	.497
200	.738
210	.897
220	.975
230	.993

The following are SPSS equivalents (the SPSS equivalent of Proc Catmod is Logistic):

USING SPSS TO SIMULATE A PSYCHOMETRIC FUNCTION/DOSE–RESPONSE CURVE

```
input program.
vector x(7).
vector y(7).
loop #i=1 to 1000.
  compute #gmn=190.
  compute #gsd=15.
  compute x1=170.
  compute x2=180.
  compute x3=190.
  compute x4=200.
  compute x5=210.
  compute x6=220.
  compute x7=230.
  loop #j=1 to 7.
    compute y(#j)=0.
    if(x(#j) gt #gmn+#gsd*normal(1)) y(#j)=1.
  end loop.
end case.
end loop.
end file.
end input program.
execute.
```

MODIFYING SPSS OUTPUT FOR INPUT INTO LOGISTIC

Highlight column x2 and cut and then paste it immediately below the last case in column x1. Then highlight column x3 and cut and paste it immediately below the new last case in column x1. Continue in this manner until all the data from variables x1 to x7 are in column x1. Do the same for y1 to y7, putting them in column y1 one after the other in the same order as the corresponding x's. This cutting and pasting adds cases with missing data to the ends of variables x1 and x2. To remove the cases with missing values, from the SPSS Data Editor menu choose DATA, SELECT CASES, if, condition is satisfied, if, x1~=missing(x1), CONTINUE, OK. SPSS has options for restructuring files in this manner.

Comparing Two Logistic Functions

In a variety of situations, two curves of the preceding form need to be compared. For example, you may wish to compare two types of psychophysical stimuli, two test items, or two drugs across a series of common dosages. The Excel extension is straightforward—simply replicate columns A to G in H to N, making sure that the random normal deviates are chosen independently. The SAS program requires a bit more work, as follows:

USING SAS TO SIMULATE DATA CREATING TWO
PSYCHOMETRIC FUNCTIONS/DOSE–RESPONSE CURVES

```
Data A;
Retain nobs 1000 err 2234
GmnA 190 GSdA 15 xA1-xA7 (170 180 190 200 210 220 230)
GmnB 200 GSdB 15 xb1-xB7 (170 180 190 200 210 220 230);
Array xas(*) xa1-xa7;
Array yas(*) ya1-ya7;
Array xbs(*) xb1-xb7;
Array ybs(*) yb1-yb7;
Do I = 1 to nobs;
Do j = 1 to 7;
If (xas(j) gt gmna + gsda*rannor(err)) then yas(j) = 1;
  else yas(j) = 0.;
If (xbs(j) gt gmnb + gsdb*rannor(err)) then ybs(j) = 1;
  else ybs(j) = 0.;
Keep xa1-xa7 ya1-ya7 xb1-xb7 yb1-yb7;
```

```
End;
Output;
End;
Proc Means Data=A;
Data A1;set A;x = xa1; y = ya1;cond = 1;keep x y
cond;output;
Data A2;set A;x = xa2; y = ya2;cond = 1;keep x y
cond;output;
Data A3;set A;x = xa3; y = ya3;cond = 1;keep x y
cond;output;
Data A4;set A;x = xa4; y = ya4;cond = 1;keep x y
cond;output;
Data A5;set A;x = xa5; y = ya5;cond = 1;keep x y
cond;output;
Data A6;set A;x = xa6; y = ya6;cond = 1;keep x y
cond;output;
Data A7;set A;x = xa7; y = ya7;cond = 1;keep x y
cond;output;
Data B1;set A;x = xb1; y = yb1;cond = 2;keep x y
cond;output;
Data B2;set A;x = xb2; y = yb2;cond = 2;keep x y
cond;output;
Data B3;set A;x = xb3; y = yb3;cond = 2;keep x y
cond;output;
Data B4;set A;x = xb4; y = yb4;cond = 2;keep x y
cond;output;
Data B5;set A;x = xb5; y = yb5;cond = 2;keep x y
cond;output;
Data B6;set A;x = xb6; y = yb6;cond = 2;keep x y
cond;output;
Data B7;set A;x = xb7; y = yb7;cond = 2;keep x y
cond;output;
Data All;Set A1 A2 A3 A4 A5 A6 A7 B1 B2 B3 B4 B5 B6 B7;
Proc Sort Data=All;by cond x;
Proc Means Data=All;by cond x;var y;
Run;
```

Following is an SPSS equivalent:

USING SPSS TO SIMULATE DATA CREATING TWO PSYCHOMETRIC FUNCTIONS/DOSE–RESPONSE CURVES

```
input program.
vector #xa(7).
vector #ya(7).
vector #xb(7).
vector #yb(7).
loop #i=1 to 1000.
  compute #gmna=190.
  compute #gsda=15.
  compute #xa1=170.
  compute #xa2=180.
  compute #xa3=190.
  compute #xa4=200.
  compute #xa5=210.
  compute #xa6=220.
  compute #xa7=230.
  compute #gmnb=200.
  compute #gsdb=15.
  loop #k=1 to 7.
    compute #xb(#k)=#xa(#k).
  end loop.
  loop #j=1 to 7.
    compute conda=1.
    compute condb=2.
    compute #ya(#j)=0.
    if(#xa(#j) gt #gmna+#gsda*normal(1)) #ya(#j)=1.
    compute xax=#xa(#j).
    compute yay=#ya(#j).
    compute #yb(#j)=0.
    if(#yb(#j) gt #gmnb+#gsdb*normal(1)) #yb(#j)=1.
    compute xbx=#xb(#j).
    compute yby=#yb(#j).
end case.
```

```
end loop.
end loop.
end file.
end input program.
execute.
```

Cut and paste column condb below the data in column conda. Cut and paste column xbx below the data in column xax and cut and paste column yby below the data in column yay. Eliminate missing values as before. Choose DATA, SPLIT FILES, organize output by groups, groups based on conda xax, sort file by grouping variables, OK. To get means of yay by conda and xax, from the SPSS Data Editor menu choose ANALYZE, SUMMARIZE, DESCRIPTIVE STATISTICS, variables yay, OK.

Psychometric Data Based on Binary Item Scoring

We will begin this section by generating binary data such as responses on ordinary classroom tests, leaving the question of recovering the parameters to Chapter 11. The basic idea is to construct a latent variable describing each individual's standing on the trait in question, which, as noted before, is usually denoted θ. Equation 6.2 defines an IOC as a three-parameter extension to the logistic function (Equation 6.1) in which θ corresponds to p. The extension also allows the curve to start at a level $c \geq 0$ instead of at 0 to reflect a base rate for chance success. When set at 0, it represents data as one might obtain from a completion item where one cannot answer correctly by guessing. It is commonly set at $1/k$ for a k-alternative multiple-choice test and at .5 for a true–false test. However, an incorrect alternative might require a lower value of c if it were attractive to those who are minimally knowledgeable and a higher value of c if it were very unattractive.

$$p(c) = c + \frac{(1-c)e^{da(\theta-b)}}{1 + e^{da(\theta-b)}}. \tag{6.2}$$

Figure 6.2 illustrates a logistic function with $a = 1$, $b = -.5$, $c = .25$, and $d = 1.7$. This might be a plausible outcome for a four-alternative multiple-choice item. We will also consider IOC curves later in this chapter in their relation to psychometric functions.

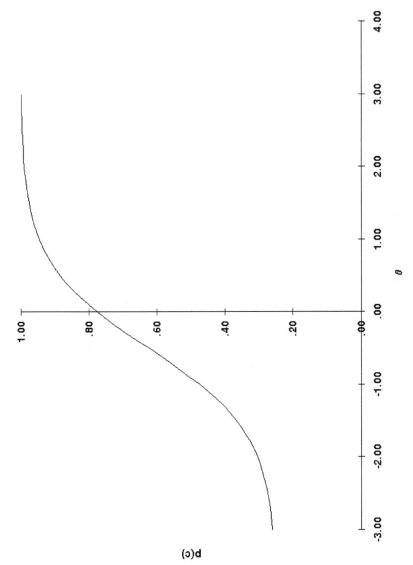

Figure 6.2. Three-parameter logistic function ($a = 1$, $b = -.50$, $c = .25$, and $d = 1.7$).

197

The following steps will simulate responses to a six-item test in Excel:

USING EXCEL TO SIMULATE BINARY TEST RESPONSES

▬ 1. Enter 100 random normal deviates in A4:A103. These represent values of θ.

▬ 2. Enter the values of a for the six items in cells B1:G1. The values chosen in the present example were –.50, –.30, –.10, .10, .30, and 50. This means the items become progressively more difficult.

▬ 3. Enter the values of b for the six items in cells B2:G2. The values chosen in the present example were all 1.0. This means that the items are equally discriminating.

▬ 4. Enter the values of c for the six items in cells B3:G3. The values chosen in the present example were all .0. This means the items are in a short-answer format. In a more general sense, these data fit a one-parameter logistic (1PL) or Rasch model because they differ only in difficulty.

▬ 5. Enter =B$3 + (1-B$3)*EXP(1.7*B$2*($A4-B$1))/ (1+EXP(1.7*B$2*($A4-B$1))) in cell B4. This represents the probability that the first respondent will answer the first item correctly. It can also be viewed as a latent variable.

▬ 6. Drag this equation through column G. This provides the remaining probabilities for the first respondent.

▬ 7. Drag B4:G4 down through row 103 to complete the data for the remaining respondents.

▬ 8. Enter =rand() in cells H4:M103. This represents the stochastic (chance) device in the model.

▬ 9. Enter =IF(B4>H4,1,0) in cell N4. This converts the latent variable $p(c)$ into an observable correct or incorrect item response for the first respondent on the first item.

10. Drag N4 through S4 to complete the item responses for the first respondent.

11. Drag N4:S4 through row 103 to complete the item responses for the remaining respondents.

12. Enter =SUM(N4:S4) in cell T4. This is the number correct obtained by the first respondent.

13. Drag T4 down through row 103.

14. Enter =AVERAGE(A4:A103) in cell A104. This is the average value of θ, which should be .0 within sampling error.

15. Drag A104 through column T to provide averages on the remaining variables. Given the values of a and c, the averages in B104:G104 should be the probabilities corresponding to b, viewing b as a z score. For example, if b were .0, the population value of the average p value would be .5. Moreover, the means in N104:S104 should parallel these means. The means in H104:M104 should be approximately .5. The average in T104 will equal the sum of the means in cells N104:S104.

16. Enter =CORREL(A4:A103,$T4:$T103) in cell A105 to obtain the correlation between θ and number correct.

17. Drag the equation in cell A105 through cell S105 to provide the remaining correlations with number correct. Note that (a) increasing a increases the corresponding correlation in cells B105:G105 and N105:S105, but (b) the correlations in H105:M105 differ from .0 only by chance.

A corresponding SAS program is as follows:

USING SAS TO SIMULATE BINARY TEST RESPONSES

```
Data MC;
Retain err 2005 nobs 1000 ni 6 c1-c6 (.0 .0 .0 .0 .0 .0)
```

```
b1-b6 (-.5 -.3 -.1 .1 .3 .5) a1-a6 (1. 1. 1. 1. 1. 1.);
Array cs (*) c1-c6;
Array bs (*) b1-b6;
Array as (*) a1-a6;
Array pcs (*) pc1-pc6;
Array irs (*) ir1-ir6;
Do I = 1 to nobs;
Theta = rannor(err);
Do j = 1 to ni;
pcs(j) = cs(j) + (1 - cs(j))*exp(1.7*as(j)*
  (theta-bs(j)))/(1 + exp(1.7*as(j)*(theta-bs(j))));
If pcs(j) ge uniform(err) then irs(j) = 1;
  else irs(j) = 0;
End;
Total = sum(of ir1-ir6);
Output;
Keep ir1-ir6 total;
End;
Proc Means Data=MC;
Proc Sort Data=MC;by ir1 ir2 ir3 ir4 ir5 ir6;
Proc Means Data=MC n;by ir1 ir2 ir3 ir4 ir5 ir6;
Run;
```

Following is an SPSS equivalent:

USING SPSS TO SIMULATE BINARY TEST RESPONSES

```
input program.
vector #c(6).
vector #b(6).
vector #a(6).
vector #pc(6).
vector ir(6).
loop #i=1 to 1000.
  compute #ni=6.
```

```
compute #b1=-.5.
compute #b2=-.3.
compute #b3=-.1.
compute #b4=.1.
compute #b5=.3.
compute #b6=.5.
loop #j=1 to 6.
  compute #a(#j)=1.
  compute #c(#j)=0.
end loop.
compute #theta=normal(1).
loop #j=1 to #ni.
  compute #pc(#j)=#c(#j)+(1-#c(#j))*exp(1.7*#a(#j)*
    (#theta-#b(#j))).
  compute #pc(#j)=#pc(#j)/(1+exp(1.7*#a(#j)*
    (#theta-#b(#j)))).
  compute ir(#j)=0.
  if(#pc(#j) ge uniform(1)) ir(#j)=1.
end loop.
compute total=sum(ir1 to ir6).
end case.
end loop.
end file.
end input program.
execute.
```

Obtain descriptive statistics for the variables ir1 to ir6 and Total as before. To get the number of cases in each combination of categories of the ir's, split the file as before with groups based on ir1 to ir6 and get descriptive statistics again on the variable Total.

Likert Data

Likert data are ratings that reflect judgmental confidence or strength of response. Assume five items are each scored on a 5-point scale. Basically, each item must contain four progressively increasing thresholds (e.g., −1,

−.5, 0, and .5). A given item is scored as a "1" if θ falls below the first cutoff, a "2" if it falls between the first and the second, a "3" if it falls between the second and the third, and so on. To keep the simulation simple, the discrimination (a) parameter is the same for all five items, and we will not have a stochastic process affect the response. The basic issue is simply to determine how many thresholds for each item a given value of θ exceeds. The Excel simulation is as follows:

USING EXCEL TO SIMULATE LIKERT TEST RESPONSES

1. Enter 100 random normal deviates in cells A2:A101, paralleling what was done in the previous simulation to generate values of θ.

2. Enter the values of the lowest thresholds in B1:K1.

3. Enter =IF($A2>B$1,1,0) in cell B2. This converts the first latent observation for the first respondent to a binary outcome for the first item.

4. Drag B2 through column F to provide the remaining observations for the first respondent with respect to the lowest threshold for the four remaining items.

5. Drag B2:F2 through row 101 to provide the data for the remaining respondents.

6. In essence, steps 3 to 5 will be performed three more times to obtain responses for the remaining thresholds. In particular, enter the second set of thresholds in cells G1:K1. Each must be numerically larger than its counterpart in B1:F1.

7. Copy cells B2:F101 to G2:K101 to obtain data for the second threshold.

8. Enter the third set of thresholds in cells L1:P1, again making sure that each value is numerically larger than its counterpart in G1:K1.

9. Copy cells B2:F101 to L2:P101 to obtain data for the third threshold.

10. Enter the final set of thresholds in cells Q1:U1, again making sure that each value is numerically larger than its counterpart in L1:P1.

11. Copy cells B2:F101 to Q2:U101 to obtain data for the final threshold.

12. Enter =SUM(B2,G2,L2,Q2)+1 in cell V2. This counts the number of thresholds exceeded by the first response made by the first respondent. Adding 1 converts the data to a 1–5 scale.

13. Drag V2 through column Z2 to obtain responses for the remaining items for this first respondent.

14. Drag cells V2:Z2 through row 101 to obtain data for the remaining respondents. This completes the actual simulation.

15. Enter =SUM(V2:Z2) in cell AA2. This provides a total score for the first respondent over the five items.

16. Drag AA2 through row 101 to obtain total scores for the remaining respondents.

17. Enter =AVERAGE(A2:A101) in cell A102. This provides the average value of θ, which should be near 0.

18. Enter =CORREL(A2:A101,$AA2:$AA101) in cell A103 to correlate θ with the obtained score.

19. Drag A102:A103 through column Z to obtain the remaining correlations with total score.

The corresponding SAS implementation is as follows:

USING SAS TO SIMULATE LIKERT TEST RESPONSES

```
Data cr;
Retain nobs 1000 ni 5 err 1998
thra1-thra5 (-1.    -.5  .0    .5   1.)
thrb1-thrb5 ( -.75 -.25 .25  .75 1.25)
thrc1-thrc5 ( -.5   .0  .5  1.00 1.5)
thrd1-thrd5 ( -.25  .25 .75 1.25 1.75);
Array thresha (*) thra1-thra5;
Array threshb (*) thrb1-thrb5;
Array threshc (*) thrc1-thrc5;
Array threshd (*) thrd1-thrd5;
Array ys (*) y1-y5;
Do I = 1 to nobs;
Theta = rannor(err);
Do j = 1 to ni;
If theta gt thresha(j) then ys(j) = 2;else ys(j) = 1;
If theta gt threshb(j) then ys(j) = ys(j) + 1;
If theta gt threshc(j) then ys(j) = ys(j) + 1;
If theta gt threshd(j) then ys(j) = ys(j) + 1;
End;
Total = sum(of y1-y5);
Keep total y1-y5;
Output;
End;
Proc Sort Data=cr;by y1 y2 y3 y4 y5;
Proc Means Data=cr n; by y1 y2 y3 y4 y5;
Run;
```

Following is an SPSS equivalent:

USING SPSS TO SIMULATE LIKERT TEST RESPONSES

```
input program.
vector #thra(5).
```

```
vector #thrb(5).
vector #thrc(5).
vector #thrd(5).
vector y(5).
loop #i=1 to 1000.
  compute #ni=5.
  compute #thra1=-1.
  compute #thra2=-.5.
  compute #thra3=0.
  compute #thra4=.5.
  compute #thra5=1.
  compute #thrb1=-.75.
  compute #thrb2=-.25.
  compute #thrb3=.25.
  compute #thrb4=.75.
  compute #thrb5=1.25.
  compute #thrc1=-.5.
  compute #thrc2=0.
  compute #thrc3=.5.
  compute #thrc4=1.
  compute #thrc5=1.5.
  compute #thrd1=-.25.
  compute #thrd2=.25.
  compute #thrd3=.75.
  compute #thrd4=1.25.
  compute #thrd5=1.75.
  compute #theta=normal(1).
  loop #j=1 to #ni.
    compute y(#j)=1.
    if(#theta gt #thra(#j)) y(#j)=2.
    if(#theta gt #thrb(#j)) y(#j)=y(#j)+1.
    if(#theta gt #thrc(#j)) y(#j)=y(#j)+1.
    if(#theta gt #thrd(#j)) y(#j)=y(#j)+1.
  end loop.
  compute total=sum(y1 to y5).
end case.
end loop.
end file.
end input program.
execute.
```

Split the file with groups based on $y1$ to $y5$ and then get descriptive statistics on the variable Total to get the number of cases in each combination of y's.

Multiple-Choice Data and Bock's Nominal Model

Bock's (1972) nominal model predicts the probability of choosing each of k alternatives on a multiple-choice test as a function of θ. The basic equation for the model is as follows:

$$p\left(X = \frac{k}{\theta}\right) = \frac{e^{a_k\theta + c_k}}{\sum e^{a_k\theta + c_k}}. \tag{6.3}$$

One constraint is that $\sum a = \sum c = 0$. A proper item is one for which the correct response has the largest value of a, which means that it has the steepest slope. For simplicity, alternative 1 will be the correct response for all items.

The following will create relevant data using Excel. A five-item, four-alternative scale with 100 participants will be assumed.

USING EXCEL TO SIMULATE MULTIPLE-CHOICE RESPONSES

1. Enter a series of 100 normal random deviates in cells A3:A102. These correspond to the values of θ.

2. Enter =rand() in cells B3:B102. This represents a stochastic component affecting the response to the first alternative.

3. Enter values for a_1 to a_4 in cells C1 to F1. The values chosen for this example are 1.90, .70, −.80, and −1.80.

4. Enter values for c_1 to c_4 in cells C2 to F2. The values chosen for this example are .60, .40, −.30, and −.70.

5. Enter =EXP(C$1*$A3+C$2) in cell C3. This represents the absolute magnitude of the tendency to choose alternative 1 for item 1 for participant 1.

6. Drag C3 through column F to obtain the absolute magnitudes of the remaining alternatives for respondent 1 for item 1.

7. Drag C3:F3 through row 102 to obtain the absolute magnitudes for the remaining participants on item 1.

8. Enter `=C3/SUM($C3:$F3)` in cell G4. This provides the probability of choosing alternative 1 for the first participant on the first item.

9. Drag G4 through to column J to provide the probabilities of choosing the remaining alternatives for the first participant on the first item.

10. Drag G4:J4 through row 102 to obtain response probabilities for the remaining respondents.

11. Enter `=IF(B3<G3,1,IF(B3<SUM(G3:H3),2, IF(B3<SUM(G3:I3),3,4)))` in cell K3. This uses the stochastic component to identify the chosen response.

12. Enter `=IF((B3<=G3),1,0)` in cell L3, which scores the item as correct or incorrect.

13. This strategy is repeated four more times to provide responses to the remaining items. Copy B3:L102 to (a) M1:M102 to obtain data for the second item, (b) X3:AH102 to obtain data for the third item, (c) AI3:AS102 to obtain data for the fourth item, and (d) AT3:BD102 to obtain data for the final item. This completes what is necessary for the simulation.

14. Enter `=SUM(L3,W3,AH3,AS3,BD3)` in cell BE3 to obtain a number correct score for the first respondent.

15. Drag BE3 through row 102 to obtain number correct scores for the remaining respondents.

16. Enter `=AVERAGE(A3:A102)` in cell A103 as a check on the random normal deviate generation process for θ. It should be approximately 0.

▬ 17. Enter =STDEV(A3:A102) in cell A104 as a further check on the random normal deviate process for θ. It should be approximately 1.0.

▬ 18. Copy A103 to B103, C103, D103, and N103. All but D103 should also be approximately .0; D103 should be approximately .5. This is a check on item 1.

▬ 19. Enter =MIN(M3:M102) in cell M103 to ensure that the correct response was chosen at least once. This is also a check on item 1 (you can also check by inspection).

▬ 20. Enter =MAX(M3:M102) in cell M104 to ensure that the last response category was chosen at least once (you can also check by inspection of column M).

▬ 21. Copy B103:N104 to O103:AA104 to obtain comparable data for item 2.

▬ 22. Copy B103:N104 to AB103:AN104 to obtain comparable data for item 3.

▬ 23. Copy B103:N104 to AO103:BA104 to obtain comparable data for item 4.

▬ 24. Copy B103:N104 to BB103:BN104 to obtain comparable data for item 5.

▬ 25. Enter =CORREL(A3:A102,$BE3:$BE102) in cell A105 to determine the correlation between θ and number correct .

▬ 26. Copy A105 to cells L105, W105, H105, AS105, and BD105 to obtain item total correlations for individual items.

The following is an SAS equivalent:

USING SAS TO SIMULATE MULTIPLE-CHOICE RESPONSES

```
Data bock;
Retain err 2001 b1-b5 (.8 .78 .8 .9 .9) nobs 1000 ni 5
  nalt 4;
Retain a1-a4 (1.9 .7 -.8 -1.8) c1-c4 (.6 .4 -.3 -.7);
Array Alts (*) alt1-alt5;
Array bs (*) b1-b5;
Array bres (*) br1-br5;
Array as (*) a1-a4;
Array cs (*) c1-c4;
Array rps (*) rp1-rp4;
Array crps (*) crp1-crp4;
Do i = 1 to ni;
bres(i) = sqrt(1 - bs(i)**2);
End;
Do I = 1 to nobs;
tot = 0;
theta = rannor(err);
Do j = 1 to NI;
rander = uniform(err);
alts(j) = 0.;
t = 0;
Do k = 1 to nalt;
rps(k) = exp(as(k)*theta + cs(k));
t = t + rps(k);
crps(k) = t;
End;
Do k = 1 to nalt;
rps(k) = rps(k)/t;
crps(k) = crps(k)/t;
If rander le crps(k) and alts(j) = 0 then alts(j) = k;
End;
If alts(j) = 1 then tot = tot + 1;
End;
Output;
Keep theta tot alt1-alt5;
End;
Proc Sort Data=bock;by tot;
```

```
Proc Means Data=bock;by tot;
Proc Freq Data=bock;tables tot*(alt1-alt5);
Proc Corr Data=bock;var theta alt1-alt5;with tot;
Proc Sort Data=bock;by theta;
Proc Print Data=bock;
Run;
```

Following is an SPSS equivalent:

USING SPSS TO SIMULATE MULTIPLE-CHOICE RESPONSES

```
input program.
vector alt(5).
vector #b(5).
vector #bres(5).
vector #a(4).
vector #c(4).
vector #rp(4).
vector #crp(4).
loop #i=1 to 1000.
  compute #b1=.8.
  compute #b2=.78.
  compute #b3=.8.
  compute #b4=.9.
  compute #b5=.9.
  compute #ni=5.
  compute #nalt=4.
  compute #a1=1.9.
  compute #a2=.7.
  compute #a3=-.8.
  compute #a4=-1.8.
  compute #c1=.6.
  compute #c2=.4.
  compute #c3=-.3.
  compute #c4=-.7.
  loop #k=1 to #ni.
    compute #bres(#k)=sqrt(1-#b(#k)**2).
  end loop.
  compute tot=0.
```

```
compute theta=normal(1).
loop #j=1 to #ni.
  compute #rander=uniform(1).
  compute #t=0.
  compute alt(#j)=0.
loop #k=1 to #nalt.
  compute #rp(#k)=exp(#a(#k)*theta+#c(#k)).
  compute #t=#t+#rp(#k).
  compute #crp(#k)=#t.
end loop.
loop #k=1 to #nalt.
  compute #rp(#k)=#rp(#k)/#t.
  compute #crp(#k)=#crp(#k)/#t.
  if(#rander le #crp(#k) and alt(#j) eq 0) alt(#j)=#k.
end loop.
  if(alt(#J) eq 1) tot=tot+1.
end loop.
end case.
end loop.
end file.
end input program.
execute.
```

Get correlations among all variables. To get cross tabulations, choose ANALYZE, DESCRIPTIVE STATISTICS, SUMMARIZE, CROSSTABS, rows alt1, alt5, columns tot, OK. Then split the file with groups based on variable tot and get descriptive statistics of theta and alt1 to alt5. To see the data sorted by theta, access the split file again and remove tot as the grouping variable. Choose analyze all cases, OK. Then choose DATA, SORT CASES, sort by theta, OK.

Chapter 11 considers an application of an item response model to these data. However, considerations of space will limit our presentation to the results of the binary data. Nonetheless, it is useful to look at the Bock model data using classical psychometrics. Figure 6.3 contains a plot of the probabilities of each alternative as a function of number correct. Note that to get five correct, one must choose alternative 1 every time, by definition. However, the main points are that alternative 1's probability increases most sharply with number correct followed by alternative 2, whereas alternative 3 and, especially, alternative 4's response probabilities decline

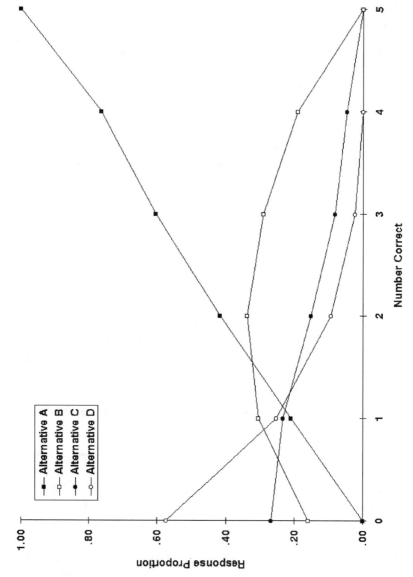

Figure 6.3. Data generated by Bock model plotted as trace line based on classical psychometrics.

with number correct. The difference between classical psychometrics and IRT is that the latter would use inference and use θ as the abscissa rather than number correct, an observed quantity.

🔲 THURSTONE'S "LAW" OF COMPARATIVE JUDGMENT AND SIGNAL DETECTION THEORY

Paired Comparisons and Thurstone Scaling

It is but a simple step from binary classification and psychometric functions to Thurstone's "law" of comparative judgment. Previously, a forced-choice task used two stimuli differing along defined physical dimensions such as intensity. The participant's task was to judge their relative magnitudes. The means and standard deviations of the two completely determined the accuracy (percentage of correct judgments).

Now, consider a task of stating preference among stimuli where there are no right or wrong answers. A series of k stimuli are presented in pairs (it can be shown that there are $k \cdot (k-1)/2$ pairs, ignoring order). If the stimulus differences are subtle or if probabilities are derived from different participants responding to a given pair, the preference probabilities will differ from .0 or 1.0. Thurstone (1927/1994) developed his law of comparative judgment to infer scale distances among the stimuli, which is discussed in a number of sources, including Nunnally and Bernstein (1994, Chapter 2). The central idea is that, in addition to being located as points in space, stimuli have a **discriminal dispersion** that causes them to form normal distributions. These discriminal dispersions cause the outcome of comparing the same stimuli to yield different results. The present discussion is limited to the case in which (a) the stimuli vary along one dimension and (b) the errors in observation of the two things being compared are unrelated. However, the simulations include both Case V, in which the discriminal dispersions are assumed to equal one another, and Case III, in which they are not. Although the simulation could proceed from individual trials, it would be fairly cumbersome. Consequently, the simulation will simply proceed from individual scale values and discriminal dispersions to generate the preference probabilities for each pair.

A spreadsheet generating the probabilities from the scale values is as follows:

USING EXCEL TO SIMULATE PAIRED-COMPARISON DATA FROM THURSTONE SCALE VALUES

■ 1. Enter the mean scale values in cells C1:G1. Test values of .00, .25, .60, .80, and 1.00 were used.

■ 2. Enter the discriminal dispersions in cells C2:G2. All initial test values were 1.0.

■ 3. Enter =C1 in cell A3. Note that you do not want to simply paste the value. By entering this as an equation, cell A3 will update automatically when C1 is changed.

■ 4. Enter =D1 in cell A4.

■ 5. Enter =E1 in cell A5.

■ 6. Enter =F1 in cell A6.

■ 7. Enter =G1 in cell A7.

■ 8. Enter =C2 in cell B3. Note that you do not want to simply paste the value. By entering this as an equation, cell B3 will update automatically when C2 is changed.

■ 9. Enter =D2 in cell B4.

■ 10. Enter =E2 in cell B5.

■ 11. Enter =F2 in cell B6.

■ 12. Enter =G3 in cell B7.

■ 13. Enter =NORMSDIST((C$1 - $A3)/SQRT(SUMSQ(C$2,$B3))) in cell C3. This formula computes the desired probability for the first pairing (NORMSDIST is an Excel function that converts a z value to a probability). Because the first pairing is between the first stimulus and itself, the result will be .5.

■ 14. Drag the equation through column G.

■ 15. Drag C3:G3 through row 7. Assuming the scale values in the first row increase and the discriminal dispersions are at least nearly equal, the preference probabilities should increase going from right to left and decrease going down.

An SAS program that will perform this analysis is as follows:

USING SAS TO SIMULATE PAIRED-COMPARISON DATA FROM THURSTONE SCALE VALUES

```
Data A;
Retain X1-X5 (.0 .25 .60 .8 1.0) sd1-sd5
  (1. 1. 1. 1. 1.);
Retain nstim 5;
Array Xs(*) X1-X5;
Array sds(*) sd1-sd5;
Do J = 1 to Nstim;
Do I = 1 to Nstim;
p = probnorm((xs(i) - xs(j))/
  sqrt(sds(i)**2 + sds(j)**2));
Output;
End;
End;
Proc Print Data=A;
Run;
```

Following is an SPSS equivalent:

USING SPSS TO SIMULATE PAIRED-COMPARISON DATA FROM THURSTONE SCALE VALUES

```
input program.
vector x(5).
vector sd(5).
loop #j=1 to 5.
loop #i=1 to 5.
```

```
compute x1=0.
compute x2=.25.
compute x3=.6.
compute x4=.8.
compute x5=1.
loop #k=1 to 5.
  compute sd(#k)=1.
end loop.
compute p=cdfnorm((x(#i)-x(#j))/
  sqrt(sd(#i)**2+sd(#j)**2)).
end case.
end loop.
end loop.
end file.
end input program.
execute.
```

▣ CONSTRUCTING A SIMILARITY MATRIX

The z-score matrix of simulated paired-comparison data is skew symmetric in that $z_{ji} = -z_{ij}$ as the data represent **dominance** relationships (**preferences**)—to the extent that stimulus A is preferred over stimulus B, the converse is not true. In contrast, similarity data are symmetric because one generally regards stimulus A's similarity to stimulus B as the same as stimulus B's similarity to stimulus A. A correlation is a form of similarity relationship, and $r_{xy} = r_{yx}$.

The **weighted Euclidean model** of multidimensional scaling is a common way to assess individual differences in the distances among stimuli for subjects (or other source of variation). Let $D_{ij \cdot k}$ denote the distance between stimuli i and j for subject k. The model assumes that individual differences in distance emerge because subjects weight dimensions differently. For example, one subject may base his or her judgments more on dimension 1 than on dimension 2 and a second subject may do the reverse. This model can be written as:

$$D_{ij \cdot k} = \sqrt{\sum w_{kr}(x_{ir} - x_{jr})^2}, \qquad (6.4)$$

where w_{kr} is the weight subject k gives dimension r and x_{ij} and x_{ik} are the distances along dimension r for the two stimuli.

▣ CLAJJIFICATION AND JIGNAL DETECTION ANALYJIJ

Yes–No Task

Classification is a general term describing a task in which the participant must assign events to categories of response. In a **signal detection task**, the task is to be as accurate as possible in choosing on the basis of experimentally defined stimulus characteristics such as whether a faint tone is or is not embedded in a noise background. The stimulus alternatives are deliberately made confusable. By contrast, classifying foods as liked or disliked would involve classification. However, it would not be signal detection because the participant is being asked to respond based on a dimension he or she uses rather than one defined by the experimenter. In addition, the various stimuli may not be confusable. For example, the participant may be asked to classify corn and string beans, which are hardly confusable.

The most basic form of classification is **binary**, for example, whether a tone was heard or not. There may be only two stimulus alternatives or there may be more than two. If there are but two stimulus alternatives, one is commonly designated **signal plus noise** $(s + n)$, and the other is designated **noise** (n). The associated responses are typically labeled "yes" (Y) and "no" (N), so this is commonly described as a **yes–no task**. The summary data appear in the form of a **confusion matrix**. One common way to present these data (among others) is as follows:

	Stimuli	
Response	$s + n$	n
Y	$p(Y/s + n) =$ Hit rate	$p(Y/n) =$ False alarm rate
N	$p(N/s + n) =$ Miss rate	$p(N/n) =$ Correct rejection rate

Note that the hit rate and miss rate are complements of one another as are the false alarm and correct rejection rates. Further discussion will center on hit and false alarm rates. The hit rate is denoted the true positive rate in some settings. Likewise, the false alarm, miss, and correct rejection rates are often denoted the false positive, false negative, and true negative rates.

Because of sensory and possibly physical variability, the stimuli vary from trial to trial. The $s+n$ falls, on average, above n, but because of this randomness, an $s+n$ trial can assume a small value and an n trial can assume a large value. Some common conceptions of the process assume that the participant adopts a criterion value and responds Y if the observation falls above the criterion and N otherwise. The hit and false alarm rates are therefore the probability that $s+n$ and n trials exceed the criterion. Specifically, the **equal-variance Gaussian** signal detection model assumes that the noise component is normally distributed with the same variance on $s+n$ and n trials. This is because the signal component is assumed constant, so its effects are simply additive. In contrast, the **unequal-variance Gaussian** signal detection model assumes that $s+n$ trials are more variable than n trials because the signal is also variable, independent of noise. In this case, the two sources of variance add. These two models have their roots in Thurstone's (1927/1994) law of comparative judgment, which is considered in more detail later. The following Excel program can illustrate either of the latter models for this binary case:

USING EXCEL TO SIMULATE YES–NO SIGNAL DETECTION DATA

1. Enter a value in cell A1 to represent the mean of the $s+n$ distribution in z-score units (the n distribution is assumed to have a mean of .0). It should be a positive number between 0 and 2.0.

2. Enter a value in cell B1 to represent the standard deviation of the $s+n$ distribution (the n distribution is assumed to have a standard deviation of 1.0.).

3. Enter a value in cell C1 to represent the criterion in z-score units. Values between ±2 are reasonable.

4. Enter 200 random normal deviates in cells A2:B101 to represent the noise components of 100 $s+n$ and 100 n trials. Note that at present these are not to be considered paired—indeed the number of these two types of trials are made equal only as a convenience

in generating the data. However, this assumption will be changed later.

▬ 5. Enter =$A$1 + B$1*A2 in cell C2 to represent the first of the $s+n$ observations.

▬ 6. Enter =B2 in cell D2 to represent the first of the n observations.

▬ 7. Enter =IF(C2 > $C$1,1,0) in cell E1 to represent the outcome of the first $s+n$ trial (i.e., whether the observation does or does not exceed the criterion).

▬ 8. Copy this equation to cell F2 to represent the outcome of the first n trial.

▬ 9. Copy row 2 through to row 101.

▬ 10. Enter =AVERAGE(E2:E101) in cell E102 to obtain the hit rate.

▬ 11. Copy this to cell F102 to obtain the false alarm rate.

Vary the parameters. In particular, note how the difference between hit and false alarm rate increases as the content of A1 is made larger, and how both increase as the content of cell C1 is made smaller.

The following is a SAS equivalent:

USING SAS TO SIMULATE YES–NO SIGNAL DETECTION DATA

```
Data A;
Retain msn .5 ssn 1.0 crit .1 nobs 1000 err 1999;
Do I = 1 to nobs;
Sn = msn + ssn*rannor(err);
n = rannor(err);
If (sn gt crit) then Rsn = 1;else Rsn = 0.;
If (n gt crit) then Rn = 1;else Rn = 0.;
Output;
Keep sn n Rsn Rn;
```

```
End;
Proc Means Data=A;
Run;
```

Paralleling the checks used in the spreadsheet, make sure that the mean stimulus magnitude on $s+n$ trials (sn) exceeds the mean stimulus magnitude on n trials and that the hit rate (Rsn) exceeds the false alarm rate (Rn).

Following is an SPSS equivalent:

USING SPSS TO SIMULATE YES–NO SIGNAL DETECTION DATA

```
input program.
loop #i=1 to 1000.
  compute #msn=.5.
  compute #ssn=1.
  compute #crit=.1.
  compute sn=#msn+#ssn*normal(1).
  compute n=normal(1).
  compute rsn=0.
  if(sn gt #crit) rsn=1.
  compute rn=0.
  if(n gt #crit) rn=1.
end case.
end loop.
end file.
end input program.
execute.
```

Confidence-Rating Task

In a multicategory **confidence-rating task**, the participant can express degrees of confidence. For example, a 7-point rating scale might be used in which 1=certain signal was absent, 2=sure signal was absent, 3=signal was probably absent, 4=uncertain, 5=signal was probably present,

6=sure signal was present, and 7=certain signal was present. The data are often expressed as follows:

	Stimulus	
Response	$s + n$	n
1	$p(1/s + n)$	$p(1/n)$
2	$p(1 \text{ or } 2/s + n)$	$p(1 \text{ or } 2/n)$
3	$p(1 \text{ or } 2 \text{ or } 3/s + n)$	$p(1 \text{ or } 2 \text{ or } 3/n)$
	$\vdots$	
7	$p(1 \text{ or } 2 \text{ or } 3 \text{ or } 4 \text{ or } 5$ $\text{or } 6 \text{ or } 7/s + n) = 1.00$	$p(1 \text{ or } 2 \text{ or } 3 \text{ or } 4 \text{ or } 5$ $\text{or } 6 \text{ or } 7/n) = 1.00$

The entries under $s+n$ can be viewed as hit rates under alternative criteria for defining a hit, and the entries under n can be viewed as corresponding false alarm rates. Note that k alternatives provide $k - 1$ usable hit rate/false alarm rate pairings because the kth alternative leads to the pairing 1.0/1.0.

The following is an adaptation of the previous Excel program:

USING EXCEL TO SIMULATE CONFIDENCE-RATING SIGNAL DETECTION DATA

1. Repeat steps 1 to 6 of the previous Excel program with one exception. Instead of placing a single criterion value in cell C1, place a series of six values in cells C1:H1. These should be in increasing magnitude.

2. Enter =IF($C2 > C$1,1,0) in cell E2. The slight modification is intended to facilitate further copying. This represents whether the first $s+n$ observation surpassed the most lenient criterion.

3. Drag E2 through column J. This represents the outcome for the first $s+n$ observation relative to the remaining criteria.

4. Enter =SUM(E2:J2) in column K2. This represents the rating of the first $s+n$ observation (i.e., the number of criteria surpassed).

▬ 5. Enter =IF($D2 > C$1,1,0) in cell L2. This represents whether the first n observation surpassed the most lenient criterion.

▬ 6. Drag L2 through column Q. This represents the outcome for the first n observation relative to the remaining criteria.

▬ 7. Enter =SUM(L2:Q2) in cell R2. This represents the rating of the first n observation (i.e., the number of criteria surpassed).

▬ 8. Drag cells E2:R2 down to row 101 to produce data for the remaining observations.

▬ 9. Enter =AVERAGE(E2:E101) in cell E103. This represents the hit rate relative to the first criterion.

▬ 10. Drag this equation through column R. Cells F103:J103 represent the hit rates relative to the remaining criteria; cell K103 is the mean rating on $s+n$ trials, cells L103:Q103 are the false alarm rates, and cell R103 is the mean rating on n trials. The proportions in cells K103:J103 should decrease as should the proportions in cells L103:Q103. In addition, the mean in cell K103 should exceed the mean in cell R103.

It is possible to determine the response using a single "If" statement, but the result is quite cumbersome and likely to contain an error. Consequently, the strategy was to compare the stimulus to each criterion in turn.

An equivalent SAS program is as follows:

USING SAS TO SIMULATE CONFIDENCE-RATING SIGNAL DETECTION DATA

```
Data A;
Retain msn .5 ssn 1.0 crit1-crit6 (-.4 -.2 0 .4 .7 1)
  nobs 1000 err 1999;
Array Crits(*) Crit1-Crit6;
Array Rsns(*) Rsn1-Rsn6;
```

```
Array Rns(*) Rn1-Rn6;
Do I = 1 to nobs;
Sn = msn + ssn*rannor(err);
n = rannor(err);
Do j = 1 to 6;
If (sn gt crits(j)) then Rsns(j) = 1;else Rsns(j) = 0.;
If (n gt crits(j)) then Rns(j) = 1;else Rns(j) = 0.;
End;
Ratsn = sum(of Rsn1-Rsn6);
Ratn = sum(of Rn1-Rn6);
Output;
Keep Rsn1-Rsn6 Rn1-Rn6 Ratsn Ratn;
End;
Proc Means Data=A;
Run;
```

Again, make sure that the values in Rsn1–Rsn6 decline, that the values in Rn1–Rn6 decline, and that Ratsn is greater than Ratn.

Following is an SPSS equivalent:

USING SPSS TO SIMULATE CONFIDENCE-RATING SIGNAL DETECTION DATA

```
input program.
vector #crit(6).
vector rsn(6).
vector rn(6).
loop #i=1 to 1000.
  compute #msn=.5.
  compute #ssn=1.
  compute #crit1=-.4.
  compute #crit2=-.2.
  compute #crit3=0.
  compute #crit4=.4.
  compute #crit5=.7.
  compute #crit6=1.
  compute #sn=#msn+#ssn*normal(1).
```

```
compute #n=normal(1).
loop #j=1 to 6.
  compute rsn(#j)=0.
  if(#sn gt #crit(#j)) rsn(#j)=1.
  compute rn(#j)=0.
  if(#n gt #crit(#j)) rn(#j)=1.
end loop.
compute ratsn=sum(rsn1 to rsn6).
compute ratn=sum(rn1 to rn6).
end case.
end loop.
end file.
end input program.
execute.
```

Forced-Choice Task

In a **forced-choice task**, the signal appears in one and only one of several observation intervals and the participant's task is to say which one. For example, the task might be auditory and the two intervals defined in time so the task is to discriminate the sequence of $s+n$ followed by n from n followed by $s+n$. Conversely, the intervals might be spatial, so the task is to say whether the stimulus is on the left or right. A commonly postulated decision rule is to choose the first interval if the magnitude of the observation in that interval exceeds the magnitude of the observation in the second interval and to choose the second interval under the converse conditions. Accuracy is commonly evaluated in terms of the percentage of correct judgments.

An Excel spreadsheet that will simulate this process is as follows:

USING EXCEL TO SIMULATE FORCED-CHOICE SIGNAL DETECTION DATA

▬ 1. Repeat steps 1, 2, and 4 from the yes–no simulation (step 3 is not necessary).

▬ 2. Enter =IF($A$1 + $B$1*A2 > B2,1,0) in cell C2. The outcome is a 1 if the magnitude of the $s+n$ observation exceeds the magnitude of the n observation and a 0 if the magnitude of the n observation exceeds the magnitude of the $s+n$ observation on the first trial.

■■■ 3. Drag the contents of cell C2 down to cell C101 to complete the observations.

■■■ 4. Enter =AVERAGE(C2:C101) to determine the accuracy.

An SAS program is as follows:

USING SAS TO SIMULATE FORCED-CHOICE SIGNAL DETECTION DATA

```
Data A;
Retain msn 1. ssn 1.0 nobs 1000 err 21999;
Do I = 1 to nobs;
If msn + ssn*rannor(err) > rannor(err) then Resp = 1;
  else Resp = 0.;
Output;
Keep Resp;
End;
Proc Means Data=A;
Run;
```

Following is an SPSS equivalent:

USING SPSS TO SIMULATE FORCED-CHOICE SIGNAL DETECTION DATA

```
input program.
loop #i=1 to 1000.
  compute #msn=1.
  compute #ssn=1.
```

```
compute resp=0.
 if(#msn+#ssn*normal(1) gt normal(1)) resp=1.
end case.
end loop.
end file.
end input program.
execute.
```

▣ PROBLEMS

6.1. After having made the modification to the multiple regression data set noted in the text, modify it again to obtain a different threshold, specifically one that divides outcomes in a 2:1 favorable manner.

6.2. Modify and generate the psychometric function/dose–response curve so that its mean is 180 and its standard deviation is 20.

6.3. Visually compare the function generated in Problem 6.2 to one having a mean of 185 and a standard deviation of 20. Repeat this exercise comparing the first function to one having a mean of 200 and a standard deviation of 20. Finally, compare the function generated in Problem 6.2 to one having a mean of 180 and a standard deviation of 15.

6.4. Generate simulated binary items for a six-item test, modifying the parameters given in the text as follows: (a) change only the guessing (c) parameters so that they are each .2, (b) change only the discrimination (b) parameters so that they are (−.7, −.2, .−1, .1, .2, and .7), and (c) change only the difficulty (a or threshold) parameters so that they are all .5. In each case, repeat the analysis performed in the text.

6.5. Generate simulated Likert items by increasing each of the threshold values by .5. Repeat the analysis performed in the text.

6.6. Generate simulated multiple-choice items, modifying the parameters given in the text as follows: (a) change only the discrimination (b) parameters to .9, (b) change only the guessing (c) parameters by reducing them .25 from the values given in the text, and (c) change only the difficulty parameters by reducing each by .3 from the values given in the text. Repeat the analysis performed in the text.

6.7. Simulate Thurstone paired-comparison data first by using scale values of .0, .50, .70, .90, and 1.30, holding the standard deviations constant at .10. Now change the five standard deviations to .10, .20, .10, .20, and .10.

6.8. Simulate signal detection data using the following parameters: (a) $\mu_{s+n} = .8$, $\sigma_{s+n} = 1$ and (b) $\mu_{s+n} = .5$, $\sigma_{s+n} = 1.4$ (assume in both cases that $\mu_n = 1$ and $\sigma_n = 1$). Obtain hit rates and false alarm rates for both sets of data using criteria of −.5, 0, and 1. A plot of false alarm rate on the ordinate and hit rate on the abscissa is known as a **receiver operating characteristic** (ROC) curve. Plot ROC curves for the data obtained in the original simulation and these two new sets of data on the same axes.

6.9. Use the two values of μ_{s+n} and σ_{s+n} from Problem 6.8 to obtain confidence-rating data. In each case, use criteria of −.7, −.5, −.3, .7, 1.0 and 1.6. Compare the response distributions to each other's and to those generated from the simulations presented in the text.

Part III

7

Exploratory Factor Analysis

1. We begin by considering a principal component analysis of a previously presented data set that consists of six variables defined by a single underlying factor plus unique error. The variables were also standard normal ($\mu = 0$, $\sigma = 1.0$) in the population.

2. Next, the common factor model is considered.

3. Factor analyses are then performed on two previously considered data sets and then examined. Each consists of six observed variables based on two underlying factors. In one analysis, variables within each cluster are correlated with each other but not with the variables in the other cluster. In the other, there is also a moderate correlation between variables in the two clusters.

4. Factor score calculation is then considered.

5. Finally, maximum likelihood exploratory factor analysis is presented.

▣ A UNIFACTOR STRUCTURE

Component Versus Common Factor Analysis

All relevant statistical packages allow the user a choice between what is known as component analysis and common factor analysis. A detailed treatment of this distinction and its implications is beyond the scope of this book but may be found in numerous sources; see, for example, Gorsuch (1983) and Nunnally and Bernstein (1994). Basically, common factor analysis seeks to partition the variance of each variable into three independent sources: (a) **common** variance (variance that is potentially or actually sharable with other variables), (b) **error** variance (unreliability), and (c) **specific** variance (variance that is systematic but unrelated to other measures). Usually, the latter two sources are left unanalyzed to form **unique** variance. This decomposition is not performed in component analysis. It can also be shown that common factors are **inferred** in the sense that they are only imperfectly related to observed variables whereas components are **exact** (**observable**) linear combinations of other variables. Component analysis is performed by using unities in the diagonal of a correlation matrix or observed variances in the diagonal of a variance–covariance matrix. Common factor analysis uses smaller values, though the exact magnitudes depend on the specific method, of which there are several. Terminologically, many authors contrast "component analysis" with "factor analysis." We prefer to contrast "component analysis" with "common factor analysis," both being forms of factor analysis, though this is an issue that generates much controversy. Empirically, their results tend to converge as either the number of variables being analyzed or the average correlation increases.

Note that this chapter is concerned with two approaches to explaining variables in terms of linear combinations. Both what we will call principal component factor analysis (some do not consider this to be "true" factor analysis) and common factor analysis are options under SAS `Proc Factor`. In contrast, component analysis per se, as exemplified by SAS `Proc Princomp`, is a somewhat different model, though both it and principal component factor analysis work with the same set of eigenvalues and eigenvectors.

Basic Factor Analysis Using SAS

We begin by analyzing the data set named "One" in Chapter 4. Recall that this data set was constructed to have six variates that were standardized in the population and that correlations between any pair were .49. The basic SAS command to perform either a component or a common factor analysis on these data is `Proc factor data=one;var s1-s6;`. The keyword `data=one` is unnecessary if this is the only or most recently created data set and the keyword `var s1-s6;` is unnecessary if these are the only variables in the data set. However, it is good practice to include both the data set name and the variables to be factored so as to minimize confusion when you later reexamine or someone else examines what you have done. Numerous other options, which can also appear as keywords, are provided to control the factoring process. Some of the more important are as follows:

1. Choosing the method of factor extraction; for example, `m=ml` means to perform maximum likelihood analysis—the default is a component analysis.

2. Defining the number of factors to be extracted; for example, `n=2` means to obtain a maximum of two factors—the default depends on the method of factor extraction.

3. Controlling the rotation process; for example, `r=v` means to perform a varimax rotation—the default is no rotation and other keywords afford further control over rotation.

4. Controlling the output, specifically the keyword `all`, which requests all output to be printed (however, a scree plot requires the additional keyword `scree`).

5. Controlling the communality estimates (values to be placed in the diagonal of the correlation matrix)—`Priors` allows the user to use reliabilities.

6. Controlling the statistics to be analyzed—normally correlations are factored, but specifying `Covariance` uses these statistics.

SAS also has a procedure specific to component analysis, not surprisingly called `Proc Princomp`, but it really adds little to the capabilities of `Proc Factor`. You will spend your time better learning the details of `Proc Factor` if you intend to perform component and/or common factor analysis extensively in SAS.

The printout begins with the means and standard deviations (for a complete listing of the output of `Proc Factor`, see the *SAS/STAT User's Guide*, Vol. 1, pp. 797–800 [SAS Institute, 1989a]). These univariate statistics can readily be checked for such elemental errors as the wrong variables (which everyone, including the authors, has specified at some time).

Means and Standard Deviations From 1000 Observations

	S1	S2	S3	S4	S5	S6
Mean	0.02856242	0.0420939	0.02083566	0.00671508	-0.0036179	0.01144711
Std Dev	0.98756626	0.97245083	0.99610162	0.99266058	0.99163399	1.03392282

The correlation matrix (commonly symbolized as **R**, but not in the SAS printout) follows:

	S1	S2	S3	S4	S5	S6
S1	1.00000	0.50205	0.51498	0.48882	0.46773	0.48153
S2	0.50205	1.00000	0.48711	0.47474	0.46698	0.51624
S3	0.51498	0.48711	1.00000	0.49391	0.48224	0.51493
S4	0.48882	0.47474	0.49391	1.00000	0.46223	0.50260
S5	0.46773	0.46698	0.48224	0.46223	1.00000	0.46896
S6	0.48153	0.51624	0.51493	0.50260	0.46896	1.00000

The next item is the inverse of the correlation matrix, commonly symbolized as $\mathbf{R}^{-1}$:

	S1	S2	S3	S4	S5	S6
S1	1.69773	-0.32350	-0.34901	-0.28399	-0.24305	-0.21408
S2	-0.32350	1.68989	-0.23665	-0.23068	-0.24683	-0.36306
S3	-0.34901	-0.23665	1.73979	-0.27916	-0.27943	-0.33430
S4	-0.28399	-0.23068	-0.27916	1.65908	-0.24105	-0.32122
S5	-0.24305	-0.24683	-0.27943	-0.24105	1.58470	-0.23365
S6	-0.21408	-0.36306	-0.33430	-0.32122	-0.23365	1.73367

This matrix has many important properties. One is that the squared multiple correlation between the variable j and the remaining variables in the set R_j^2 is given by

$$R_j^2 = 1 - \frac{1}{r^{jj}}, \tag{7.1}$$

where r^{jj} is the value of this variable on the diagonal of $\mathbf{R}^{-1}$. Thus, given that $r^{33} = 1.73979$, $R_3^2 = 1 - 1/1.73979 = .425$. If R_j^2 is small, that variable has little in common with the remaining variables (a related index follows). Conversely, if R_j^2 is close to 1.0, it is difficult to compute the inverse for models that require it, a situation known as **multicollinearity**. An even more severe computational problem is that a variable that is linearly dependent on others in the matrix, such as an average of several variables where the individual variables are themselves included, precludes computation of the inverse. It is difficult to think of circumstances in which including a variable that has a linear dependency is wise. Component solutions do not require an inverse but variables with either a large or a small R_j^2 should be considered for deletion from the analysis.

Partial correlations controlling all other variables likewise are also obtainable from $\mathbf{R}^{-1}$ and are available for printout. For example, the partial correlation between variables 3 and 4, partialling out variables 1, 2, 5, and 6, can be written

$$(r_{34.1256}) = -\frac{r^{34}}{\sqrt{r^{33}r^{44}}} = -\frac{-.27916}{\sqrt{1.73979 \cdot 1.65908}} = -.164.$$

A low value means that the covariance between the two variables is also shared with the other variables in the data set.

```
Partial Correlations Controlling All Other Variables
           S1         S2         S3         S4         S5         S6
S1    1.00000    0.19099    0.20308    0.16921    0.14818    0.12478
S2    0.19099    1.00000    0.13802    0.13777    0.15083    0.21211
S3    0.20308    0.13802    1.00000    0.16431    0.16829    0.19249
S4    0.16921    0.13777    0.16431    1.00000    0.14866    0.18940
S5    0.14818    0.15083    0.16829    0.14866    1.00000    0.14096
S6    0.12478    0.21211    0.19249    0.18940    0.14096    1.00000
```

Kaiser's measure of sampling adequacy is a function of the difference between these values and the ordinary (Pearson) correlations. As noted in the *SAS/STAT User's Guide*, Vol. 1 (SAS Institute, 1989a) and elsewhere, values above .8 for individual variables indicate that the variable in question is satisfactory for inclusion in the analysis, and values below .5 suggest deletion or other modification.

```
Kaiser's Measure of Sampling Adequacy: Overall MSA = 0.89508027
       S1         S2         S3         S4         S5         S6
0.893482   0.893967   0.890707   0.898565   0.905580   0.889623
```

Next appears the eigenanalysis (see Appendix A for a detailed examination of its mechanics). The default in SAS is to use values of 1.0 in the diagonals of $\mathbf{R}$ (component analysis). This is explicitly stated in the printout and is a check that the intended factoring method was actually used. This statement is followed by the sum of the eigenvalues (which equals the number of variables in a component analysis, i.e., 6 in the present case) and their average (1.0 in the present case). Next are the specific eigenvalues (which, following standard notation, will be denoted λ_i), the difference between successive values (which is useful for users of the scree criterion to define the number of factors), the proportion of variance accounted for ($\lambda_I/\sum\lambda$), and, finally, the cumulative proportion. In a component analysis, the total variance $\sum\lambda = k =$ the number of variables, so the proportion of variance accounted for by a given eigenvector is λ_I/k.

	1	2	3	4	5	6
Eigenvalue	3.4425	0.5503	0.5277	0.5227	0.4999	0.4569
Difference	2.8922	0.0226	0.0050	0.0228	0.0430	
Proportion	0.5737	0.0917	0.0880	0.0871	0.0833	0.0761
Cumulative	0.5737	0.6655	0.7534	0.8405	0.9239	1.0000

These eigenvalues are plotted in Figure 7.1. They are used by default to determine the number of factors SAS retains in several (but not all) methods of extracting factors. The most common form this criterion takes is the Kaiser–Guttman rule (Guttman, 1954; Kaiser, 1960, 1970), which is to retain as many factors as there are eigenvalues greater than or equal to 1.0. However, this method is not necessarily the best one. Gorsuch (1983), Nunnally and Bernstein (1994), and Zwick and Velicer (1982, 1986) are among those who have discussed this topic at length. The printout states how many factors meet whatever criterion is used and identify the chosen method of initial factor extraction.

1 factors will be retained by the MINEIGEN criterion.
Initial Factor Method: Principal Components

The eigenvector(s) corresponding to each factor (one in the present case) are then presented. These are standardized, meaning that the sum of the elements squared is 1.0. They are of relatively limited interest.

S1 0.41010
S2 0.40901
S3 0.41535
S4 0.40554
S5 0.39500
S6 0.41415

Note that $.41010^2 + .40901^2 + \cdots + .41415^2 = 1$.

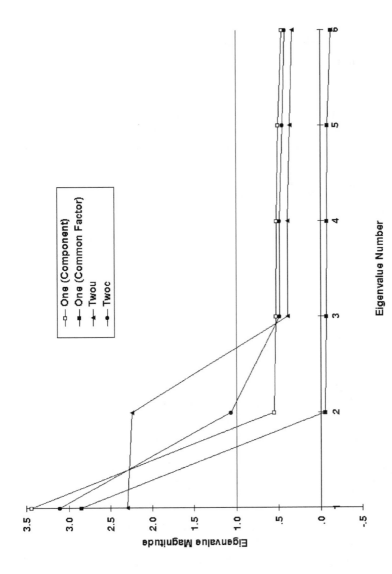

Figure 7.1. Scree plot for (a) component solution of data containing a single factor, (b) common factor solution of data containing a single factor, (c) component solution of data containing two factors in which the clusters are independent of one another (Twou), and (d) component solution of data containing two factors in which the clusters are correlated (Twoc).

The factor pattern is of more interest. Each element of the pattern (b_i) equals the square root of the eigenvalue (λ) times the corresponding element of the normalized eigenvector. In the present case,

$$b_1 = \sqrt{3.4425} \text{ (the square root of } \lambda_1)$$
$$\times 0.41010 \text{ (the first element of the normalized eigenvector)}$$
$$= 0.76090.$$

Pattern elements explain variables in terms of components/factors. An initial solution or an orthogonal rotated solution is a right-angled solution (if there is more than one factor) in which the pattern element equals the structure element or correlation between variable and component/factor, so the structure elements need not be reported. However, they are reported in an oblique solution. The following are the pattern elements:

S1   0.76090
S2   0.75887
S3   0.77063
S4   0.75243
S5   0.73289
S6   0.76842

Note that the magnitudes of the pattern elements should all be .7 in the population; all obtained values are slightly overestimated. Snook and Gorsuch (1989) have noted this tendency for a component solution to overestimate the magnitude of pattern weights.

Squaring and summing the pattern elements for each factor, that is, computing $\sum b_i^2 = 0.76090^2 + 0.75887^2 + \cdots + .76842$, gives the total variance explained by that factor, in this case, 3.442475. Because only one factor was extracted, this is also the **total** variance explained by all of the factors, but in a multifactor solution, the total is the sum of these individual values. Other programs divide by the number of variables to obtain the proportion of variance explained, which is .573 here (for some reason, SAS has resisted making this simple calculation). Others find it more convenient to express the result as a percentage (57.3%).

SAS then provides what it calls `Communality estimates for each variable`, which are commonly symbolized as h^2. These are the sums of the squared pattern elements across factors. Because there is only one factor, these equal the squared pattern elements (e.g., $0.76090^2 = .578963$ for S1). We prefer to call these values "communalities" as we reserve the term "communality estimates" for the quantities placed in the diagonal of the correlation matrix (R). Regardless, the full set of h^2 values is as follows:

S1	S2	S3	S4	S5	S6
0.578963	0.575891	0.593875	0.566154	0.537124	0.590468

Scoring coefficients (also known as factor score weights) describe (in a component solution) or estimate (in a common factor solution) component/factor scores from observed variables. Computational methods are not and cannot be unique. Gorsuch (1983), Harman (1976), and Mulaik (1972) discuss this topic in depth. The output notifies the user that the regression method has been used. Also, because this is a component solution, there is a perfect multiple correlation between component/factor scores and the observed variables that give rise to them.

```
Scoring Coefficients Estimated by Regression
Squared Multiple Correlations of the Variables With
Each Factor
   FACTOR1
   1.000000
Initial Factor Method: Principal Components
```

Principal component weights (w_{ij}) are defined as

$$\frac{b_{ij}}{\lambda_j} = \frac{e_{ij}}{\sqrt{\lambda_j}},$$

where w_{ij} is the weight to be applied to variable j in estimating factor i, b_{ij} is the corresponding pattern weight, λ_i is the ith eigenvalue, and e_{ij} is the corresponding weight on the normalized eigenvector. In the present case, there is only one factor to consider so $\lambda_1 = \lambda = 3.442475$, and $w_{16} = .76090/\sqrt{3.442475} = 0.22103$. The complete vector of scoring coefficients is as follows:

```
S1  0.22103
S2  0.22044
S3  0.22386
S4  0.21857
S5  0.21290
S6  0.22322
```

What is known as the **residual correlation matrix** follows. This is obtained by subtracting the estimated correlation matrix (in an orthogonal solution such as the present, $\hat{\mathbf{R}} = \mathbf{BB'}$, where $\mathbf{B}$ is the factor pattern) from the actual correlation matrix ($\mathbf{R}$). The term "residual correlation" is a misnomer. Differences between correlations are covariances rather than correlations. Note that in a component solution the off-diagonal values tend to be negative, implying that the values in $\mathbf{B}$ are too large (Snook and Gorsuch, 1989). The values on the diagonal are equal to 1 minus the communality estimates, defined previously. It is desirable that these values be small and without pattern, but the matrix of partial correlations controlling other factors, which is defined below, should be consulted as well to this end.

	S1	S2	S3	S4	S5	S6
S1	0.42104	-0.07538	-0.07139	-0.08370	-0.08992	-0.10316
S2	-0.07538	0.42411	-0.09770	-0.09626	-0.08919	-0.06690
S3	-0.07139	-0.09770	0.40613	-0.08594	-0.08254	-0.07723
S4	-0.08370	-0.09626	-0.08594	0.43385	-0.08922	-0.07558
S5	-0.08992	-0.08919	-0.08254	-0.08922	0.46288	-0.09421
S6	-0.10316	-0.06690	-0.07723	-0.07558	-0.09421	0.40953

The **root mean square off-diagonal residuals** are obtained by taking the square root of the average of the squares of these "residual correlations," ignoring entries on the diagonal. Note that the square root of an average of squares appears in a variety of statistical contexts and is referred to as a root mean square (RMS). This is done both for individual variables (i.e., the five terms within a given column) and for all terms. Taking S1 as an example,

$$\{[(-0.07538)^2 + (-0.07139)^2 + \cdots + (-0.10316)^2]/2\}^{1/2} = 0.085453.$$

The set of all terms is as follows:

```
Root Mean Square Off-Diagonal Residuals: Overall = 0.08581600
        S1          S2          S3          S4          S5          S6
0.085453    0.085934    0.083435    0.086408    0.089094    0.084463
```

The root mean square off-diagonal residuals may be converted to what SAS denotes as partial correlations controlling other factors quite simply (these are often called residual correlations). Let $\mathbf{C}$ denote any covariance matrix, c_{ij} denote any off-diagonal entry in this matrix, and c_{ii} and c_{jj}

denote the diagonal entries associated with c_{ij}. The correlation between any two variables, r_{ij}, is given by

$$r_{ij} = \frac{c_{ij}}{\sqrt{c_{ii}c_{jj}}}. \tag{7.2}$$

For example, -0.08594, 0.40613, and 0.43385 are c_{12}, c_{11}, and c_{22}, respectively. Then $r_{12} = -0.08594/(0.40613 \cdot 0.43385)^{1/2} = -0.204$. Velicer (1976; Zwick and Velicer, 1982, 1986), in particular, has argued that such residuals are important in determining the number of factors to retain, and few would argue about the importance of examining clusters of correlations present in residual correlation matrices.

The complete residual correlation matrix is as follows:

	S1	S2	S3	S4	S5	S6
S1	1.00000	-0.17838	-0.17264	-0.19585	-0.20368	-0.24842
S2	-0.17838	1.00000	-0.23541	-0.22441	-0.20130	-0.16051
S3	-0.17264	-0.23541	1.00000	-0.20474	-0.19038	-0.18938
S4	-0.19585	-0.22441	-0.20474	1.00000	-0.19910	-0.17932
S5	-0.20368	-0.20130	-0.19038	-0.19910	1.00000	-0.21638
S6	-0.24842	-0.16051	-0.18938	-0.17932	-0.21638	1.00000

Finally, these correlations are converted to root mean square terms, as were the covariances.

```
Root Mean Square Off-Diagonal Partials: Overall = 0.20131074
       S1        S2        S3        S4        S5        S6
 0.201584  0.201933  0.199627  0.201212  0.202342  0.201156
```

Basic Factor Analysis Using SPSS

To perform a factor analysis in SPSS, select **Data Reduction** from the Analyze menu followed by **Factor** ([Alt]-A,D,F). This brings down a box in which you identify the variables to be factored and select, in turn, five further selections. You click on a list of names, either individually or in groups, which then appear under **Variables**.

The five selections primarily control what is to be output and are organized as follows:

1. **Descriptives.** This gives the option of providing univariate statistics, the initial solution, the correlation matrix, and several of its properties, such as its determinant and its inverse.

2. **Extraction**. This provides alternative ways to extract factors (principal components, principal axis, maximum likelihood, etc.), as well as allowing the user to define the criterion for factor extraction (minimum eigenvalue, e.g., Kaiser–Guttman vs. set number of factors) and to obtain a scree plot. The maximum number of iterations is set here, when applicable (e.g., in maximum likelihood factoring).

3. **Rotation**. No rotation, three orthogonal rotations (varimax, equamax, and quartimax), and two oblique rotations (promax and direct oblimin) are provided as options. Also provided are options to select the form of printout.

4. **Scores**. This provides an option to save factor scores and to compute them by three different methods.

5. **Options**. This controls how missing values are handled and how printout of coefficients is ordered.

There is not much different about the way SPSS reports basic factor analytic output. The quantities are labeled well in both cases (a minor point is that what SAS refers to as `Kaiser's Measure of Sampling Adequacy` is called `Kaiser-Meyer-Olkin Measure of Sampling Adequacy` in SPSS. Other than that, it is rather simple to make menu choices that parallel SAS commands in performing factor analysis. Unfortunately, going in the reverse direction is a bit more difficult because of the specifics of the SAS language. Consequently, we will focus on SAS commands and output for the remainder of this chapter, though we will note points of relevance to SPSS users, where appropriate.

Classical Common Factor Analysis

We will use the term "classical common factor analysis" to refer to the factoring of a correlation matrix using some form of communality estimation in which interest is focused on accounting for variance. This is in contrast to the use of unity communalities in component analysis, discussed previously, and inferentially based procedures such as maximum likelihood exploratory factor analysis, discussed later in this chapter, as well as with confirmatory factor analysis and the factoring of other than correlation matrices, also discussed later in this chapter.

SAS offers several options to perform classical common factor analysis. Three communality estimation algorithms that are especially worthy of note are as follows:

1. **Squared multiple correlations (SMCs)**. This may be accomplished by modifying the preceding principal component example to read `Proc factor method = principal all priors = smc data=one;`. An m can be used as an alias for method; `prin` or p can be used as an alias for principal, and the keyword `all` again provides all output. The relationship between the squared multiple correlations and the diagonal elements of the inverse of the correlation matrix was noted before.

2. **Reliabilities**. This is a possibility when the variables being factored are themselves scale scores whose reliabilities can be computed by such means as the Alpha option of `Proc Corr`. This option is chosen as `Proc factor method = principal data=one; Priors xx xx xx xx;` (where each `xx` denotes a value and the order corresponds to that given in the `var` statement or the order in the file—there must be exactly as many values given in the `Priors` statement as there are variables in the analysis). Although this option is not chosen very often, it has some interesting features when used to analyze unique variance (i.e., specific variance and error variance). This was discussed in Harman (1976). Bernstein, Jaremko, and Hinkley (1994) provide a recent example.

3. **Iteration**. This involves using a preliminary value, typically an SMC, as communality estimates, factoring, obtaining communalities (h^2 values), comparing them, using the communalities as new communality estimates, and so on, until the communality estimates and communalities converge, which is not guaranteed.

The preferred method is a matter of historic debate; ideally, your results should not depend greatly on which method has been chosen. Differences between component and common factor analysis will tend to stand out most with the use of SMC communality estimates, so they will be used in the demonstration.

The initial output (means, standard deviations, correlation matrix, inverse of correlation matrix, partial correlations controlling other factors, and Kaiser's measures of sampling adequacy) are identical to those noted in a component analysis. The first indication that the analysis is any different is that the values of R_j^2 (the SMC) are presented. These values, which can be obtained from the $\mathbf{R}^{-1}$ presented previously and Equation 7.1, are

S1	S2	S3	S4	S5	S6
0.410979	0.408245	0.425218	0.397256	0.368965	0.423190

The current version of SAS only presents the eigenvalues obtained using these communality estimates in the diagonal of $\mathbf{R}$, though older versions and other programs present both the component eigenvalues and those derived from communality estimates. The eigenanalysis is as follows:

```
Eigenvalues of the Reduced Correlation Matrix: Total = 2.43385161 Average = 0.40564193
```

	1	2	3	4	5	6
Eigenvalue	2.8489	−0.0578	−0.0682	−0.0767	−0.0853	−0.1270
Difference	2.9066	0.0105	0.0084	0.0087	0.0417	
Proportion	1.1705	−0.0237	−0.0280	−0.0315	−0.0351	−0.0522
Cumulative	1.1705	1.1468	1.1188	1.0873	1.0522	1.0000

An essential point is that the first eigenvalue (λ_1) is smaller (2.8489) than it was in the component analysis (3.4425) due to the reduction in magnitude of the communality estimates. This difference in magnitude has implications throughout the analysis. SPSS provides a somewhat more abbreviated listing of the eigenanalysis.

In addition, instead of using the Kaiser–Guttman $(\lambda > 1)$ criterion for default factor retention when communalities are specified, SAS uses a proportion of variance criterion by default. SAS does provide preliminary (component) eigenvalues in some classical common factor approaches, specifically, when iterations are used. Still, these data provide only a single factor, as they should. SPSS, however, uses the Kaiser–Guttman criterion by default and also allows users to specify the number of factors, but does not employ a proportion of variance criterion.

The next point to note is that the eigenvector elements, pattern elements, variance(s) explained, communality (estimates, h^2), and related terms are smaller than their component counterparts, reflecting the smaller value of λ from which they were derived. Moreover, they are nearly equal in magnitude to one another. This is because a common factor analysis attempts to eliminate measurement error. In the present case, there was no difference among the variables once this component was eliminated. The elements of the normalized eigenvector are as follows:

```
S1  0.41075
S2  0.40929
S3  0.41802
S4  0.40429
S5  0.38996
S6  0.41655
```

The pattern elements, which can be derived from the eigenvector elements, are as follows. Note that they are smaller than their component counterparts; indeed, they are all nearly equal to .70:

```
S1  0.69328
S2  0.69082
S3  0.70556
S4  0.68238
S5  0.65819
S6  0.70307
```

Accordingly, the variance explained by the factor and the individual communality estimates are

```
Final Communality Estimates: Total = 2.848853
      S1        S2        S3        S4        S5        S6
 0.480644  0.477230  0.497810  0.465645  0.433219  0.494306
```

Because this is a common factor solution, the squared multiple correlations of the variables with the factor is less than 1.0 at 0.827742. Similarly, the scoring coefficients are

```
S1  0.20301
S2  0.20103
S3  0.21262
S4  0.19442
S5  0.17810
S6  0.21081
```

These values are smaller than their counterparts in the component solution for the same reason that the pattern elements are smaller. Their lesser magnitudes cause the variance of factor scores to be less than the 1.0 value, which is the variance of component scores. In fact, their variance in the original sample will be precisely 0.827742.

The residual "correlations" are as follows:

	S1	S2	S3	S4	S5	S6
S1	0.51936	0.02311	0.02583	0.01573	0.01142	-0.00590
S2	0.02311	0.52277	-0.00030	0.00334	0.01229	0.03055
S3	0.02583	-0.00030	0.50219	0.01245	0.01785	0.01888
S4	0.01573	0.00334	0.01245	0.53436	0.01309	0.02284
S5	0.01142	0.01229	0.01785	0.01309	0.56678	0.00620
S6	-0.00590	0.03055	0.01888	0.02284	0.00620	0.50569

Note that these values are both smaller in absolute value and more symmetrically distributed about .0 than were the component values. Consequently, the RMS residual values are smaller, both overall and for the individual variables:

```
Root Mean Square Off-Diagonal Residuals: Overall=0.01687349
      S1        S2        S3        S4        S5        S6
0.017967  0.018053  0.017305  0.014876  0.012726  0.019413
```

The same is true for the partial correlations controlling other factors (the actual residual correlations), which will not be presented.

▣ MULTIFACTOR STRUCTURES USING DATA WITH INDEPENDENT CLUSTERS

Chapter 4 had an SAS data step called "Two," which generated a two-factor solution. This was run twice. In the first case, the data step was left as originally presented, except that the name of the data set was changed to "Twou." In the second case, denoted "Twoc," the first three lines were changed as follows:

```
Data Twoc;
Retain bI1-bI6 (.7 .7 .7 .25 .25 .25);
Retain bII1-bII6 (.25 .25 .25 .7 .7 .7);
```

The population correlation between any two variables is the sum of the cross products in the two retain statements. Thus, in data Twou, any correlations among y_1 to y_3 will equal $.8^2 + .0^2 = .64$, and any correlations among y_4 to y_6 will equal $.0^2 + .8^2 = .64$. However, other population correlations will be 0 (e.g., $r_{14} = .8 \cdot 0 + .0 \cdot .8$. In contrast, population correlations within y_1 to y_3 and within y_4 to y_6 will equal .55 ($.7^2 + .25^2$ or the reverse order), whereas correlations between variables in the two sets will equal .35 ($.7 \cdot .35 + .35 \cdot .7$) in data set Twoc.

The means and standard deviations for the two data sets were as follows:

Variable	Twou Mean	Twou S.D.	Twoc Mean	Twoc S.D.
Y1	−.06	.99	−.06	.98
Y2	−.04	.98	−.03	.97
Y3	−.01	1.01	.00	1.00
Y4	.01	.98	.00	.98
Y5	.01	1.00	.00	1.00
Y6	.01	1.02	.00	1.02

The following matrix contains correlations from data Twou above the diagonal and data Twoc below the diagonal:

	Y1	Y2	Y3	Y4	Y5	Y6
Y1	1.00	.65	.63	.00	−.01	.02
Y2	.56	1.00	.62	−.01	.01	.00
Y3	.54	.52	1.00	−.04	−.03	−.02
Y4	.36	.34	.30	1.00	.62	.63
Y5	.35	.37	.31	.52	1.00	.64
Y6	.38	.35	.33	.54	.56	1.00

Each set was factored using the following commands:

```
Proc factor all data=twou rotate=promax;var y1-y6;
Proc factor all data=twoc rotate=promax;var y1-y6;
```

In other words, component solutions were generated (what is to be generated does not depend on the choice of model) and, assuming a multifactor solution, an oblique promax rotation was applied. By default, a promax rotation uses an orthogonal prerotation, varimax, so this option allows for the exploration of both correlated and uncorrelated rotations. The "u" and "c" at the end of the two data set names are intended to denote "uncorrelated" and "correlated," respectively. However, the factors they produce should not be thought of as uncorrelated or correlated. What will happen is that the uncorrelated data will produce factors that are relatively uncorrelated with any reasonable rotation strategy. In contrast, the correlated data set can also be described in terms of an uncorrelated solution, but this solution will have some shortcomings compared to an oblique rotation.

As with other simulations, the sample values for the statistics of basic interest (the correlations) were similar to the population value. For example, within-cluster correlations for Twou ranged from .62 to .65 and between-cluster correlations ranged from .00 to .02. These ranges were from .52 to .56 and from .30 to .38 for Twoc.

The two sets of eigenvalues appear in Figure 7.1. Note that, in both cases, the first two eigenvalues exceed the Kaiser–Guttman $\lambda > 1$ criterion. However, there is an important difference between the two. In the case of data set Twou, these two eigenvalues are nearly equal (and would be in the population), whereas, in the case of Twoc, they are not. In the former case, the independence of the two clusters means they contribute independently; in the latter case, the fact that they are correlated means that part of their contribution to the explanation of the total is redundant.

It is useful to consider the factor pattern, the variances explained by each factor, and what SAS calls the communality estimates in a two-factor solution. These are as follows for Twou:

```
Factor Pattern
       FACTOR1  FACTOR2
Y1     0.66787  0.56805
Y2     0.67012  0.55817
Y3     0.68765  0.52081
Y4    -0.56230  0.65352
Y5    -0.55690  0.66550
Y6    -0.54955  0.67959
Variance Explained by Each Factor
   FACTOR1    FACTOR2
   2.296293   2.237304
Final Communality Estimates: Total = 4.533598
        Y1         Y2         Y3         Y4         Y5         Y6
   0.768724   0.760618   0.744110   0.743269   0.753027   0.763851
```

The formula describing variance explained becomes a bit complicated because there are now two factors, so we will refer to it as $\sum b_{ij}^2$, where the subscript i denotes the variable and j denotes the factor. Consequently, $0.66787^2 + 0.67012^2 + \cdots - 0.54955^2 = 2.296293$ and

$0.56805^2 + 0.55817 + \cdots + 0.67959^2 = 2.237304$; that is, the variances accounted for reflect summing and squaring columns. Because these initial factors are independent, the total (4.533598) is the sum of the two individual values (2.296293 and 2.237304). The communality "estimates" (h^2 values) are obtained by summing and squaring along rows so $0.66787^2 + 0.56805^2 = 0.768724$, and so on.

Figure 7.2 is a scatterplot of the factor pattern. The point to note is that the six variables "float free" of the axes (factors) so that one factor does not define a variable any better than another. In other words, the values are always intermediate—they are never close to 1.0 (which would mean that the variable is well defined by the factor) or .0 (which would mean that the variable has nothing to do with the factor). However, if you were to draw lines from the origin of the graph through the middle of each of the two sets of points, the lines would be nearly at right angles to one another. This is a subjective method of defining a **rotation**; we will now turn to a consideration of analytic methods.

Orthogonal Rotation

If a promax rotation is chosen, SAS identifies the next section as a "prerotation," as it is a step toward an ultimate promax rotation in this problem, but this can be regarded as an end in itself. The default, which is usually a reasonable choice, is varimax, an orthogonal solution whose properties are of interest.

The first thing to note is the transformation matrix, which has the following form when there are two factors:

$\cos(\theta) \quad -\sin(\phi)$
$\sin(\theta) \quad \cos(\phi)$

where θ is the amount by which factor I is rotated and ϕ is the amount by which factor II is rotated. However, in an orthogonal rotation, $\theta = \phi$. In the present case, this matrix is:

```
            1           2
1    0.77178    -0.63588
2    0.63588     0.77178
```

$\theta = \cos^{-1}(.77178) = 39.5°$. Note that $.77178^2 + .63588^2 = 1$.

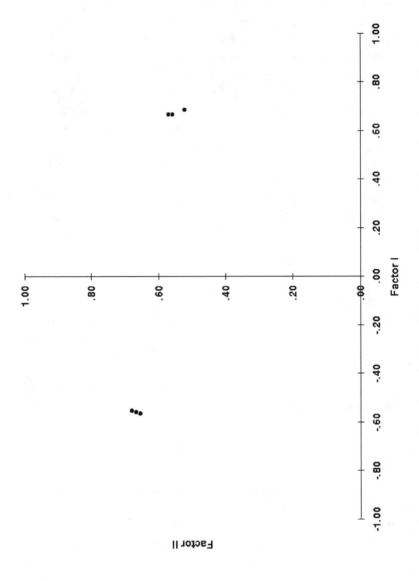

Figure 7.2. Initial factor pattern derived from data reflecting two independent clusters (Twou).

Applying the transformation matrix to the initial (unrotated) factors provides the rotated factor loadings (pattern elements), as follows:

	FACTOR1	FACTOR2
Y1	0.87666	0.01372
Y2	0.87212	0.00467
Y3	0.86189	-0.03532
Y4	-0.01841	0.86193
Y5	-0.00662	0.86775
Y6	0.00801	0.87395

These data are plotted in Figure 7.3. Note that the elements are either much larger (between .86 and .88) or smaller (between –.03 and .02). They fall very near the axes so the axes describe the variables well; these axes reflect what the variables have in common. The factor variances remain nearly equal at 2.272441 and 2.261157; the total remains unchanged at 4.533598. SAS then provides individual and total communality estimates, but these are identical to what was obtained with the initial

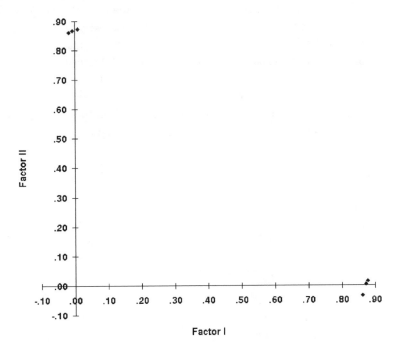

Figure 7.3. Varimax rotated factor pattern derived from data reflecting two independent clusters (Twou).

factors—rotation does not add to or subtract from the variance explained. Next, SAS provides the scoring coefficients. Because this is a component solution, the squared multiple correlations of the variables with each factor are 1.0.

The standardized scoring coefficients are as follows:

	FACTOR1	FACTOR2
Y1	0.38592	0.01101
Y2	0.38387	0.00698
Y3	0.37914	-0.01076
Y4	-0.00325	0.38115
Y5	0.00198	0.38379
Y6	0.00845	0.38661

Note that whereas the pattern loadings for variables Y_1 to Y_3 on rotated factor I and for variables Y_4 to Y_6 on rotated factor II were nearly .8, they are roughly .4 here. This is an artifact of scaling. It is more important that the converse elements are still nearly .0. This means that Y_1 to Y_3 will not have anything to do with defining scores on rotated factor I, and Y_4 to Y_6 will not have anything to do with defining scores on rotated factor II.

A promax rotation takes the prerotation pattern (varimax in this case) and defines a **target matrix** by raising the pattern elements to a power. The factors are then rotated separately in an attempt to make them fall as close as possible to this target matrix. The matrix providing this best fit is called a **Procrustean transformation matrix**. It is specific to promax so you will not find it if you choose some other oblique rotation such as Harris–Kaiser or direct oblimin.

	1	2
1	1.14787	0.00716
2	0.00724	1.15131

This matrix is then normalized so that its sums of squares of elements in columns (but not necessarily rows) are 1.0. This is called the **normalized oblique transformation matrix**.

	1	2
1	0.76782	-0.63112
2	0.64079	0.77578

Factor I is rotated by $\cos^{-1}(.76782) = 39.8°$, and factor II is rotated by $\cos^{-1}(.77178) = 39.1°$. The correlation between the two factors is

–0.01253. A real problem that led to an angle and correlation this small would dictate retaining the simpler geometry of the varimax rotation. There is still vigorous debate about whether to employ orthogonal or oblique rotations in exploratory factor analysis, but even the most die-hard advocate of oblique rotations would not use one here. Although we could continue with the description of the data in Twou, it will be far more instructive to turn to the data in Twoc.

▣ MULTIFACTOR STRUCTURES USING DATA WITH RELATED CLUSTERS

Recall that the within-cluster correlations ranged from .52 to .56 and the between-cluster correlations ranged from .30 to .38 for data Twoc. The initial factor pattern is as follows:

```
Factor Pattern
      FACTOR1    FACTOR2
Y1    0.74046    0.39112
Y2    0.73051    0.39874
Y3    0.68899    0.46971
Y4    0.70777   -0.42705
Y5    0.71981   -0.41752
Y6    0.73121   -0.41265
```

Figure 7.4 is a graphical representation of these data. The point to note here is that lines passed through each of the two clusters would form an oblique angle.

The factor variances are as follows:

```
FACTOR1    FACTOR2
3.110375   1.059580
Final Communality Estimates: Total = 4.169955
```

Note that factor I accounts for roughly three times the variance as does factor II.

The communality "estimates" are as follows:

```
    Y1         Y2         Y3         Y4         Y5         Y6
0.701266   0.692636   0.695337   0.683317   0.692451   0.704948
```

The residual covariances and correlations do not offer any additional information so they will be ignored.

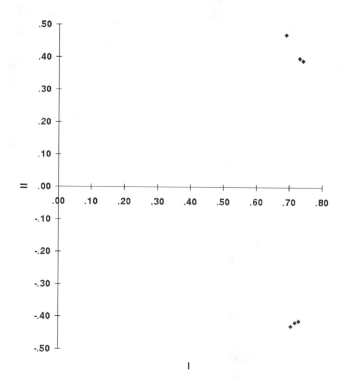

Figure 7.4. Initial factor pattern derived from data reflecting related clusters (Twoc).

Despite the oblique structure of these data, the varimax prerotation preserves the orthogonality of the factors. The transformation matrix is as follows:

```
          1          2
1     0.70867   0.70554
2    -0.70554   0.70867
```

This corresponds to a rotation angle of 44.87°. The resulting factor pattern is as follows:

```
      FACTOR1   FACTOR2
Y1    0.24879   0.79961
Y2    0.23636   0.79798
Y3    0.15686   0.81898
Y4    0.80288   0.19673
Y5    0.80469   0.21197
Y6    0.80933   0.22347
```

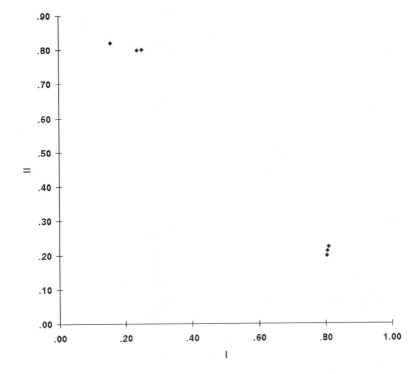

Figure 7.5. Varimax rotated factor pattern derived from data reflecting two related clusters (Twoc).

Figure 7.5 represents these data graphically. Note that although they fall closer to the axes than their unrotated counterparts, these data do not fit the two factors as well as data Twou fit them.

The variances explained are as follows:

```
The Variance explained by Each Factor
 FACTOR1   FACTOR2
2.089513  2.080442
Final Communality Estimates: Total = 4.169955
```

Note that the two factors now roughly account for the same amount of variance.

The communality estimates, however, remain unchanged:

Y1	Y2	Y3	Y4	Y5	Y6
0.701266	0.692636	0.695337	0.683317	0.692451	0.704948

The squared multiple correlations between the variables with each factor are again 1.00 because this is a component solution.

The following are the standardized scoring coefficients:

	FACTOR1	FACTOR2
Y1	−0.09173	0.42955
Y2	−0.09907	0.43239
Y3	−0.15579	0.47044
Y4	0.44562	−0.12507
Y5	0.44202	−0.11597
Y6	0.44137	−0.11012

Again, a target matrix, a Procrustean transformation matrix, and a normalized oblique transformation matrix are provided for the promax rotation.

Target Matrix for Procrustean Transformation

	FACTOR1	FACTOR2
Y1	0.02862	0.91891
Y2	0.02500	0.93044
Y3	0.00727	1.00000
Y4	1.00000	0.01423
Y5	0.98691	0.01745
Y6	0.97750	0.01990

Procrustean Transformation Matrix

	1	2
1	1.30706	−0.30942
2	−0.32065	1.26148

Normalized Oblique Transformation Matrix

	1	2
1	0.58677	0.58254
2	−0.96343	0.96600

The resulting interfactor correlation matrix appears next. The single relevant value is .46276. The $\cos^{-1}$ of this value is 62.43°. This certainly is far removed from being nearly orthogonal as in data Twou!

The factor pattern, reference structure, and factor structure are three matrices, which are identical in an orthogonal solution but quite distinct here. First, the **rotated factor pattern** consists, as before, of standardized regression coefficients that explain variables in terms of factors (they are identified as such):

```
Rotated Factor Pattern (Std Reg Coefs)
        FACTOR1    FACTOR2
Y1      0.05766    0.80917
Y2      0.04448    0.81073
Y3     -0.04826    0.85510
Y4      0.82674   -0.00023
Y5      0.82462    0.01599
Y6      0.82661    0.02734
```

Reference axes are axes that are orthogonal to all but one of the rotated axes. They play an important role in computing some rotations, and were very important conceptually to Thurstone (1935, 1947; also see Yates, 1987), but they are of little explanatory value. With two factors, reference axis I is a vector that is orthogonal to rotated axis II and vice versa. Consequently, the correlation between the two reference axes is the negative complement of the correlation between the two rotated axes or −0.46276.

Any structure in factor analysis denotes the correlation between a variable and a factor. The **reference structure** matrix consists of semipartial correlations among variables and the reference structure.

```
        FACTOR1    FACTOR2
Y1      0.05111    0.71732
Y2      0.03943    0.71870
Y3     -0.04278    0.75803
Y4      0.73289   -0.00020
Y5      0.73101    0.01417
Y6      0.73278    0.02423
```

The sum of squared structure elements defines the variance explained by a given factor controlling for the other factor. In the present case, there are two such sums: 1.614458 and 1.606479 (i.e., $0.05111^2 + 0.03943^2 + \cdots + 0.73278^2 = 1.614458$). Note that this sum is less than the total variance explained of 4.169955 because that which is common to the two factors is eliminated; this is not trivial because of the high factor correlation.

The third matrix consists of correlations between the variables and the factors or **factor structure**:

```
Factor Structure (Correlations)
     FACTOR1  FACTOR2
Y1   0.43211  0.83585
Y2   0.41965  0.83131
Y3   0.34745  0.83277
Y4   0.82663  0.38235
Y5   0.83202  0.39759
Y6   0.83926  0.40986
```

The sums of squares of these elements define the variance explained by each factor, ignoring other factors so that $0.43211^2 + 0.41965^2 + \cdots + 0.83926^2 = 2.563476$ and $0.83585^2 + 0.83131^2 + \cdots + 0.40986^2 = 2.555498$. Note that these add to more than the total because the variance they share appears in each of the sums so it is counted twice. However, note that the variance accounted for by factor I controlling for factor II (1.614458) plus the variance accounted for by factor II ignoring factor I (2.563476) equals the total variance accounted for (4.169955). Similarly, the variance accounted for by factor II controlling for factor I (1.606479) plus the variance accounted for by factor I ignoring factor II (2.555498) also equals the total variance accounted for.

The analysis concludes with the individual h^2 values, and data concerning the scoring coefficients complete the analysis. Again, variables not contributing to the definition of the factors have negligible weights (−.03 to .01), but, other than that, these data provide nothing new.

▣ CALCULATING FACTOR SCORES

Although the coefficients needed to generate factor scores are readily available, they are relatively unimportant to factor analysis for the following reasons:

1. Factor scores per se are rarely needed as most applications are concerned with the structure of the variables and not with individuals' scores on these variables.

2. Scoring coefficients have only a pseudo-precision because they are subject to sampling error. In fact, each weight depends complexly on the variables in the analysis and their structure. Sampling error is the cumulative effect of these influences.

3. Only in component analysis are the coefficients unique. It is readily demonstrated that alternative sets of coefficients can produce scores that are imperfectly correlated with one another yet consistent with the underlying model (Gorsuch, 1983; Harman, 1976; Mulaik, 1972). This is because common factors are only partially determined, one point, we feel, in favor of critics of the common factor model (but, we also add, not the only one). The term "calculating" was used as an umbrella term to include both the "exact" procedures used in component analysis and the estimations of common factor analysis.

4. Simpler algorithms such as unit weights often generalize better to **new** populations than the pseudo-precise weights provided by the algorithm. Moreover, they can often help one better understand what comprises the factors. Whenever possible, one should explore simplifications.

Nonetheless, it is not wrong to use these weights to generate factor scores. If you are new to factor analysis, you should examine the properties of the resulting scores (i.e., their means, variances, and intercorrelations). However, do not simply copy the scoring coefficients to a new data step because of the likelihood of error; there are simpler ways:

1. In SAS, specify `out=` with a file name and `nfactors=` with a specified number of factors as part of the `Proc Factor` statement. For example, `Proc Factor all out=scores nfactors=1 data= one;var s1-s6;` will create a data set named "Scores" based on the data set named "One" studied earlier in this chapter. Successive factor scores will be identified as "Factor1," "Factor2," and so on. These scores are "exact" if a component model is used and "estimated" if a common factor model is used. The data set will also contain the original variables used in the data step. These can be dropped quite simply; for example, adding the statements `Data scores; set scores; keep Factor1;` will limit the data set to the (single) set of factor scores generated by the preceding run.

2. SAS provides three useful scoring algorithms. One of these is `Proc Rank` (SAS Institute, 1990, pp. 493–502), whose name describes its function—it is not of major import here. Another is `Proc Standard` (SAS Institute, 1990, pp. 551–558), which converts raw scores to distributions having a desired mean and

standard deviation, including z scores. For example, suppose you had one set named "Olddata" and wished to standardize variables named X and Y, but not Z. You then wished to include all three in a set named "Newdata." The following code can be used: `Proc Standard Data=olddata out=newdata mean=0 std=1;var x y;`. The defaults would have been to use the most recently created data set.

3. `Proc Score` (SAS Institute, 1990, pp. 1479–1492) forms linear combinations of variables by combining one data set containing coefficients, which may be any kind of weights, and a second data set that consists of the variables to be combined. For example, the following program factors the data set previously identified as Twou to obtain its factor weights and then applies these weights to a parallel data set, Newtwo, in a cross-replication design:

```
Data Twou;
Retain bI1-bI6 (.8 .8 .8 .0 .0 .0);
Retain bII1-bII6 (.0 .0 .0 .8 .8 .8);
Retain nobs 1000 nitems 6 err 1998;
Array Ys(*) y1-y6;
Array BIs(*) bI1-bI6;
Array BIIs(*) bII1-bII6;
Array Bress(*) bres1-bres6;
Do I = 1 to nitems;bress(I) = sqrt(1 - bIs(I)**2 -
  bIIs(I)**2);
End;
Do I = 1 to nobs;
X1 = rannor(err);
X2 = rannor(err);
Do j = 1 to nitems;
Ys(J) = BIs(J)*X1 + BIIs(J)*X2 + Bress(J)*rannor(err);
End;
Keep y1-y6;
Output;
End;
Proc Factor All Data=Twou outstat=oneout;var y1-y6;
Data Newtwo;
Title "Parallel Data Set";
Retain bI1-bI6 (.8 .8 .8 .0 .0 .0);
Retain bII1-bII6 (.0 .0 .0 .8 .8 .8);
Retain nobs 1000 nitems 6 err 1999;
Array Ys(*) y1-y6;
```

```
Array BIs(*) bI1-bI6;
Array BIIs(*) bII1-bII6;
Array Bress(*) bres1-bres6;
Do I = 1 to nitems;bress(I) = sqrt(1 - bIs(I)**2 -
  bIIs(I)**2);
End;
Do I = 1 to nobs;
X1 = rannor(err);
X2 = rannor(err);
Do j = 1 to nitems;
Ys(J) = BIs(J)*X1 + BIIs(J)*X2 + Bress(J)*rannor(err);
End;
Keep y1-y6;
Output;
End;
Proc Score Data=Newtwo score=oneout out=fscore;
  var y1-y6;
Proc Corr Data=fscore;var factor1-factor2;
Run;
```

It is of interest to note that the weights originally obtained in Twou would have produced perfectly orthogonal and standardized factor scores by definition, because the default of a component solution was chosen. When applied to a parallel set, the results were as follows:

```
Variable    N      Mean    Std Dev      Sum    Minimum    Maximum
FACTOR1   1000  0.012251  1.003624  12.250944  -3.670275  3.410819
FACTOR2   1000  0.021710  1.009290  21.709507  -2.793440  3.639442
Pearson Correlation Coefficients / Prob > |R| under Ho: Rho = 0 / N = 1000
            FACTOR1    FACTOR2
FACTOR1     1.00000   -0.00724
              0.0       0.8191
FACTOR2    -0.00724    1.00000
              0.8191    0.0
```

Note that deleting `var factor1-factor2` from the next to the last line would generate correlations between the original variables and the two factors in addition to the data on the two factors. These correlations formed the factor structure in the original analysis that, in this orthogonal solution, was also the factor pattern. The reader is encouraged to see how similar the original and cross-replicated results are.

4. SPSS provides factor scores as an option in the Scores menu. This provides three different methods for obtaining factor scores: regression (as in SAS), Bartlett's (1937) minimization of unique factor variance, and Anderson and Rubin's (1956) generation of uncorrelated scores. The scores can be saved as new variables and/or the scoring matrix can be saved.

▣ EXPLORATORY MAXIMUM LIKELIHOOD FACTOR ANALYSIS

Any maximum likelihood (ML) analysis involves estimating the most probable parameters given the data. In this particular case, the parameters are pattern elements, and the data are correlations. More specifically, the six-variable problem provides 15 (6 · 5/2) distinct known quantities. Estimating one factor involves six unknown pattern elements, leaving 15 − 6 or 9 quantities that are free to vary—degrees of freedom (df). Estimating two orthogonal factors involves 12 unknowns, but the orthogonality constraint (requirement) leaves 15 − 11 = 4 df. Note that attempting to estimate more than two factors would require more unknowns than known values. The basic idea is that the estimated pattern matrix ($\hat{\mathbf{B}}$) gives rise to an estimated correlation matrix, $\hat{\mathbf{R}} = \hat{\mathbf{B}}\hat{\mathbf{B}}'$. The estimation process makes $\hat{\mathbf{R}}$ as close to the actual correlation matrix, $\mathbf{R}$, as possible in a maximum likelihood sense. Specifically, the goal is to minimize the ML fitting function, F, defined as

$$F = \mathrm{tr}(\hat{\mathbf{R}}^{-1}) + \ln|\hat{\mathbf{R}}| - \ln|\mathbf{R}| - q, \tag{7.3}$$

where tr is the trace function (sum of the diagonal elements), $|\mathbf{R}|$ is the determinant of $\mathbf{R}$, $|\hat{\mathbf{R}}|$ is the determinant of $\hat{\mathbf{R}}$, q is the number of variables, and ln is the natural-log function. It is possible, and indeed, mandatory, in some settings to fit the variance–covariance matrix ($\mathbf{C}$) rather than $\mathbf{R}$. Similar, but distinct, fitting functions give rise to unweighted least squares and generalized least squares factor analysis. Both SAS and SPSS can factor either $\mathbf{C}$ or $\mathbf{R}$.

The fact that the value of this fitting function is proportional to chi-square is basic to ML analysis. This likelihood ratio chi-square (G^2) is computed somewhat differently from the Pearson chi-square (χ^2) taught in basic statistics; they are defined as follows:

$$G^2 = 2\sum o \cdot \ln\left(\frac{o}{e}\right),$$ (7.4a)

$$\chi^2 = \sum\left[\frac{(o-e)^2}{e}\right].$$ (7.4b)

Exploratory maximum likelihood and related forms of factor analysis evaluate the fit of a given number of factors in which the only basic requirement is that successive factors be orthogonal prior to rotation. As in any other form of exploratory factor analysis, subsequent rotation may be oblique. Two additional models are also usually evaluated in addition to the one specified. One is the **saturated** model in which all of the *df* are exhausted, achieving the best possible (but not necessarily perfect) fit. The other is the **null** model in which there are no common factors so the fit is as poor as it can be.

The SAS command to conduct an exploratory ML analysis takes the following form. Note that it is necessary to specify the number of factors, unlike other methods:

```
Proc Factor all data=one method=ml n=1;var s1-s6;
```

Some anomaly can occur in estimation. There may be multicollinearity so that the computation of $\mathbf{R}^{-1}$ is unstable. If one attempts to extract too many factors, $\hat{\mathbf{R}}^{-1}$ will become unstable. SAS may or may not provide an error message. For example, it was not possible to obtain a two-factor solution for data set one—SAS simply substituted a one-factor solution though SPSS did note an error condition.

The SAS analysis begins as if one had conducted a common factor analysis with SMC communality estimates. One obtains the means, variances, correlation matrix, correlation matrix inverse, partial correlations controlling others factors, Kaiser measures of sampling adequacy, and eigenvalues obtained using the SMCs. The next item to appear is an eigenanalysis. This result is numerically different from both results presented previously.

	1	2	3	4	5	6
Eigenvalue	4.8045	−0.0983	−0.1150	−0.1238	−0.1453	−0.2172
Difference	4.9027	0.0167	0.0088	0.0215	0.0718	
Proportion	1.1704	−0.0239	−0.0280	−0.0302	−0.0354	−0.0529
Cumulative	1.1704	1.1465	1.1185	1.0883	1.0529	1.0000

Assuming convergence takes place, the next item to be presented is an iteration history. This output contains a criterion value, a ridge value, and a change, followed by the six estimates. In the present case, convergence took place after the second iteration:

```
Iter   Criterion   Ridge    Change   Communalities
1        0.00563   0.000   0.08916   0.49480 0.49108 0.51438 0.47867 0.44307 0.51009
2        0.00563   0.000   0.00006   0.49476 0.49107 0.51436 0.47869 0.44309 0.51015
```

The criterion is the value obtained from Equation 7.3. **Ridge** is a quantity used in estimation that need not concern us. **Change** is an index of conversion as the process stops when this falls below a criterion value (it is possible to adjust this, but there is rarely any reason to do so). Finally, communalities are the h^2 values estimated by the model.

The ML output begins with a chi-square (G^2) value described as

```
Test of HO: No common factors.
        vs HA: At least one common factor.
   Chi-square = 2117.529 df = 15 Prob>chi**2 = 0.0001
```

This is obtained from the fitting function using a null matrix (matrix of zeros) for $\hat{\mathbf{B}}$, which would be the case if there were no common factors. The large value of G^2 and correspondingly small probability means that this null model can be rejected so at least one factor is present.

The next G^2 value is described as

```
Test of HO: 1 Factors are sufficient.
        vs HA: More factors are needed.
   Chi-square = 5.602 df = 9 Prob>chi**2 = 0.7790
```

This value is a slightly modified product of the fitting function and the sample size. Part of the modification is to use what is called *Bartlett's correction* (Bartlett, 1950). The small value of G^2 means that one can tentatively accept the one-factor model.

The output further provides several measures of fit:

1. The unmodified chi-square (G^2) Bartlett correction, which equals 5.6219235054.

2. The Akaike information criterion (AIC, Akaike, 1974, 1987), which equals the uncorrected $G^2 - 2 \cdot df$ or $5.6219235054 - 2 \cdot 9 =$

−12.37807649. This is a descriptive statistic designed to favor models that estimate relatively few parameters, though there is debate as to how well it accomplishes this end.

3. Schwarz's Bayesian criterion (Schwarz, 1978; Sclove, 1987), a similar descriptive measure that others favor. It equals −56.54787401 in the present case (the goal is to have both the AIC and the Schwarz criterion be as small as possible).

4. The Tucker–Lewis (1972) reliability coefficient, a measure of the difference in fit between the obtained and null model. The value in the present case is 1.0026934013. Good fit implies a value near 1.0, as here.

5. The squared canonical correlation between the factor and the variables, which equals 0.852040.

The following output consists of the eigenvector elements, namely, the eigenvectors and what are titled in the output as `Eigenvalues of the Weighted Reduced Correlation Matrix`, which represent the final estimates:

```
Total = 5.75859845 Average = 0.95976641
```

	1	2	3	4	5	6
Eigenvalue	5.7586	0.0547	0.0299	0.0002	−0.0009	−0.0840
Difference	5.7039	0.0248	0.0297	0.0011	0.0832	
Proportion	1.0000	0.0095	0.0052	0.0000	−0.0001	−0.0146
Cumulative	1.0000	1.0095	1.0147	1.0147	1.0146	1.0000

Next are the pattern elements:

```
FACTOR1
S1        0.70339
S2        0.70077
S3        0.71719
S4        0.69188
S5        0.66565
S6        0.71425
```

The variance explained by the single factor appears in both weighted and unweighted form

```
Weighted     5.758598
Unweighted   2.932116
Final Communality Estimates and Variable Weights
Total Communality: Weighted = 5.758598 Unweighted = 2.932116
```

as do the communalities

	S1	S2	S3	S4	S5	S6
Communality	0.494754	0.491074	0.514360	0.478693	0.443085	0.510150
Weight	1.979240	1.964920	2.059140	1.918252	1.795608	2.041438

The scoring coefficients and residual statistics then appear. They are obtained as noted before, so they will not be presented.

One- and two-factor solutions were obtained for data set Twou. The basic points to note is that the G^2 value for the presence of at least one factor was, by definition, the same in both cases (2376.24 on 15 df, $p < .0001$). However, the G^2 value for the residual decreased from a very large value (1176.01 on 9 df, $p < .0001$) to a very small value (3.77 on 4 df, ns). In other words, the data indicated that there was a significant residual past the first factor but not the second. In a like manner, the AIC, Schwarz, and Tucker–Lewis measures went from 1180.15, 1162.15, and .18 to –4.21, –23.84, and 1.00. The canonical correlation changed from a single value of .84 to two such values. Similar results held for analyses conducted on data set Twoc. The disparity between one- and two-factor solutions was not nearly as great because of the redundancy of the second factor. A related difference between the two pairs of analyses is that the canonical correlation values differed substantially (.85 and .57) instead of being nearly equal in the Twoc solution.

▣ PROBLEMƒ

7.1. Factor data set "One" from Chapter 4, exploring the various methods of communality estimation available in your statistical program.

7.2. Change the communality (value assigned b in the third line) from .7 to .9 and factor the data again. Compare the outcomes of a principal component analysis and a common factor analysis of your choice other than maximum likelihood to those obtained previously.

7.3. Take data "Two" and use it to create a one-factor data set. First, change the values of bI1 to bI6 from .8, .8, .8, .0, .0, and .0 to .8, .8, .8 .4, .4, and .4. Then, change all of the values of bII1 to bII6 to .0 in the second and third lines. Repeat the analyses performed in Problem 7.2.

7.4. Generate data with two independent clusters. Explore the rotation options specific to your package. Be sure to include options not to rotate, an orthogonal and an oblique selection.

7.5. Repeat Problem 7.4 generating data with two related clusters.

7.6. Take the data sets generated in Problems 7.2 and 7.3 and subject them to both component and common factor options. Specifically, use maximum likelihood solutions specifying one and two factors and examine the inferential (G^2) values from the various solutions.

7.8. Take the data generated in Problems 7.4 and 7.5 and subject them to both component and common factor options. Specifically, use maximum likelihood solutions specifying two and three factors. Were you able to obtain a solution with three factors? Why or why not?

8

Confirmatory Factor Analysis and Related Procedures

CHAPTER OBJECTIVES

The following topics are considered in this chapter:

1. Confirmatory factor analysis

2. Clustering

3. Multidimensional scaling

▣ CONFIRMATORY FACTOR ANALYSIS

The basic constraint upon successive initial factors beyond satisfying the ML or other criterion in **exploratory** factor analysis is that these factors be orthogonal, a constraint that may be dropped in the process of rotating them. **Confirmatory** factor analysis allows the user to impose a wide variety of other constraints and/or fix (set) parameter estimates to values dictated by theory. A model can be evaluated independently of any other model in terms of either an inferential (G^2) or a descriptive criterion. However, the more usual strategy is to compare the fit of the model to a more general model, specifically, one that includes all the constraints in the model of interest save those specific to the theory of note. The former model is thus **nested** within the more general model. The difference between the two **model** G^2 values is itself a G^2 statistic known as a **difference** G^2. If this G^2 is significantly greater than chance, one may conclude that the constraint(s) degraded the fit so they are not appropriate (assuming, of course, that one wishes to rely solely on an inferential criterion, which has been a matter of some debate for some time; see Bentler & Bonnett, 1980).

Perhaps the three most common and important comparisons of this form are as follows:

1. A model in which certain pattern elements are set to 0 versus a model in which pattern elements are allowed to assume nonzero values

2. A model in which factors are made to be orthogonal versus a model in which factors are allowed to be correlated

3. A model in which pattern elements that are not constrained to 0 are allowed to vary freely among themselves versus one in which they are constrained to equality

If, as has been assumed thus far, the correlation matrix (terms of the form $r = s_{xy}/s_x s_y$) rather than the variance–covariance matrix (terms of the form s_{xy}) is to be analyzed, constraining the pattern elements to equality also constrains the unique variance to equality because the two add to 1.0. However, analyzing the variance–covariance matrix allows one to constrain the pattern elements to equality while leaving the uniquenesses varying freely (a **tau-equivalent** model) rather than constraining both the pattern elements to equality and the uniquenesses to equality (a **parallel** model). Moreover, if there are separate groups (e.g., males and females), one can analyze the sum-of-products matrix (terms of the form $\sum XY$) to test hypotheses about the equality of group means and/or standard deviations on the factor(s).

One well-known program to test models in confirmatory factor analysis is Karl Jöreskog's (Jöreskog & Sörbom, 1993) linear structural relations (LISREL) program. LISREL was the only program for so long that the abstract model that it implemented was also referred to as LISREL, and Jöreskog played an important role in developing the model as well as the program (Jöreskog, 1969). LISREL was originally a standalone program.

Later, it became an add-on to SPSS as well as a standalone, but more recently is again a standalone program. AMOS is the SPSS current product, but it may not be found in all installations, as it is a separate, standalone product. Peter Bentler's EQS program (1985), though newer than LISREL, has also been available for a long time, both as a standalone and as part of BMD (which, like EQS, originated at the University of California at Los Angeles). It has a more graphic interface and is the clearest rival to LISREL.

Although just becoming known, perhaps the most widely available program is based on the work of Roderick McDonald, who is now at the University of Illinois. This is `Proc Calis`, which is a standard part of SAS. In our opinion, it is by far the most poorly documented. However, it is extremely powerful and, because of the flexibility of its syntax, can be acquired easily by anyone who has used competing programs. Specifically, it allows models to be defined in what are in effect four different ways. However, in our view, two (`Ram` and `Cosan`) are cumbersome for problems that are limited to confirmatory factor analysis. However, as will be discussed in the next chapter, `Proc Calis` can be used for structural equation modeling, which includes confirmatory factor analysis (also known as a measurement model). The other two, `Lineqs` (linear equations) and `Factor`, are, in our opinion, much easier to use (the following examples will use `Lineqs`). We will first use `Lineqs` to compare a general one-factor solution of data One, which is of the form $X_i = b_i x_i + e_i$, and a parallel model, which is of the form $X_i = bx_i + e$, as all b_i are equal to one another as are all e_i.

The steps needed for the two solutions using `Lineqs` are as follows:

```
Title 'Lineqs General';
Proc Calis Corr Data=one method=ml;var s1-s6;
Lineqs s1=beta1 f1 + e1,
       s2=beta2 f1 + e2,
       s3=beta3 f1 + e3,
       s4=beta4 f1 + e4,
       s5=beta5 f1 + e5,
       s6=beta6 f1 + e6;
std f1 = 1,e1-e6=ve1-ve6;
Title 'Lineqs Parallel';
Proc Calis Corr Data=one method=ml;var s1-s6;
Lineqs s1=beta f1 + e1,
```

```
      s2=beta f1 + e2,
      s3=beta f1 + e3,
      s4=beta f1 + e4,
      s5=beta f1 + e5,
      s6=beta f1 + e6;
std f1 = 1,e1-e6=ve1;
Run;
```

This general model is equivalent to the exploratory maximum likelihood model. This will always be the case when (a) pattern elements are allowed to vary freely and (b) orthogonal factors are specified (or, as here, there is only a single factor to consider). However, differences in the values of statistics can arise because of differences in the computing algorithm and the criterion used for convergence. In addition, although we are using data set One, it should not be considered proper to base hypotheses for a confirmatory factor analysis on a previous exploratory factor analysis of that same data set—the hypotheses applied to a given data set should be derived prior to that analysis.

The first line of **Proc Calis** is very much like its exploratory counterpart, **Proc Factor**. This is true whichever model is chosen. **Lineqs** defines the model in terms of a series of linear equations of the form $X = bf + e$, which, except for the lack of a multiplication sign, is a direct statement of the relationships involved. In the case of the first (general one-factor) model, using six different names for the pattern elements (beta1–beta6) generates as many distinct values, whereas in the second (parallel one-factor) model, using the same name (beta) constrains these values to equality. Unfortunately, things are not done the same way as regards the error terms in the current version of **Proc Calis** because using the same name for the error terms in the parallel model will not work. Instead, one uses six different names (e1–e6) in both models. The constraint of equal error terms is introduced in the **std** statement, which defines the standard deviations of the latent variables. In both cases, it is desired to have the factor assume unit variance (the $f = 1$ part). The second part is critical: **e1-e6=ve1-ve6** in the general model frees these terms, whereas **e1-e6=ve1** constrains them to a common value (ve1), which is estimated from the data.

The log messages are a bit foreboding, although examination of the results indicates that the analysis was conducted as intended. The general model gave rise to the following statements:

```
NOTE: Type of model matrix _PHI_ changed to: Diagonal Matrix.
NOTE: Type of model matrix _BETA_ changed to: Identity Matrix.
NOTE: Due to a sparse Jacobian the Hessian algorithm 11 will be used.
NOTE: ABSGCONV convergence criterion satisfied.
NOTE: The PROCEDURE CALIS used 1.1 seconds.
```

The parallel model gave rise to these even more abstruse statements:

```
WARNING: Shorter parameter list than variable list in STD statement.
         Parameter list filled up with 5 entries VE1 . .
NOTE: Type of model matrix _PHI_ changed to: Diagonal Matrix.
NOTE: Type of model matrix _BETA_ changed to: Identity Matrix.
NOTE: ABSGCONV convergence criterion satisfied.
NOTE: The PROCEDURE CALIS used 1.2 seconds.
```

The output for the general model is as follows:

1. The various matrices used in the analysis are described.

```
Covariance Structure Analysis: Pattern and Initial Values
     LINEQS Model Statement
    -------------------------------
       Matrix          Rows & Cols       Matrix Type
    TERM 1----------------------------------------------------------
    1   _SEL_           6     13         SELECTION
    2   _BETA_         13     13         EQSBETA        IMINUSINV
    3   _GAMMA_        13      7         EQSGAMMA
    4   _PHI_           7      7         SYMMETRIC
```

2. More critically, the structure of the problem is described. The term "endogenous variables" refers to the variables manifest in the problem (i.e., s1–s6 generated by data One). This appears as

```
    Number of endogenous variables = 6
    Manifest:      S1      S2      S3      S4      S5      S6
```

3. Conversely, "exogenous variables" are variables that are not generated by the model, the single factor (F1) and the six error terms (E1–E6).

```
    Number of exogenous variables = 7
    Latent:        F1
    Error:         E1      E2      E3      E4      E5      E6
```

4. The estimation equations are then presented. Be sure to verify that these equations are correct, which they are.

```
Manifest Variable Equations
Initial Estimates
S1 = . *F1 + 1.0000 E1
        BETA1
S2 = . *F1 + 1.0000 E2
        BETA2
...
S6 = . *F1 + 1.0000 E6
        BETA6
   Variances of Exogenous Variables
- - - - - - - - - - - - - - - - - - - - - - - - - - - - - - - - - - - -
Variable    Parameter    Estimate
```

Variable	Parameter	Estimate
F1		1.000000
E1	VE1	.
E2	VE2	.
E3	VE3	.
E4	VE4	.
E5	VE5	.
E6	VE6	.

5. The univariate and bivariate data follow. These are the same as those presented in Chapter 4. A few additional items, such as the determinant of the correlation matrix and its natural log, which are 0.1194 and -2.126, respectively, also appear.

6. Early versions of structural modeling programs required the user to supply initial estimates of the parameters. This is no longer necessary as the program provides these initial estimates. These estimates are only of interest to the extent that they are similar but distinct from the SMC solution given previously and reveal the correct number of unknowns being estimated (12 for beta1–beta6 and e1–e6), which they do.

7. The iteration history is then described:

```
Levenberg-Marquardt Optimization
Scaling Update of More (1978)
Number of Parameter Estimates 12
Number of Functions (Observations) 21
Optimization Start: Active Constraints = 0 Criterion = 0.006 Maximum
Gradient Element = 0.007
Radius = 1.000
Iter  rest  nfun  act   optcrit   difcrit  maxgrad  lambda    rho
   1     0     2    0   0.005628  0.000051  0.0005        0  0.940
   2     0     3    0   0.005628  2.095E-7  0.00003       0  0.934
   3     0     4    0   0.005628  9.28E-10  2.05E-6       0  0.934
Optimization Results: Iterations = 3 Function Calls = 5 Jacobian Calls = 4
Active Constraints = 0
Criterion = 0.0056275511 Maximum Gradient Element = 2.05412E-6 Lambda = 0
Rho = 0.9337
Radius = 0.0001025
NOTE: ABSGCONV convergence criterion satisfied.
```

8. The results of interest then begin to appear. These begin with the predicted model matrix. This matrix is similar to the matrix of partial correlations controlling other factors. The difference is that the predicted model matrix is $\beta\beta'$ and the partial correlations controlling other factors matrix is $\mathbf{R} - \beta\beta'$. As a result, this matrix will not be presented.

9. A series of fit statistics then appears.

```
Fit Criterion . . . . . . . . . . . . . . . . . . .      0.0056
Goodness of Fit Index (GFI) . . . . . . . . . . . .      0.9982
GFI Adjusted for Degrees of Freedom (AGFI). . . . .      0.9957
Root Mean Square Residual (RMR) . . . . . . . . . .      0.0081
Parsimonious GFI (Mulaik, 1989) . . . . . . . . . .      0.5989
Chi-square = 5.6219        df = 9        Prob>chi**2 = 0.7771
Null Model Chi-square:     df = 15                    2123.5519
RMSEA Estimate . . . . . . . . . 0.0000  90%C.I.[., 0.0241]
Probability of Close Fit . . . . . . . . . . . . .       0.9998
ECVI Estimate . . . . . . . . . 0.0298  90%C.I.[., 0.0385]
Bentler's Comparative Fit Index . . . . . . . . . .      1.0000
Normal Theory Reweighted LS Chi-square . . . . . .       5.4721
Akaike's Information Criterion. . . . . . . . . . .    -12.3781
Bozdogan's (1987) CAIC. . . . . . . . . . . . . . .    -65.5479
Schwarz's Bayesian Criterion. . . . . . . . . . . .    -56.5479
McDonald's (1989) Centrality. . . . . . . . . . . .      1.0017
Bentler & Bonett's (1980) Non-normed Index . . . .      1.0027
Bentler & Bonett's (1980) NFI . . . . . . . . . . .      0.9974
James, Mulaik, & Brett's (1982) Parsimonious NFI. .      0.5984
Z-Test of Wilson & Hilferty (1931). . . . . . . . .     -0.7667
Bollen's (1986) Normed Index Rho1 . . . . . . . . .      0.9956
Bollen's (1988) Non-normed Index Delta2 . . . . . .      1.0016
Hoelter's (1983) Critical N . . . . . . . . . . . .        3008
```

The chi-square and null model chi-square presented here are nearly identical to the values obtained in the exploratory maximum likelihood algorithm where they were 5.602 and 2117.529, respectively. These are simply differences due to algorithm; a suitable alteration of the convergence criterion should make these two results identical to any desired degree of precision.

10. The raw score equations are then presented:

```
S1       =    0.7034*F1   +   1.0000 E1
Std Err       0.0295 BETA1
t Value    23.8088
S2       =    0.7008*F1   +   1.0000 E2
Std Err       0.0296 BETA2
t Value    23.6915
S3       =    0.7172*F1   +   1.0000 E3
Std Err       0.0294 BETA3
t Value    24.4330
S4       =    0.6919*F1   +   1.0000 E4
Std Err       0.0297 BETA4
t Value    23.2961
S5       =    0.6656*F1   +   1.0000 E5
Std Err       0.0300 BETA5
t Value    22.1534
S6       =    0.7142*F1   +   1.0000 E6
Std Err       0.0294 BETA6
t Value    24.2991
```

The t values test the null hypothesis that the parameter estimate is 0 and equal the estimate divided by the standard error (e.g., the value for s1 [23.8088] is .7034/.0295).

11. Next are the variances of exogenous variables. Recall that the factor variance was constrained to 1.0, but the error terms were each estimated from the data. Again, t values are computed as the ratio of parameter estimates to standard errors.

Variable	Parameter	Standard Estimate	Error	t Value
F1		1.000000	0	0.000
E1	VE1	0.505247	0.027338	18.481
E2	VE2	0.508925	0.027452	18.538
E3	VE3	0.485639	0.026740	18.161
E4	VE4	0.521307	0.027840	18.725
E5	VE5	0.556915	0.028982	19.216
E6	VE6	0.489850	0.026868	18.232

12. The standardized parameter estimates, which are the ones of interest, are presented. Slightly edited, these are as follows:

```
S1 = 0.7034*F1 + 0.7108 E1
S2 = 0.7008*F1 + 0.7134 E2
S3 = 0.7172*F1 + 0.6969 E3
S4 = 0.6919*F1 + 0.7220 E4
S5 = 0.6656*F1 + 0.7463 E5
S6 = 0.7142*F1 + 0.6999 E6
```

Note that the pattern elements are identical to those obtained with the exploratory maximum likelihood algorithm within rounding error.

13. Finally, the error variance, total variance, and R^2 for each variable are presented. The error variance for s1 is 0.505247 (.7108²), the total variance is 1.00000 (a necessity, as the variable is standardized), and the R^2 is 0.494753 (.7034²).

Fit of Parallel Model

Having considered the general structure of the printout of the general model, it is of interest to look at the second model, which defines the measures as **parallel** because they are constrained to the same pattern weight and (because this is a solution derived from a correlation matrix) uniqueness. The G^2 increased to 11.1847 on 19 *df* for this parallel model. However, compared to the value of 5.6219 on 9 *df* obtained with the general model, the increase of 5.5638 was nonsignificant (*df* = 10). Measures such as Akaike's information criterion (–26.8153) also decreased. One should conclude that the measures are parallel, which, of course, they were.

Respecification

Models tested on real data typically require respecification. It is all to easy to fool oneself into believing that a respecification results in a better model when, in fact, all one has done is to capitalize upon chance. For a discussion of techniques used to respecify, see the April 1990 issue of **Multivariate Behavioral Research**.

Clustering

Clustering is a collection of methods to aggregate profiles (vectors) into groups based on their similarity. In this sense, it shares certain similarities with factor analysis and multidimensional scaling. Most forms of clustering, like most forms of multidimensional scaling, place fewer assumptions upon the data than does factor analysis. However, it is not as well equipped to test well-formulated hypotheses as is factor analysis, and in this sense it is also like multidimensional scaling. It differs from both in that the emphasis is upon determining similarities among stimuli rather than in generating a space to be dimensionalized.

A cluster may be **disjoint** in that a profile can be in one and only one cluster or it may be **overlapping** and allow a profile to be placed in multiple clusters. It may be deterministic (not a standard term) in that a profile can be in one and only one cluster or it can be **fuzzy** (a standard term) so that membership is expressed in degrees (probabilistically). Clustering is often **hierarchical** in that one starts with a set of k profiles (vectors). The two most similar (using various mathematical definitions) are placed into a cluster (grouped or averaged), leading to a set of $k - 1$ profiles. The next most similar pair, which could either be two of the original profiles or an original profile and the profile recently formed by averaging, are then grouped and averaged, leading to a set of $k - 2$ profiles, and so on. The process continues until there is but a single cluster. The data include the history of grouping and various measures of error (misfit) based on the heterogeneity of the clusters. Typically, one focuses at a given level of clustering even when the process is hierarchical. The cluster is chosen when there is a disproportionately large increase in error or small increase in fit past this point.

Major Options

The time required for `Proc Cluster` is an exponential function of the number of observations. Although the radical increases in computer speed have partially offset the problem of analyzing very large data sets, SAS includes two procedures that are effectively preprocessors. These are `Proc Fastclus` and `Proc Aceclus`. The latter assumes that the data meet normal curve assumptions.

The default method of input is as discussed previously—each record consists of a profile with perhaps other data. An alternative is to input a **distance** matrix in which element d_{ij} represents the distance between profiles i and j. This matrix need not be derived from Euclidean metric and is the way that more advanced users can cluster based on such other metrics as the City Block (see Tversky & Gati, 1982, for a relevant discussion).

The most extensive set of SAS options deals with how dissimilarity is defined, denoted the `method=` option in `Proc Cluster`. An important thing to keep in mind is that data with a well-defined structure tend to provide highly similar results with the various methods, as is also true in factor analysis. Of course, important results may contain noise, so it is important for someone who uses clustering to understand something about these alternative methods.

The following are some of the major methods, as presented in SAS `Proc Cluster`:

1. **Average linkage**. The average linkage between two clusters is the distance between a member of one cluster and a member of the second cluster, averaged over all combinations obtained from each cluster. This method tends to join clusters that are homogeneous and, conversely, is sensitive to outliers.

2. **Centroid**. This method defines the distance between two clusters as the squared distance from their centroids (averages). Consequently, the homogeneity of the cluster is irrelevant, so that it is relatively insensitive to outliers.

3. **Complete linkage**. The complete linkage between two clusters is the largest distance between any pair of profiles. It, too, is sensitive to outliers.

4. **Density linkage**. This denotes a series of methods that do not use distances to compute (dis)similarity. Instead, they compute distances in terms of the probability density of the profiles.

5. **EML**. This method assumes that profiles within clusters have a multivariate normal distribution and maximizes the likelihood of the hierarchy.

6. **Ward's**. This method basically maximizes the F ratio of variance between clusters relative to variance within clusters. It tends to join clusters with relatively few observations and to form clusters of equal size. It is also highly sensitive to outliers.

For a complete list of their output, see Chapter 2 of *SAS/STAT Software: Changes and Enhancements Through Release 6.12* (SAS Institute, 1997). Also see Anderberg (1973), Blashfeld and Aldenderfer (1978), and Everitt (1980).

For example, consider the four profiles in Figure 8.1. Profiles 1 and 2 are similar in increasing from measures A to D, and profiles 3 and 4 are similar in decreasing from measures A to D. Visually, profiles 3 and 4 appear to have a greater difference than profiles 1 and 2 because of the former's disparity on measure D.

The following SAS program performs a hierarchical clustering of the four profiles and looks in more detail at the two-cluster solution:

```
Data Cluster;
Input ID A B C D;
Output;
Cards;
1   1   2   9 10
2   2   3 10 11
3 11 10   1   2
4 12 11   2   5
;
Proc Cluster Data=Cluster method=centroid outtree=tree
Pseudo Rmsstd Rsquare Simple;
Var A B C D;
Copy ID;
Proc Tree Data=tree horizontal out=outer nclusters=2;
Copy ID A B C D;
Proc Print Data=Outer;
```

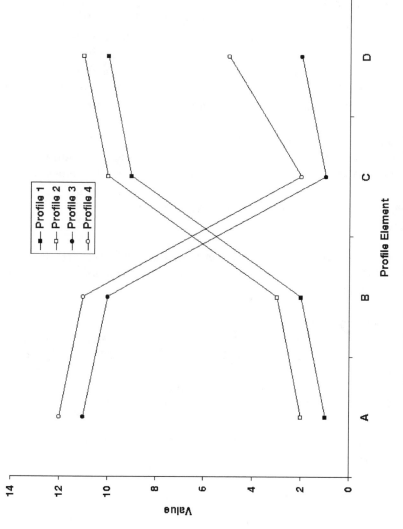

Figure 8.1. Four profiles used in cluster analysis.

```
Proc Sort Data=Outer;by Clusname;
Proc Means Data=Outer;by Clusname;
Run;
```

The data step merely inputs the four profiles depicted in Figure 8.1. Proc Cluster is then invoked using the centroid. For example, the squared distance between profiles 1 and 2 is $(1-2)^2 + (2-3)^2 + (9-10)^2 + (10-11)^2 = 4$. The output may be viewed graphically as a tree in which the trunk is the final (single) cluster; the first level of branching is the two-category solution and so on. A variety of statistics are requested. Another form of clustering, not applicable here but useful in other circumstances, is known as the cubic clustering criterion (CCC, Sarle, 1983). The Copy statement includes the ID variable as an identifier.

Proc Tree then prints the tree and creates a file named Outer. We have chosen to use a horizontal format for the two-cluster solution. The contents of Outer include the cluster name as well as the values of the individual profile elements. The values are then averaged within clusters using Proc Means, which defines the cluster **centroids**.

The results of the clustering are as follows:

1. Although the results are somewhat method dependent, the first results to appear in the present case are simple statistics on the four profile elements.

```
Simple Statistics
             Mean    Std Dev   Skewness   Kurtosis   Bimodality
A          6.50000   5.80230   0.00000    -5.70591    0.12830
B          6.50000   4.65475   0.00000    -5.54556    0.12572
C          5.50000   4.65475   0.00000    -5.54556    0.12572
D          7.00000   4.24264  -0.36665    -3.43827    0.11275
```

2. The next results are an eigenanalysis of the variance–covariance matrix formed by the profiles. In this case, the four profiles clearly differ along a single dimension with 97% of the variance explained by the first eigenvector.

```
     Eigenvalue  Difference  Proportion  Cumulative
1      92.4127    89.9193     0.972765    0.97277
2       2.4934     2.3996     0.026247    0.99901
3       0.0939     0.0939     0.000988    1.00000
4       0.0000        .       0.000000    1.00000
```

3. The root mean square total-sample standard deviation, which indexes variability pooled over profile elements, and the root mean square distance between observations, which indexes the dissimilarity of the profiles, then appear, the values of which are 4.873397 and 13.78405, respectively. The large value of the latter relative to the former indicates that differences among profiles are large.

4. The history of cluster formation appears next. It indicates that the first step was to join observations 1 and 2 because their distance (.145096) is less than any other pair of profiles. This new cluster contains two profiles. Observations 3 and 4 are then joined to form a second cluster, again with two profiles. Their normalized centroid distance (**Norm Cent Dist**) of .251312 is less than the sum of the two distances between each of them and the previously formed cluster. Finally, the two clusters are joined. The distance between them (1.207433) is the only distance left. All four initial observations are thus placed into one large cluster. The other statistics are (a) the **root mean square standard deviation** (**RMS STD**), which is obtained by pooling the variances among profiles within clusters and taking the square root of the result; (b) the **SPRSQ**, which represents the change in the squared multiple correlation from the previous number of clusters; (c) the squared multiple correlation (**RSQ**), which is 1 – the ratio of the pooled variance within clusters to the total variance pooled across profile elements; and (d) the **Pseudo F** and **Pseudo t**2** statistics. The latter are approximate tests of the null hypothesis that the profiles at each level do not differ. Note that the pooled **RMS STD** increases as the number of clusters decreases because clusters are made to be increasingly heterogeneous. The **RSQ** decreases because of this increasing heterogeneity. The approximate **F** and **t**2** tests indicate that the differences among profiles at each stage are large.

NCL	-Clusters Joined-		FREQ	RMS STD	SPRSQ	RSQ	Pseudo F	Pseudo t**2	Norm Cent Dist	Tie
3	OB1	OB2	2	0.70711	0.007018	0.992982	70.8	.	0.145095	
2	OB3	OB4	2	1.22474	0.021053	0.971930	69.3	.	0.251312	
1	CL3	CL2	4	4.87340	0.971930	0.000000	.	69.3	1.207433	

5. The data for the two-cluster solution appear graphically as the output of **Proc Tree**.

```
Distance Between Cluster Centroids
          1.2  1.1   1   0.9  0.8  0.7  0.6  0.5  0.4  0.3  0.2  0.1   0
          +----+----+----+----+----+----+----+----+----+----+----+----+
N  OB1  XXXXXXXXXXXXXXXXXXXXXXXXXXXXXXXXXXXXXXXXXXXXXXXXXXXXXXXXXX........
a       XXXXXXXXXXXXXXXXXXXXXXXXXXXXXXXXXXXXXXXXXXXXXXXXXXXXXXXXXXXXXX
m       XXXXXXXXXXXXXXXXXXXXXXXXXXXXXXXXXXXXXXXXXXXXXXXXXXXXXXXXXXXXXX
e       XXXXXXXXXXXXXXXXXXXXXXXXXXXXXXXXXXXXXXXXXXXXXXXXXXXXXXXXXXXXXX
        XXXXXXXXXXXXXXXXXXXXXXXXXXXXXXXXXXXXXXXXXXXXXXXXXXXXXXXXXXXXXX
o       XXXXXXXXXXXXXXXXXXXXXXXXXXXXXXXXXXXXXXXXXXXXXXXXXXXXXXXXXXXXXX
f  OB2  XXXXXXXXXXXXXXXXXXXXXXXXXXXXXXXXXXXXXXXXXXXXXXXXXXXXXXXXXXXXXX........
        X
o       X
b       X
s       X
e       X
r  OB3  XXXXXXXXXXXXXXXXXXXXXXXXXXXXXXXXXXXXXXXXXXXXXXXXXXXXXXXXXX........
v       XXXXXXXXXXXXXXXXXXXXXXXXXXXXXXXXXXXXXXXXXXXXXXXXXXXXXXXX
a       XXXXXXXXXXXXXXXXXXXXXXXXXXXXXXXXXXXXXXXXXXXXXXXXXXXXXXXX
t       XXXXXXXXXXXXXXXXXXXXXXXXXXXXXXXXXXXXXXXXXXXXXXXXXXXXXXXX
i       XXXXXXXXXXXXXXXXXXXXXXXXXXXXXXXXXXXXXXXXXXXXXXXXXXXXXXXX
o       XXXXXXXXXXXXXXXXXXXXXXXXXXXXXXXXXXXXXXXXXXXXXXXXXXXXXXXX
n  OB4  XXXXXXXXXXXXXXXXXXXXXXXXXXXXXXXXXXXXXXXXXXXXXXXXXXXXXXXX........
```

The data output from Proc Tree to form data set Outer consists of the contents of the two-factor solution. This includes the value of ID and the values for each profile element, which were included by using the Copy command. Note that sorting and using either the variable named "Cluster" or the variable named "Clusname" will produce the centroids.

OBS	_NAME_	ID	A	B	C	D	CLUSTER	CLUSNAME
1	OB1	1	1	2	9	10	1	CL3
2	OB2	2	2	3	10	11	1	CL3
3	OB3	3	11	10	1	2	2	CL2
4	OB4	4	12	11	2	5	2	CL2

6. Finally, the results of Proc Means provide the centroids for the two clusters plus other statistics such as the standard deviation within groups.

```
CLUSNAME=CL2
Variable   N        Mean      Std Dev     Minimum       Maximum
-----------------------------------------------------------------
A          2    11.5000000    0.7071068   11.0000000    12.0000000
B          2    10.5000000    0.7071068   10.0000000    11.0000000
C          2     1.5000000    0.7071068    1.0000000     2.0000000
D          2     3.5000000    2.1213203    2.0000000     5.0000000
-----------------------------------------------------------------

CLUSNAME=CL3
Variable   N        Mean      Std Dev     Minimum       Maximum
-----------------------------------------------------------------
A          2     1.5000000    0.7071068    1.0000000     2.0000000
B          2     2.5000000    0.7071068    2.0000000     3.0000000
C          2     9.5000000    0.7071068    9.0000000    10.0000000
D          2    10.5000000    0.7071068   10.0000000    11.0000000
-----------------------------------------------------------------
```

As a second example, consider the data set named Cluster presented in Chapter 4. This data set generates profiles defined by a single factor within groups that are based on prototypes. The results duplicate what has already been discussed. A question of major interest is how accurately the three-cluster solution recovers the prototypes.

The program code is as follows:

```
Proc Corr Data=Clusters;
Var A B C;By Group;
Proc ANOVA Data=Clusters;
Class Group;
Model A B C = Group;
Manova H = Group;
Means Group;
Proc Cluster Data=Clusters method=Ward outtree=
    tree;var A B C;
Copy Group;
Proc Tree Data=tree horizontal nclusters=3 out=outer;
Copy Group;
Proc Print Data=tree;
Proc Sort Data=Outer;by _Name_;
Data Outer;set outer;
Proc Print Data=Outer;
Proc Freq;Tables Cluster*Group;
Run;
```

The table of frequencies relating actual cluster membership (group) to assigned cluster (cluster) is as follows:

		Group		
Cluster	1	2	3	Total
1	18	8	6	32
2	2	8	7	17
3	0	4	7	11
Total	20	20	20	60

Note that group 1 does not have to correspond to cluster 1 although the individual groups do correspond to the individual clusters in this particular example. The simplest definition of accuracy is the percentage of correct classifications, which is $(18 + 8 + 7)/60$ or 55%, where 33% is chance. Choice of clustering algorithm will have an impact on this figure, but so does the amount of separation of the prototypes in the latent data. Moreover, one algorithm may work better in one situation than another, but the reverse may hold true in other situations. By providing the reader with the code, both of these variables can be studied rather easily.

▣ MULTIDIMENJIONAL JCALING

Multidimensional scaling (MDS) is concerned with transforming input measures of dissimilarity to output measures of distance and inserting vectors to dimensionalize these distances. The SAS implementation (`Proc MDS`) provides the user with several ways of analyzing the dissimilarities:

1. As is—no transformations are made on the dissimilarities. The data are cast in a space having a dimensionality defined by the user and the misfit of the data to that space is determined. This is known as **absolute** in SAS.

2. By transforming the dissimilarities to distances via a **ratio** (multiplicative) transformation of the form $d' = ad$, where d' is a distance measure, a is a (multiplicative) constant, and d is the original dissimilarity. This transformation differs minimally from the absolute transformation.

3. By subjecting them to an **interval** (linear) transformation of the form $d' = ad + b$, where b is an (additive) constant and the remaining terms are as before.

4. By subjecting them to a **log interval** (power) transformation, which is of the form $d' = ad^b$, where b is an exponent.

5. By subjecting them to an **ordinal** (nonmetric) transformation, which can be of any form subject to the requirement that rank orderings of distances are preserved. The ordinal model is perhaps the most popular among behavioral scientists because it makes the fewest assumptions about the input.

If you do not have access to SAS, Alscal, a very similar program, can be downloaded free—see http://forrest.psych.unc.edu/research/ALSCAL.html.

MDS dimensionalizes data in a manner that is similar to factor analysis, but it differs in that the dissimilarities that are input need not themselves be **Gramian**. A Gramian matrix (**B**) is the result of multiplying another matrix (**A**) by its transpose—**B** = **AA′** or **A′A**. The eigenvalues of **B** will always be nonnegative. Ordinary sum-of-products, covariance, and correlation matrices (with ones in the diagonal) are all Gramian and therefore suitable for factor analysis. However, interval, log interval, and ordinal MDS procedures all allow non-Gramian data to be analyzed. Moreover, the nonmetric option provides a better fit to a correlation matrix than can be obtained with factor analysis. However, this does not mean that MDS is superior to factor analysis, as the transformation is generally ad hoc. In addition, there is a difference of one dimension that is effectively an artifact of method (Davison, 1985). Both factor analysis and MDS have their advantages and disadvantages. In particular, factor analysis can be exploratory or confirmatory; those versions of MDS that are generally available are strictly exploratory.

Perhaps the simplest way to introduce MDS is to scale distances between cities, a demonstration used by Kruskal and Wish (1978) in their well-known introduction to the topic. Depending on which program you use, you may or may not be limited to Euclidean distance measures (distances on a flat surface, which obey the Pythagorean theorem that the squared distance along the hypotenuse of a right triangle equals the sum of the squared distances along the other sides). Strictly speaking, distances on the earth's surface are non-Euclidean due to the approximate sphericity of the earth, but this has little effect over relatively small distances. Our example uses actual distances among eight cities in Texas rather than a simulation. The cities are Austin, Dallas, El Paso, Fort Worth, Houston, Lubbock, San Antonio, and Texarkana.

Following is the SAS program:

```
Data Cities;
Input City $ Dallas Ft_Worth Texark San_Ant Austin El_Paso Houston
Lubbock;
Cards;
Dallas     0.      .      .      .      .      .      .      .
Ft_Worth  31.1    0.      .      .      .      .      .      .
Texark   176.1  207.1    0.      .      .      .      .      .
San_Ant  272.1  261.1  415.1    0.      .      .      .      .
Austin   195.1  188.1  338.1   78.1    0.      .      .      .
El_Paso  611.1  581.1  787.1  546.1  571.1    0.      .      .
Houston  240.1  260.1  286.1  195.1  161.1  728.1    0.      .
Lubbock  322.1  291.1  471.1  384.1  372.1  339.1  511.1    0.
Proc Print Data=Cities;
Proc mds Data=Cities level=absolute dim=1 pfinal;
id city;
Proc mds Data=Cities level=absolute dim=1 pfinal;
id city;
Proc mds Data=Cities level=absolute out=Cityout Outfit=Cityfit
Outres=Cityres dim=2 pfinal;
id city;
Proc Print Data=Cityout;
Proc Print Data=Cityfit;
Proc Print Data=Cityres;
Proc Plot Data=cityd vtoh=1.7;
plot dim2 * dim1 $ city / haxis=by 500 vaxis=by 500;
where _type_ = 'CONFIG';
Proc mds Data=Cities level=absolute dim=3 pfinal;
id city;
Run;
```

Proc MDS was invoked separately for one-, two-, and three-dimensional solutions in order to concentrate upon the two-dimensional results, which were known to be correct. The basic data could have been generated in one procedure using "Dimension 1 to 3." The procedure itself generates very little output. The key data are in three files that, by default, contain information about (a) the iteration history and parameter estimates (the Out data set here named Cityout), (b) the fit measures (the Outfit data set, here named Cityfit), and (c) the residual analysis and transformation (Outres, here named Cityres). These data sets may be printed using **Proc Print** or plotted using **Proc Plot**. The material generated by MDS itself is as follows. First, the properties of the data are given: (a) Shape, the shape of the matrix; (b) the conditionality of the data; (c) Level, the level of analysis as given by the **Proc** MDS statement; and (d) Coef, the type of matrix, which determines the interpretation of the distance measures. In the present case, the shape is triangular, as all values on or above the diagonal are missing. This is standard for a basic MDS solution. The data are matrix conditional because the various distances are to be viewed as forming a single set, as opposed, say, to having different matrices reflect different subjects, as will be illustrated later. The level of scaling was specified as absolute in the **Proc** statement.

A series of parameters used to control the estimation process appears next. These are (a) Dimension, the number of dimensions; (b) Formula, the formula used to define misfit; (c) Fit, the data used for the formula for misfit; (d) Gconverge, the program's tolerance for misfit; (e) Maxiter, the maximum number of iterations performed; (f) Over, the **overrelaxation factor** part of the criterion defining fit; and (g) Ridge, also used in the estimation process. One-, two-, and three-dimensional solutions were obtained in the present case. The intent was to demonstrate the effects of choosing both too few and too many dimensions. Obviously, the two-dimensional solution should fit perfectly within limits imposed by rounding error and the slight deviation from Euclidean distances within relatively small areas (Texans would never admit that the distances are ever small). The default formula for misfit depends on the transformation used. In general, it is some variant on $\sqrt{1-R^2}$, where R is the multiple correlation between disparities (inputs) and distances (outputs). The value assigned to Fit is the exponent to which distances are raised, if Fit is a number; Fit = 0 or Fit = Log performs an analysis on log distances. The default is Fit = 1, which means that fit is defined in terms of actual distances. Of the remaining quantities, the one most likely to be important to a user is Maxiter, although the default value of 100 is liberal.

The fit data for the one-dimensional solution was

Iteration	Type	Badness-of-Fit Criterion	Change in Criterion	Convergence Measure
0	Initial	0.258439	.	0.552388
1	Lev-Mar	0.215427	0.043012	0.000057982

Likewise, the fit data for the two-dimensional solution was

Iteration	Type	Badness-of-Fit Criterion	Change in Criterion	Convergence Measure
0	Initial	0.009461	.	0.554125
1	Lev-Mar	0.007879	0.001582	0.005372

Finally, the fit data for the three-dimensional solution was

Iteration	Type	Badness-of-Fit Criterion	Change in Criterion	Convergence Measure
0	Initial	0.010898	.	0.834240
1	Lev-Mar	0.008489	0.002409	0.735001
2	Lev-Mar	0.008349	0.000140	0.755353
3	Lev-Mar	0.007277	0.001072	0.398546
4	Gau-New	0.006968	0.000308	0.382422
5	Lev-Mar	0.006842	0.000126	0.202284
6	Lev-Mar	0.006828	0.000013952	0.233469
7	Lev-Mar	0.006786	0.000041919	0.115184
8	Lev-Mar	0.006769	0.000016693	0.053574
9	Lev-Mar	0.006767	0.000002541	0.023481
10	Lev-Mar	0.006766	0.000000566	0.011403
11	Lev-Mar	0.006766	0.000000113	0.005123

The one- and two-dimensional solutions converged as rapidly as could be the case. Although the three-dimensional solution did take longer, note that, after step 4, all changes in the criterion after the fourth iteration occurred in the fourth decimal place. The two-dimensional solution clearly fit better than the one-dimensional solution: $\sqrt{1-R^2} = 0.215427$ versus 0.007879, but the fit of the three-dimensional solution was only marginally better than the fit of the two-dimensional solution: 0.006766 versus 0.007879. Always keep in mind that any solution in a higher dimensionity must fit better than one in a lower dimensionality, but that one pays a penalty by having an additional set of parameter estimates. It is clear that MDS recognized the dimensionality of the space.

Proc Print contains the Criterion values for the three solutions along with the coordinate(s). Because this is an absolute solution, the coordinate values, like the input disparities, are in miles. Figure 8.2 contains the results of a separate application of Proc Plot in which only the two-dimensional solution was obtained. Note that the location of the cities roughly corresponds to the way they would appear on a map with north presented at the top. This is fortuitous. It arose because the plot presents the major dimension of variation along the abscissa. In this case,

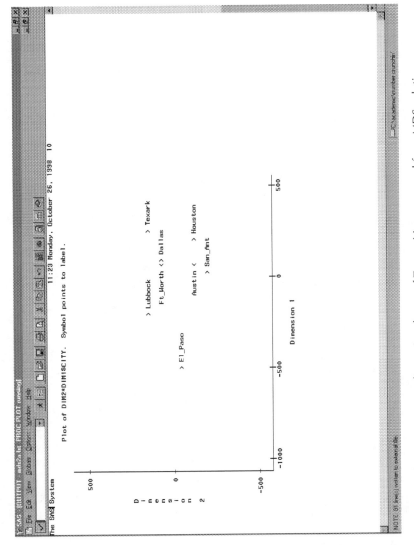

Figure 8.2. Two-dimensional map of Texas cities recovered from MDS solution.

the major dimension was defined by the distance from Texarkana, on the far east of the state, to El Paso, at the far west. In another application, this major dimension could just as well be north–south, which would rotate the contents by 90° (in actuality, the map is rotated slightly as Texarkana is also farther north than El Paso, although this disparity is very slight compared to the east–west difference).

▣ PROBLEMſ

8.1. Take the data set as modified in Problem 7.2 and use it to test unifactor hypotheses that (a) do not make any further assumptions about pattern weights, (b) assume each variable has the same pattern weight (tau equivalence), (c) assume all variables have the same pattern weights and the same uniquenesses (parallelism), and (d) assume each variable has a pattern loading of exactly .9 and a uniqueness of $\sqrt{1 - .9^2}$.

8.2. Take the data set as modified in Problem 7.3 and use it to test the same unifactor hypotheses.

8.3. Modify the data step that produces data set 2 so that the first set of pattern weights is .7, .7, .7, .0, .0, and .0 and the second set of pattern weights is .0, .0, .0, .7, .7, and .7. Use these to test the following two-factor hypotheses: (a) there are two orthogonal factors; (b) factor I is defined by the first three variables and factor II is defined by the second three variables, but no further assumptions are made about the contributions of the defining variables; (c) factor I is defined equally by the first three variables and factor II is defined equally by the second three variables, but the pattern loadings on factor I are not necessarily the same as on factor II; (d) factor I is defined equally by the first three variables and factor II is defined equally by the second three variables, and all pattern loadings are the same; and (e) all six measures are parallel.

8.4. Generate four profiles of your choice, paralleling what was done in the text, and explore the various clustering options.

8.5. Take the correlation matrix used in Problem 8.3 and subject it to clustering, appropriately taking into account the fact that correlations are measures of similarity rather than distance.

8.6. Take the following matrix and perform a multidimensional scaling using each of the various options (absolute, multiplicative, interval, log interval [power], and nonmetric) in one and two dimensions. Note that this matrix should fit an absolute or multiplicative model in two dimensions exactly because it is of the form $\mathbf{XX'}$, where $\mathbf{X}$ is a set of two-dimensional coordinates.

13	8	17	13	27	28
8	5	10	7	16	16
17	10	25	23	39	44
13	7	23	26	35	44
27	16	39	35	61	68
28	16	44	44	68	80

8.7. Repeat the exercise with the following matrix. Note that it should fit a linear but not an absolute or multiplicative two-dimensional model because entries in this matrix, $\mathbf{Y}$, are of the form $y_{ij} = 2x_{ij} + 3$.

29	19	37	29	57	59
19	13	23	17	35	35
37	23	53	49	81	91
29	17	49	55	73	91
57	35	81	73	125	139
59	35	91	91	139	163

8.8. Repeat the exercise with the following matrix. Note that it should fit a log interval (power) two-dimensional model because entries in this matrix, $\mathbf{Y}$, are of the form $y_{ij} = x_{ij}^2$.

169	64	289	169	729	784
64	25	100	49	256	256
289	100	625	529	1521	1936
169	49	529	676	1225	1936
729	256	1521	1225	3721	4624
784	256	1936	1936	4624	6400

8.9. Repeat the exercise with the following data. Note that this should fit a two-dimensional nonmetric model because entries are of the form $y_{ij} = x_{ij}^2 + 5$.

174	69	294	174	734	789
69	30	105	54	261	261
294	105	630	534	1526	1941
174	54	534	681	1230	1941
734	261	1526	1230	3726	4629
789	261	1941	1941	4629	6405

8.10. Repeat the exercise with the following data. Note that it will not fit a two-dimensional nonmetric model because entries were obtained by adding a random number from 0 to 10 to values in the previous matrix.

180	75	298	174	735	794
70	32	113	57	265	264
299	108	636	539	1527	1947
178	61	534	690	1234	1946
743	262	1534	1232	3735	4633
795	263	1949	1947	4638	6414

9

Regression, Canonical Correlation, and Covariance Structure Analysis

CHAPTER OBJECTIVES

The two main purposes of this chapter are (a) to explore regression models and (b) to combine structural models of this form with measurement (factor-analytic) models of the previous chapter to form a covariance structure. In a regression model, one or more criteria are predicted from variables external to them. In contrast, predictors in factor-analytic models are linear combinations of the variables to be predicted.

The specific major topic areas are as follows:

1. Basic multiple regression, including the analysis of residuals, hierarchical regression (including moderated multiple regression), stepwise regression, and multiple regression based on criteria other than ordinary least squares (OLS), for example, maximum likelihood (ML)

2. Analysis of variance (ANOVA), which is a special case of multiple regression in which the predictors are categorical—this includes the ANOVA for unequal N (nonorthogonal ANOVA) and the analysis of covariance

3. Canonical correlation analysis

4. Covariance structure analysis

▣ ORDINARY LINEAR REGREJJION

In linear regression, the problem is how to best linearly combine a series of predictor variables, $X_1, X_2, \ldots, X_k$, to make the resulting linear combination most closely approximate a criterion variable Y. Historically, "most closely approximates" means "in a least squares sense." This means that if we apply the weights $b_1, b_2, \ldots, b_k$, to the predictors and produce a linear combination (composite variable), $\hat{Y}$, the sum of the squared discrepancies between the original Y variables and this composite, $\sum (Y - \hat{Y})^2$, is a minimum. In other words, this discrepancy is smaller than it can be with any other set of weights. If each observation is given equal weight, the method is known as **ordinary least squares** (OLS) or **unweighted least squares** (ULS). One alternative, known as **generalized least squares** (GLS) or **weighted least squares** (WLS), consists of a related class of strategies that weight observations differentially, perhaps based on their reliability. In all forms of regression, including ML, b weights are applied to unscaled observations (raw scores), whereas beta (β) weights are applied to standard scores. Both are important as will be discussed later. The collective term **regression weight** applies to them jointly. Note that some statisticians treat WLS as a subset of GLS (Weisberg, 1985).

Items of major interest to the analysis, in addition to the raw score (b) and standardized (β) regression weights, are

1. The linear correlation between $\hat{Y}$ and Y or **multiple correlation** (R). Note that this symbol is in italics, following convention. Its square (R^2) likewise will always appear in italics. Conversely, **R** in boldface and similar expressions denote **matrices** of correlations (which are sometimes used to compute R).

2. The standard errors of the b and β weights, which describe the lack of precision in their estimation.

3. Various inferential tests, for example, the null hypothesis that the population value of R is .0 and the null hypothesis that the individual population values of b (and therefore β) are .0.

Writing an OLS program for a fixed set of predictors is not very difficult though it is more challenging to write one for stepwise selection (but see the following discussion); indeed, there are several distinct algorithms that will produce the vector of weights (**b** or β). Three will be discussed here.

Method 1

1. Let **X** denote a matrix containing 1s in the first column and the k predictors in the remaining k columns. The elements of **X** are normally thought of as values of continuous variables. In contrast, they are discrete in the ANOVA, discussed later. In fact, nearly all of the mathematics is the same in these two cases.

2. Let **y** denote the known vector criterion measures as a column vector.

3. Let **b** denote the unknown vector of b weights similarly expressed as a column vector. Then, b_0 is the first element of this vector, which describes the intercept term of the equation. The remaining elements relate to their corresponding predictor (e.g., b_3 is applied to variable X_3).

4. The basic regression equation can be stated as

$$\mathbf{y} = \mathbf{Xb}. \tag{9.1}$$

5. Standard matrix operations can be used to solve for the unknown **b** matrix, as follows

$$\mathbf{b} = (\mathbf{X'X})^{-1}(\mathbf{X'y}). \tag{9.2}$$

6. It is cumbersome to describe how to obtain the vector of standardized weights, β, from **b** in matrix notation. However, the individual elements are simply

$$\beta_i = \frac{s_y}{s_i} b_i \tag{9.3a}$$

and, conversely,

$$b_i = \frac{s_i}{s_y} \beta_j. \tag{9.3b}$$

Note that $\mathbf{X'X}$ contains terms of the form $\sum X^2$ and $\mathbf{X'y}$ contains terms of the form $\sum XY$. Moreover, the first diagonal element of $\mathbf{X'X}$ contains the number of observations.

Method 2

In this case, one needs to compute the matrix of correlations among the predictors $(\overline{\mathbf{R}}_\rho)$ and the vector of correlations between the criterion (Y) and each predictor in turn $(X_1, X_2, \ldots)$. This column vector is known as the validity vector, $\mathbf{v}$. We will denote the elements of this matrix as $r_{y1}, r_{y2} \ldots$.

The result is that

$$\beta = \overline{\mathbf{R}}_\rho \mathbf{v}. \tag{9.4}$$

One can obtain $\mathbf{b}$ instead of β by replacing the correlations in $\mathbf{R}_\rho$ and $\mathbf{v}$ with covariances. These covariance matrices may be identified as $\mathbf{C}_{xx}$ and $\mathbf{c}_{xy}$, respectively. Note that $\mathbf{C}_{xx}$ and $\mathbf{c}_{xy}$ equal $\mathbf{X'X}$ and $\mathbf{X'y}$ when the latter are divided by the number of subjects, N. However, this latter term cancels because one premultiplies $\mathbf{v}$ by $\overline{\mathbf{R}}_\rho$, making these two methods equivalent. More important is that $\overline{\mathbf{R}}_\rho$ becomes an identity matrix when the predictors are mutually independent so that it can be eliminated from the equation. In that case, the β weights are simply the validity coefficients, that is, the zero-order correlations between each predictor in turn and the criterion.

Method 3

Let $\mathbf{R}^*$ define an augmented matrix containing both the predictors (in the first k positions) and the criterion (in the last or $k + 1$st position). Then, $R^2 = 1 - 1/r^{*(k+1)(k+1)}$, where $r^{*(k+1)(k+1)}$ is the last element in $\mathbf{R}^{*-1}$ (the criterion need not physically be placed as the last element; this has been done to simplify visualization). Recall that the relationship between the diagonal elements of the inverse of a correlation matrix and R^2 was noted in the previous chapter. SAS uses a variation of this procedure in which the augmented variance–covariance matrix $(\mathbf{C}^*)$ is inverted instead of $\mathbf{R}^*$.

The ith β weight, β_i, can be calculated as follows:

$$\beta_i = -\frac{r^{*i,(k+1)}}{r^{*(k+1)(k+1)}}. \tag{9.5}$$

These do not exhaust the possible ways the regression weights may be calculated. Once you have β, $\mathbf{R}_\rho^{-1}$, and $\mathbf{v}$ (all of which should be included in any results), R^2 may be obtained using

$$R^2 = \beta'\mathbf{v} \tag{9.6a}$$

$$R^2 = \mathbf{v'R}_\rho^{-1}\mathbf{v}. \tag{9.6b}$$

Equation 9.6a states that R^2 is the product of the β weights and the validities. Equation 9.6b states that R^2 is the sum of the squared products of the validities using $\mathbf{R}_\rho^{-1}$ to correct for the correlations among the predictors.

The standard error of a regression weight, s_{b_i}, is another important quantity. It describes the lack of precision in estimation, and dividing it into b provides a t statistic to test the null hypothesis that the parameter's true value is .0. The equation for s_{b_i} can be written as

$$s_{b_i} = \sqrt{\frac{(r^{ii})(1 - R^2)}{N - k - 1}}. \qquad (9.7)$$

It is especially easy to recalculate the quantities necessary when one or more predictors are to be eliminated from the equation. Although one could simply start from scratch, given the speed of present computers, only two basic steps are needed to recompute the basic terms. Say that predictor X_p is to be eliminated:

1. Recompute $\mathbf{v}$ by eliminating r_{jy}.

2. Recompute $\mathbf{R}^{-1}$ as follows:

$$r^{ij*} = r^{ij} - \frac{r^{ip} r^{ip}}{r^{pp}}. \qquad (9.8)$$

In this equation, r^{ij*} is the element in row i and column j of the revised matrix, r^{ij} is the corresponding element in the original matrix, and so on.

We will now turn to a consideration of two related regression problems using a slight modification of a simulation presented in Chapter 5. The first involves three predictors that are uncorrelated in the population (X1–X3), as in the original simulation, and the second involves predictors derived from the first as linear combinations (XN1–XN3).

Basic Multiple Regression in SAS

SAS has several distinct procedures for OLS multiple regression. The most important of these are

1. **Proc GLM**. This is basically a superset of all SAS OLS regression procedures as it can perform the widest variety of analyses. These include complex analysis of variance (ANOVA) designs with unequal Ns. However, it uses the most resources and must be invoked separately for each model tested.

2. **Proc ANOVA**. This is limited to ANOVAs with either equal or balanced designs though it can handle designs that are extremely complex.

3. **Proc RSReg**. This is basically designed for all subset regression in which it is desired to obtain values of R^2 and the regression weights for various combinations of predictors.

4. **Proc Reg**. This is the basic regression routine used for most OLS regression. One of its advantages is that several models can be tested within one procedure statement.

In addition, **Proc Calis** performs ML multiple regression, and **Proc Catmod** analyzes categorical data. Both will be demonstrated later. **Procs Lifereg, Logistic, Nlin, Orthoreg, Probit, Transreg, Autoreg, Pdlreg, Syslin**, and **Mode** are also available, but will not be discussed. The last four require an SAS add-on called ETS.

Proc Reg will be used for the following analysis:

```
Data A;
Retain nobs 1000 b1-b3 (.4 .5 .6) err 3213;
Bres = sqrt(1 - b1**2 - b2**2 - b3**2);
Array Xs(*) X1-X3;
Do I = 1 to nobs;
Do J = 1 to 3;
Xs(J) = rannor(err);
End;
Yp = b1*x1 + b2*x2 + b3*x3;
Y = Yp + bres*rannor(err);
X1 = 50 + X1*10;
X2 = 50 + X2*10;
X3 = 50 + X3*10;
Y = 100 + 15*Y;
XN1 = X1 + X2;
XN2 = X1 + X3;
XN3 = X2 + X3;
```

```
Output;
End;
Proc Reg Data=A all;
Model Y = X1-X3/all;
Model Y = XN1-XN3/all;
Run;
```

Whereas X1 to X3 were originally standardized in the population, they are scaled here as T scores ($\mu = 50$, $\sigma = 10$). Likewise, Y was originally scaled to a population mean of 0. It can be shown that its population standard deviation is R^2, but it is rescaled to $\mu = 100$ and $\sigma = 15$, known as a **deviation IQ**. The `Proc` statement requests all output available from the set of variables used in the analysis—the two sets of predictors (X1–X3 and XN1–XN3) and the criterion Y, as noted later. The `Model` statements define what is to be evaluated and are of the form `Model Criterion = Predictors`. The `All` statement used here provides the widest variety of outputs relevant to the specific model. The result is two OLS regression analyses respectively using X1 to X3 and XN1 to XN3 as predictors of Y. Part of the output requested in each model is an analysis of the residuals, to be considered in a later section of this chapter. This particular part of the output is rather voluminous.

1. The output begins with **simple descriptive statistics**. Note that these data include variables employed in either model.

Variables	Sum	Mean	Uncorrected SS	Variance	Std Deviation
INTERCEP	1000	1	1000	0	0
X1	49853.997236	49.853997236	2589390.8362	104.0738696	10.20166014
X2	50280.101956	50.280101956	2627430.1836	99.44097183	9.9720094179
X3	50569.704714	50.569704714	2654917.8997	97.720585372	9.885372293
Y	100457.13761	100.45713761	10312489.682	221.07425935	14.86856615
XN1	100134.09919	100.13409919	10224655.039	198.01523326	14.071788559
XN2	100423.70195	100.42370195	10292015.818	207.30320807	14.398027923
XN3	100849.80667	100.84980667	10369706.752	199.2224689	14.114618978

2. Next is a matrix identified as USSCP (**uncorrected sums of squares and cross products**), which contains terms of the form $\sum X^2$ and $\sum XY$. Note their relationship to the matrix terms discussed previously in Method 1. This matrix contains all the data necessary for the analysis other than the individual observations used in the residual analysis.

USSCP	INTERCEP	X1	X2	X3
INTERCEP	1000	49853.997236	50280.101956	50569.704714
X1	49853.997236	2589390.8362	2503917.0097	2523853.5412
X2	50280.101956	2503917.0097	2627430.1836	2543679.3343
X3	50569.704714	2523853.5412	2543679.3343	2654917.8997
Y	100457.13761	5069303.6505	5122688.8901	5169946.2318
XN1	100134.09919	5093307.8459	5131347.1933	5067532.8756
XN2	100423.70195	5113244.3774	5047596.344	5178771.4409
XN3	100849.80667	5027770.5509	5171109.5179	5198597.234

USSCP	Y	XN1	XN2	XN3
INTERCEP	100457.13761	100134.09919	100423.70195	100849.80667
X1	5069303.6505	5093307.8459	5113244.3774	5027770.5509
X2	5122688.8901	5131347.1933	5047596.344	5171109.5179
X3	5169946.2318	5067532.8756	5178771.4409	5198597.234
Y	10312489.682	10191992.541	10239249.882	10292635.122
XN1	10191992.541	10224655.039	10160840.721	10198880.069
XN2	10239249.882	10160840.721	10292015.818	10226367.785
XN3	10292635.122	10198880.069	10226367.785	10369706.752

3. The last set of data based on all seven variables is CORR (the **correlation matrix**). Compare this matrix with the output of `Proc Corr` in Chapter 5. In particular, recall that X1 to X3 are independent of one another in the population, but XN1 to XN3 are not. Because the latter are based on sums of variables that individually correlate with the criterion, their correlations with the criterion are higher than those of X1 to X3 (.64 to .77 vs. .40 to .61).

CORR	X1	X2	X3	Y
X1	1.0000	–0.0270	0.0273	0.4033
X2	–0.0270	1.0000	0.0105	0.4840
X3	0.0273	0.0105	1.0000	0.6120
Y	0.4033	0.4840	0.6120	1.0000
XN1	0.7058	0.6891	0.0272	0.6354
XN2	0.7273	–0.0120	0.7059	0.7059
XN3	0.0000	0.7138	0.7077	0.7706

CORR	XN1	XN2	XN3
X1	0.7058	0.7273	0.0000
X2	0.6891	–0.0120	0.7138
X3	0.0272	0.7059	0.7077
Y	0.6354	0.7059	0.7706
XN1	1.0000	0.5188	0.5059
XN2	0.5188	1.0000	0.4859
XN3	0.5059	0.4859	1.0000

4. The output of Model 1 follows, starting with X′X X′Y Y′Y (the **model cross products**):

X'X	INTERCEP	X1	X2	X3	Y
INTERCEP	1000	49853.997236	50280.101956	50569.704714	100457.13761
X1	49853.997236	2589390.8362	2503917.0097	2523853.5412	5069303.6505
X2	50280.101956	2503917.0097	2627430.1836	2543679.3343	5122688.8901
X3	50569.704714	2523853.5412	2543679.3343	2654917.8997	5169946.2318
Y	100457.13761	5069303.6505	5122688.8901	5169946.2318	10312489.682

5. Next, output identified as X′X inverse, parameter estimates, and SSE are presented. The first $k - 1$ rows and columns contain $(\mathbf{X'X})^{-1}$. The last column of the first $k - 1$ rows and the last row of the first $k - 1$ columns contain $(\mathbf{X'X})^{-1}(\mathbf{X'y})$, the raw-score regression weights, as one can see. Finally, the last column and row contains the sum of squared residuals.

	INTERCEP	X1	X2	X3	Y
INTERCEP	0.0760171907	-0.000479884	-0.000514232	-0.000499062	-10.58256674
X1	-0.000479884	9.6325516E-6	2.6920824E-7	-2.743443E-7	0.5833301381
X2	-0.000514232	2.6920824E-7	0.0000100749	-1.138271E-7	0.7285320813
X3	-0.000499062	-2.743443E-7	-1.138271E-7	0.0000102524	0.8963409651
Y	-10.58256674	0.5833301381	0.7285320813	0.8963409651	52428.652462

6. The next quantities to be presented are familiar from any prior exposure you may have had to the ANOVA—the **summary table** partitioning of the criterion sums of squares, $\sum y^2$.

Source	DF	Sum of Squares	Mean Square	F Value	Prob>F
Model	3	168424.53263	56141.51088	1066.534	0.0001
Error	996	52428.65246	52.63921		
C Total	999	220853.18509			

Note that the degrees of freedom (df) for the total are the number of observations $(1000) - 1$, the degrees of freedom for the model are the number of predictors (3—X1–X3), and the degrees of freedom for error are the difference between the two. The remaining 1 df associated with estimating the intercept plays no role at this point, but will shortly.

7. Next presented are (a) the **square root of the mean square error** (Root MSE), (b) the **criterion mean** (Dep Mean), (c) the **coefficient of variation** (C.V.), (d) the R^2 (R-square), and (e) an estimate R^2 **adjusting** for the number of predictors and observations (Adj R-Sq).

The root MSE is simply the square root of the mean square error ($\sqrt{52.63921}$). The coefficient of variation = the criterion standard deviation/the criterion mean. The adjusted R^2 corrects for the fact that the R^2 has an upward bias because it capitalizes on the sampling error in the correlations. Specifically, even when the validity is .0 in the population, it will not be exactly .0 in the sample. This will tend to inflate the observed R^2. The most common form of correction uses the following:

$$R_a^2 = 1 - (1 - R^2) \cdot \frac{n-1}{n-p}, \qquad (9.9)$$

where R_a^2 is the adjusted value, R^2 is the observed (unadjusted) value, n is the number of observations, and p is the number of predictors. For reasons kept a corporate secret at the SAS home office in Cary, NC, SAS does not present R itself despite the fact that anyone with access to the program could make this trivial modification.

In this example, the observed R^2 of .7626 and the adjusted value of .7619 are both very close to the expected value of .77 = $.4^2 + .5^2 + .6^2$. Actually, the observed value will usually be at least as large as the expected value because of the bias. The bias increases with the number of predictors and decreases with the sample size. The reason for this bias is discussed in many sources such as Nunnally and Bernstein (1994).

Root MSE	7.25529	R-square	0.7626
Dep Mean	100.45714	Adj R-sq	0.7619
C.V.	7.22227		

8. For each predictor variable (including the intercept), next appear (a) its df, (b) **its parameter estimate** (b weight), (c) its **standard error**, (d) its t **value**, (e) the **probability** associated with the null hypothesis that b is 0, and (f) the **Type I sum of squares** (SS).

There are several things to note here. First, note that the df will always be 1 in Proc Reg because the model estimates the linear regression of the predictor on the criterion, which, in turn, means that only a slope needs be estimated. Next, note that t is the parameter estimate divided by its standard error. In the present case, the parameter estimates for the three predictors are all significantly different from 0, but the parameter estimate for the intercept is not. Finally, the Type I SS is the sum of squares for each predictor associated with the criterion **at the time it is entered**. In other words, it adjusts for terms that have previously been entered in the model but ignores the contribution of those to be entered. The sum of squares is not of inherent interest, but it is used to calculate other terms of interest, most specifically the various semi-partial and partial correlations.

Variable	DF	Parameter Estimate	Standard Error	T for H0: Parameter=0	Prob > |T|	Type I SS
INTERCEP	1	-10.582567	2.00037117	-5.290	0.0001	10091636
X1	1	0.583330	0.02251777	25.905	0.0001	35923
X2	1	0.728532	0.02302901	31.635	0.0001	54137
X3	1	0.896341	0.02323102	38.584	0.0001	78365

9. The printout continues with (a) the **Type II** sum of squares, (b) the **standardized estimates** (β weights), (c) the **squared semipartial Type I correlation**, (d) the **squared partial Type I correlation**, and (e) the **squared semipartial Type II correlation**.

The Type II SS contrasts with the Type I SS in that it is the sum of squares for each predictor associated with the criterion, **adjusting for all other "appropriate" terms**. These are the other predictors in the model except for interactions with the term under consideration. For example, suppose a study includes the effects of A, B, C, and their interaction. The Type II sum of squares for the main effect of C would define its magnitude after A, B, and $A * B$ were incorporated into the model. Similar to its Type I counterpart, the Type II sum of squares is not of intrinsic interest but is used to calculate other terms of interest (SAS also uses Type III and Type IV sums of squares, which are discussed later in the context of the analysis of variance). Note that the standardized estimate of the intercept (β_0) is always .0 because the regression line passes through the criterion mean, which is 0 when standardized. The two squared semipartial correlations relate to the rest of the printout in a straightforward way. They are obtained by dividing the respective sums of squares by the total sum of squares (220853.18509). For example, the squared semipartial correlation between X_1 and Y, controlling for the other predictors but not Y (the semipartial correlation), is 35923/220853.18509 = .16265. Note that the partial correlations control for Y as well as the other predictors. The partial correlations are of more interest as they describe the correlation between predictor and criterion, fully controlling for the other predictors. However, both squared semi-partial correlations describe the increment in R^2, accounting for the added predictors, so it is of more than trivial value.

Variable	DF	Type II SS	Standardized Estimate	Squared Semi-partial Corr Type I	Squared Partial Corr Type I	Squared Semi-partial Corr Type II
INTERCEP	1	1473.228853	0.00000000	.	.	.
X1	1	35325	0.40023603	0.16265506	0.16265506	0.15994984
X2	1	52681	0.48860991	0.24512722	0.29274341	0.23853529
X3	1	78365	0.59593266	0.35482630	0.59914840	0.35482630

10. Next appear (a) the **Type II squared partial correlation**, (b) the **tolerance**, and (c) the **variance inflation factor**. The tolerance and variance inflation factors are related, extremely important measures. The tolerance is equal to $1/r^{ii}$ and the variance inflation equals r^{ii} itself, as derived from the inverse of the correlations among the predictors (not the augmented matrix, which also contains the criterion). A predictor with high tolerance and low variable inflation (in both cases, 1.0 is the limit) can contribute independently to the criterion. Of course, this is the way the predictors were created to begin with.

Variable	DF	Squared Partial Corr Type II	Tolerance	Variance Inflation
INTERCEP	1	.	.	0.00000000
X1	1	0.40255029	0.99850781	1.00149442
X2	1	0.50120172	0.99914403	1.00085671
X3	1	0.59914840	0.99912869	1.00087207

11. COVB (the **covariance of the estimates**) describes how independent these estimates are in absolute terms. The formula to calculate the terms of this matrix simply involves removing the square root sign from Equation 9.7 and substituting r^{ij} for r^{ii} so that all elements of $\mathbf{R}^{-1}$ are used, not just the diagonals. Conversely, the diagonal terms of this matrix are the variances of the estimate so that their square roots are the standard errors noted previously in Equation 9.7, letting $i = j$.

COVB	INTERCEP	X1	X2	X3
INTERCEP	4.0014848142	–0.025260689	–0.02706878	–0.02627024
X1	–0.025260689	0.0005070499	0.0000141709	–0.000014441
X2	–0.02706878	0.0000141709	0.0005303351	–5.991767E-6
X3	–0.02627024	–0.000014441	–5.991767E-6	0.0005396801

12. CORRB (the **correlation of the estimates**) is obtained by converting the covariances and variances to correlations using the standard relationship, Equation 6.2. Diagonal elements will, of course, be 1.0. Again note that the correlations (off-diagonal elements) are very small because of the independence of the predictors.

CORRB	INTERCEP	X1	X2	X3
INTERCEP	1.0000	-0.5608	-0.5876	-0.5653
X1	-0.5608	1.0000	0.0273	-0.0276
X2	-0.5876	0.0273	1.0000	-0.0112
X3	-0.5653	-0.0276	-0.0112	1.0000

13. The **sequential parameter** estimates describe how the models evolve as predictors are added. Note that by specifying `Model Y = X1-X3/all;` we told SAS to enter the variables in the order X1, X2, and X3. Had we chosen `Model Y = X3 X2 X1/all;`, the variables would have been entered in the reverse order. The first model therefore incorporates only the mean, so it is of the form $\hat{Y} = \overline{X}$. In this model, the regression coefficients for the three actual predictors are set to 0. X1 is then entered so that its b weight now differs from 0, but those for X2 and X3 do not, and so on. Finally, the terms in the bottom row are those for the model as fully specified.

INTERCEP	X1	X2	X3
100.45713761	0	0	0
71.152793146	0.5878033074	0	0
33.049010368	0.6073152693	0.7384836554	0
-10.58256674	0.5833301381	0.7285320813	0.8963409651

14. The **consistent covariances of the estimates** (ACOV) are defined in White (1980). They define what the covariances among estimates would be if one rejected the notion of homoscedasticity (i.e., if error variance were assumed to be heterogeneous).

ACOV	INTERCEP	X1	X2	X3
INTERCEP	4.1459639619	-0.027025334	-0.026542277	-0.028196083
X1	-0.027025334	0.0004996545	0.0000225208	0.0000174468
X2	-0.026542277	0.0000225208	0.0005165867	-5.976085E-6
X3	-0.028196083	0.0000174468	-5.976085E-6	0.0005488711

15. **Tests of first- and second-moment specification** also follow from White (1980), and test the null hypothesis that the data are, in

fact, homoscedastic. In the present case, the small value of χ^2 implies that they are homoscedastic.

```
DF:  9  Chisq Value:  4.6987  Prob>Chisq:  0.8597
```

This completes the output for the first model. The form of the output for the second model parallels this, so it need not be repeated in all its detail. However, it is very instructive to compare certain key parameters (see Table 9.1). The reader is encouraged to make a full comparison.

The important points to note from Table 9.1 are the following:

1. Recombining the predictors linearly as we did left the overall R^2 unaffected because we neither added to nor subtracted from the original predictability.

2. The β weights for independent predictors are generally larger than the β weights for correlated predictors. This is because β weights define what predictors share with the criterion independently of the other predictors. X1 to X3 were constructed to contribute more independently than XN1 to XN3, even though the latter had higher validities (see before).

3. The various partial and semipartial correlations were lower for this same reason. This is particularly true of the Type II correlations because they adjust for more of the shared variance.

TABLE 9.1 **Comparison of Selected Parameter Estimates for Model 1 (Uncorrelated Predictors) and Model 2 (Correlated Predictors)**

Parameter	Model 1	Model 2
R^2	.76	.76
β_1	.40	.20
β_2	.49	.36
β_3	.60	.49
$r_{1y.23}$	.16	.02
$r_{1y.23}$	.40	.10
r_{b1b3} (correlation between estimates of b_1 and b_3)	−.03	−.34
Tolerance of X_1	1.00	.65

4. Whereas the parameter estimates derived from independent predictors are uncorrelated, parameter estimates derived from correlated predictors are not.

5. Tolerances for uncorrelated predictors are higher than tolerances for correlated predictors, again because of the independence of the former.

Residual Analysis

Residuals (Y^*) are based on the disparity between observed criterion measures (Y) and predicted criterion measures ($\hat{Y} = b_0 + \sum b_i X_i$). Were all relevant predictors to be included and the disparity simply due to normally distributed error, these residuals would have a normal distribution with a mean of .0 and a standard deviation equal to the square root of $1 - R^2$. A residual analysis examines these residuals using a variety of indexes (because the two models make identical predictions, only one set of residuals is needed for purposes of discussion).

Part of the analysis simply consists of displaying each criterion value and a series of somewhat self-explanatory statistics once the abbreviations are deciphered. Table 9.2 contains these data using the first observation as an example (the table is transposed relative to the SAS output).

The first Y observation is 92.9, and its corresponding predictor values are 62.527, 40.985, and 45.005. Its associated predicted value therefore equals $-10.582567 + 0.583330 \cdot (62.527) + 0.728532 \cdot (40.985) +$

TABLE 9.2   Residual Statistics

Quantity	Value
Dependent variable, Y	92.90
Predicted, $\hat{Y}$	96.09
Standard error of $\hat{Y}$	.44
Lower 95% C.I. for Y	95.22
Upper 95% C.I. for Y	96.96
Lower 95% C.I. for $\hat{Y}$	81.83
Upper 95% C.I. for $\hat{Y}$	110.40
Residual, Y^*	−3.20
Standard error of Y^*	7.24

$0.896341 \cdot (45.005) = 96.09$. Note that Y (92.9) $- \hat{Y}$ (96.09) $= Y^*$ (−3.20). The standard errors of $\hat{Y}$ and Y^* for the case of multiple predictors are given by formulas that are ill suited to hand calculation (Pedhazur, 1982, pp. 143–146). These standard errors are then used to obtain confidence intervals by ordinary means ($\hat{Y} \pm 1.96^*s_{y'}$ and $Y^* \pm 1.96^*s_{y*}$, where $s_{y'}$ and s_{y*} are the standard errors of the predicted scores and residuals, respectively.

The **Studentized residuals** appear next. These are obtained by standardizing $\hat{Y}$ and dividing the result by the standard error ($-.21/.48 = -.44$ for the first observation). **Studentizing** casts the residual into the metric of a normal distribution. These values are plotted to the nearest half z-score unit to facilitate the detection of **outliers**. Outliers are residuals that are so large in absolute value that it is unlikely they emerged by pure chance. They often suggest that one or more important predictors have been omitted from the model. Cook's D statistic is then presented for each value. This is a measure of the influence that the data point has on the analysis based on the change in regression weights following its deletion. Observations with large values of D are thus often outliers. It is also possible to obtain Durbin–Watson statistics, which detect autocorrelations (lack of independence in successive values). This is normally only of interest when the observations are ordered in time and specifically not of interest when the observations simply consist of responses from a sample of individuals.

Three final terms appear:

```
Sum of Residuals                 -104E-12
Sum of Squared Residuals      52428.6525
Predicted Resid SS (Press)    52850.7430
```

The `Sum of Residuals` ($\sum Y^*$) is a check on the adequacy of the calculation. It should effectively be 0, as it is. The `Sum of Squared Residuals` is the actual error sum of squares (see before). The `Predicted Resid SS` (often abbreviated as PRESS) is the sum of squared **predicted residuals**, where a predicted residual is a residual obtained from dropping that observation from the parameter estimate. The actual and predicted sums should be similar in value with, as here, the PRESS being slightly larger. This is a global check on how well assumptions such as homoscedasticity have been met.

Basic Multiple Regression in SPSS

After a data set is defined in SPSS, linear regression is invoked by choosing **Regression** from the Analyze menu, followed by **Linear**, Alt - A,R,L. This opens a window (see Figure 9.1). As can be seen, this is where the user specifies the independent and dependent variables and chooses options for output. The Block option allows multiple models to be fit. The options include

1. Use of weighted least squares

2. Statistics, which is concerned with which statistics to compute

3. Plots, which is concerned with graphical evaluation of residuals

4. Save, which allows such data sets as residuals to be saved

5. Options, which is concerned with significance levels and the handling of missing data

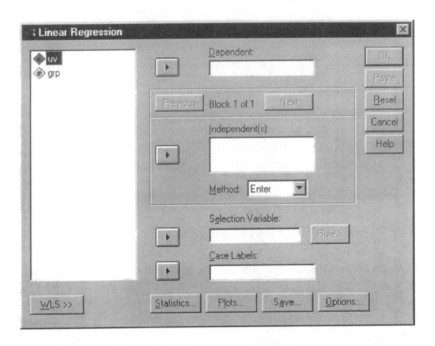

Figure 9.1. Basic SPSS window used in linear regression.

The output contains many of the same statistics as SAS:

1. Basic univariate statistics

2. Correlations

3. R^2

4. Adjusted R^2

5. Standard error of R^2

6. ANOVA table and associated statistics such as F

7. Covariance and correlations among the estimates, b

8. Standard error of b

9. β weights

10. Tolerances

11. Variable inflation factors

12. t for the null hypothesis that $b = 0$

13. Significance level associated with t (unlike SAS, SPSS also provides confidence intervals on b)

14. R itself

15. Component structure of the predictors

16. Summary statistics on residuals

The SPSS output can be displayed graphically using `Plot` and has gotten better in recent editions, but Excel's graphic capabilities are at least as good.

Basic Multiple Regression in Excel

Excel's Data Analysis package can perform a rudimentary multiple regression analysis if the package previously has been installed (you may have already used it to perform the simulations in the earlier chapters). You can tell if this package has been installed by looking to see if it is an option under Tools (`Alt`-T). If it is not, choose Tools, **Add-Ins** (`Alt`- T, I). Click on Analysis ToolPak (**not** Analysis ToolPak—VBA) and then press Enter. Once installed, choose **Data Analysis** (Alt-T, D) and then choose **Regression** from the Tools window. However, Excel sometime leads to anomalous results, and it is suggested for multiple regression applications only under "emergency" conditions. Use SAS, SPSS, or some comparable package for this and any other more complex statistical analysis.

The Excel regression dialogue asks where the predictors and criterion are and the format in which you want the output. The statistics that are output are somewhat limited, particularly with regard to collinearity. However, the basic statistics related to R, the ANOVA table, b (but not β), and the residuals are presented. Despite criticisms of Excel's accuracy, its results agreed with both SAS and SPSS in our test runs. Excel provides excellent graphic output of residual data, which is a valuable supplement to SAS if you have not installed the latter's graphic module.

For example, Figure 9.2 is a scatterplot of a predictor (X_1) (abscissa) against the residuals, Y^* (ordinate). Assuming that the relationship is linear and that error is bivariate normal, this plot should be an ellipse with its major axis parallel to the abscissa, denoting $r \approx 0$. Similar plots were generated for X_2 and X_3 but are not shown.

Figure 9.3 is a line-fit scatterplot of X_1 against both Y (filled circles) and $\hat{Y}$ (open circles). If the model fits as specified, each of the two rela-

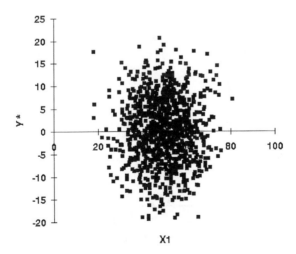

Figure 9.2. Excel scatterplot of variable X_1 (abscissa) against Y^* (residual).

tionships will be linear. Again, similar plots were generated for X_2 and X_3 but are not shown.

Finally, a normal probability plot is a plot of Y against the cumulative normal distribution (see Figure 9.4). This plot should be linear when Y is normally distributed.

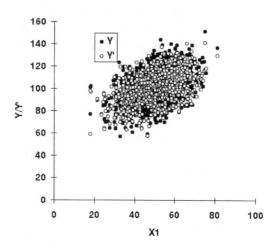

Figure 9.3. Excel scatterplot of variable X_1 against both Y (filled circles) and $\hat{Y}$ (open circles).

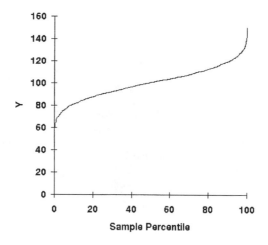

Figure 9.4. Excel normal probability plot.

Hierarchical (Incremental) Regression

Although it is not extremely difficult to write a program that performs basic multiple regression for a predesignated set of predictors, more complex issues arise when a major purpose of the analysis is to determine which predictors to retain and which to ignore. There are two basic strategies. One is to formulate hypotheses and to let these hypotheses dictate the regression strategy. Basically, the strategy is to test the hypothesis that predictor set *A* relates to the criterion. Once that has been evaluated, the next hypothesis is that predictor set *B* adds to the information provided by *A*, and so on. The first predictor set (*A*) typically has known properties, but additional set *B* (*C*, *D*, etc.) is (are) less understood. For example, one does not need to investigate the hypothesis that a measure of academic ability (*A*) relates to academic performance, as that is well known. What may be more controversial is whether some personality variable (*B*) contributes to performance. Simply relating *B* to performance is not very informative. It may simply be that people high in *B* are also high in *A*. That would probably be of little interest. Testing the increment in R^2 for predictors *A* + *B* relative to the R^2 for *A* alone tests the hypothesis that *B* relates to the criterion **independently** of *A*. In other words, this increment in R^2 controls for the effects of *A*. The *F* test for an increment in R^2 when predictor set *B* is added to predictor set *A* can be written as

$$F = \frac{(R^2_{a+b} - R^2_a)(N - k_{a+b} - 1)}{(1 - R^2_{a+b})(k_{a+b} - k_a)},$$
(9.10)

where R^2_{a+b} is the R^2 obtained with both predictor sets *A* and *B* as predictors, R^2_a is the R^2 obtained with predictor set *A*, k_{a+b} is the total number of predictors in sets *A* and *B*, k_a is the number of predictors in set *A*, and *N* is the total number of observations.

A hierarchical approach can be used in various ways. A typical example is when one predictor set is relatively inexpensive (such as objective personality test data) and a second is relatively expensive (such as a mental health professional's judgment). The question is not whether the judgments relate to the criterion—it is whether they improve prediction over the cheaper set. A second example involves the use of a hierarchical analysis as a test of **moderation**. Moderation is said to occur whenever the relationship between two variables is affected by a third, such as whether the relationship between academic ability and performance differs by gender. This issue will be examined later in this chapter after we have had an opportunity to consider multiple regression based on ML methods.

Stepwise Regression

Although there is some inconsistency in use of the term "stepwise," we follow what perhaps most do in referring to it as a set of techniques in which data rather than theory are responsible for the final choice of predictors. This includes **forward selection** in which one begins with a **mean model** $(\hat{Y} = \overline{X})$. Predictors are added, with the effect of their increment tested as in Equation 9.10. The process stops when there are no longer any predictors to significantly increment R^2. Conversely, **backward selection** begins with a **saturated model** (i.e., a model that uses as many predictors as there are independently varying quantities). As a result, it has no residual df. Effects of decrements are evaluated, and the process stops when no predictors can be deleted. What some (e.g., SAS) specifically call "stepwise" combines backward and forward selection by adding and deleting variables although perhaps most use this in the generic sense of this section.

Unfortunately, Wilkinson (1979) noted many years ago that there is a major problem with stepwise approaches. Equation 9.10 assumes that a fixed set of predictors is to be evaluated, whereas the various stepwise algorithms make multiple comparisons. The probability that at least one comparison will be significant at a given α level is much greater than α. Specifically, it is given by the relationship $p = 1 - (1 - \alpha)^q$, where p is the probability that at least one comparison will be significant, α is the nominal significance level for a single comparison, and q is the number of comparisons. For example, if q is 13, then p is .48 or nearly 10 times the level for a single comparison. The actual correction needed for stepwise regression is more complex than this, but, unfortunately, no correction has been incorporated into any stepwise regression algorithm to our knowledge. Thus, the user is not informed of the error in the values used.

The following simulation illustrates this point:

```
Data x;
Array xs (*) x1-x20;
Retain err 1996 nobs 200 ni 20;
Do i = 1 to nobs;
y = rannor(err);
Do j= 1 to 20;
xs(j) = rannor(err);
End;
```

```
Output;
Keep x1-x20 y;
End;
Proc Reg Data=x;
Model y = x1-x20/selection=stepwise;
Run;
```

You should verify that this program creates a series of 20 "predictors" that are uncorrelated in the population with both one another and the criterion. However, the sample correlations are not exactly 0 so the stepwise solution, given 20 chances, "finds" at least one significant predictor (five in the case of this random number seed). Of course, different seeds produce both different numbers of "predictors" and different specific "predictors."

Maximum Likelihood and Other Alternatives to Ordinary Least Squares Regression

We have already mentioned generalized (weighted) least squares and ML as alternatives to OLS regression. Still a third option is minimum chi-square, which can be used when the criterion is dichotomous. However, the only widely used alternative to OLS at present is ML. ML estimation can be used to cover a wide range of problems, including those involving categorical criteria to be considered in Chapter 11. Any covariance structure analysis program such as EQS or LISREL can perform ML regression, but we will limit demonstration to Proc Calis because it is almost certain to be available at university computer centers.

The problem is again to predict Y from orthogonal predictors X1–X3. The code was as follows:

```
Proc Calis cov data = A all;var x1-x3 y;
lineqs y = beta1 x1 + beta2 x2 + beta3 x3 + e1;
std e1 = ve1;
```

This specification calls for the analysis of the variance–covariance matrix rather than the correlation matrix and provides the b weights, β weights, and R, among many other statistics for analyzing data from a single group of subjects. There are three alternative specifications. The default is to analyze the correlation matrix, which provides the β weights and R but not the b weights because it analyzes standardized data. One may also choose ucov instead of cov and the additional option aug to add in a vector of 1s. These are used jointly to test group differences, which we did not do because the simulation only involves one group.

The output for the first application of Proc Calis consisted of the following (see the section on confirmatory ML factor analysis in the previous chapter):

1. A description of the model's specifications in terms of the various matrices and variables employed. This application uses the correlation matrix, so most statistics refer to standardized values. The model estimates four values: three β weights and an error variance, denoted BETA1 to BETA3 and VE1.

2. Univariate and bivariate statistics, including the variance–covariance matrix. Because the results are sensitive to kurtosis differing from normality, several statistics of this form appear, as follows:

```
Mardia's Multivariate Kurtosis . . . . . . . .        -0.2609
Relative Multivariate Kurtosis . . . . . . . .         0.9891
Normalized Multivariate Kurtosis . . . . . . .        -0.5954
Mardia-Based Kappa (Browne, 1982). . . . . . .        -0.0109
Mean Scaled Univariate Kurtosis . . . . . . .         -0.0258
Adjusted Mean Scaled Univariate Kurtosis . . .        -0.0258
Observation numbers with largest contribution to kurtosis
       855      818      430        6        272
      1164  641.4325  610.0944  546.0081  514.7496
```

3. Initial estimates

```
BETA1   1     0.58333    Matrix Entry: _GAMMA_[1:1]
BETA2   2     0.72853    Matrix Entry: _GAMMA_[1:2]
BETA3   3     0.89634    Matrix Entry: _GAMMA_[1:3]
VE1     4    52.48113    Matrix Entry: _PHI_[4:4]
```

4. Notification that estimation had been performed, leading to inferential and descriptive measures of fit, as discussed in the previous chapter.

5. The regression equations, referred to as the Manifest Variable Equations. These are nearly, but not exactly identical to the parameters of the OLS equation—the t values can be seen to be marginally greater for the data of the first analysis, based on the correlation matrix.

```
Y      =    0.5833*X1   +   0.7285*X2   +   0.8963*X3   +   1.0000 E1
Std Err     0.0225 BETA1    0.0230 BETA2    0.0232 BETA3
t Value   25.9443          31.6830          38.6419
```

6. **Variances of Exogenous Variables** present these statistics on both the observed variables (X1–X3) and the error term (described as E1 in the model but assigned the symbolic name VE1 for purposes of estimation).

```
                                     Standard
Variable  Parameter    Estimate      Error   t Value
-----------------------------------------------------

X1                    104.073870         0   0.000
X2                     99.440972         0   0.000
X3                     97.720585         0   0.000
E1           VE1       52.481134  2.348202  22.349
```

7. The covariances among the predictor estimates appear next, followed by the regression equation in standard score form (note that the error term appears as the square root of the estimated variance [$\sqrt{.237391} = .4872$]). The squared error and R^2 follow :

```
                              Standard
Parameter     Estimate         Error          t Value
-----------------------------------------------------

X2  X1       -2.749804            0           0.000
X3  X1        2.754377            0           0.000
X3  X2        1.030456            0           0.000
```

8. The standardized regression equation (i.e., the β weights) follows:

```
Y = 0.4002*X1 + 0.4886*X2 + 0.5959*X3 + 0.4872 E1
```

9. The error variance, total variance, and R^2 appear next:

```
                Error          Total
Variable      Variance        Variance      R-squared
             52.481134       221.074259      0.762609
  1  Y
```

10. **Correlations Among Exogenous Variables**, which are actually the correlations among the estimates, are then presented:

```
Parameter              Estimate
- - - - - - - - - - - - - - - - - - - - - - - - - - -
X2    X1              -0.027030
X3    X1               0.027312
X3    X2               0.010453
```

11. Next to appear are the Total Effects of Exogenous on Endogenous Variables. Total effects are correlations.

	X1	X2	X3
Y	0.5833301381	0.7285320813	0.8963409651

12. The Indirect Effects of Exogenous on Endogenous Variables follow. Indirect effects are pattern (β) weights. However, because the predictors are all exogenous, the model contains no indirect effects. Indirect effects are effects of endogenous variables being acted upon by exogenous or other endogenous variables. An endogenous variable cannot act upon an exogenous variable, because the latter is defined as being beyond the model's ability to analyze. Exogenous variables X1, X2, and X3 thus are all assumed to act directly upon Y, causing the resulting values to be 0.

	X1	X2	X3
Y	0	0	0

13. Next, the Lagrange Multiplier and Wald Test Indexes are presented (Bentler, 1986; Buse, 1982; Lee, 1985; MacCallum, 1986). A Wald statistic is the squared difference between a parameter estimate and 1.00 divided by its squared standard error. A Lagrange multiplier has a more complex definition (see SAS Institute, 1990, p. 1009). These statistics evaluate the effects of deleting each parameter, in turn, from the model relative to including it. Both are approximately distributed as χ^2. These data first appear for the exogenous variables and then for the endogenous variables. The results begin with the relationships among the predictors, identified as _PHI_[4:4], which are not informative as the indexes are all .0. These are followed by the relationships between each predictor, in turn, and the criterion, identified as _GAMMA_[1:3], which is of more interest.

```
Lagrange Multiplier and Wald Test Indexes _PHI_[4:4]
Symmetric Matrix
Univariate Tests for Constant Constraints
-----------------------------------------
|  Lagrange Multiplier or Wald Index   |
-----------------------------------------
|  Probability | Approx Change of Value  |
-----------------------------------------
              X1            X2           X3        E1
X1  0.000         0.000         0.000        SING
    1.000  0.000  1.000 -0.000  1.000 -0.000   .          .
X2  0.000         0.000         0.000        SING
    1.000 -0.000  1.000  0.000  1.000  0.000   .          .
X3  0.000         0.000         0.000        SING
    1.000 -0.000  1.000  0.000  1.000  0.000   .          .
E1  SING          SING          SING       499.500  [VE1]
     .      .       .      .       .      .

Lagrange Multiplier and Wald Test Indexes _GAMMA_[1:3]
General Matrix
Univariate Tests for Constant Constraints
-----------------------------------------
|  Lagrange Multiplier or Wald Index   |
-----------------------------------------
|  Probability | Approx Change of Value  |
-----------------------------------------
              X1            X2           X3
Y  673.107[BETA1]  1003.814[BETA2]  1493.194[BETA3]
```

The overall results hardly suggest the need for ML regression as the parameter estimates are indistinguishable from those obtained with OLS regression. However, the utility of ML regression is better seen in some form of hierarchical analysis, as will be shown. By its very nature, moderated multiple regression illustrates such a hierarchical scheme so we will now turn to it.

Moderated Multiple Regression

Previously, it was noted that OLS could be used to evaluate moderator effects—situations in which a third variable influences the relationship between two others, that is, an interacting variable (Darlington, 1990, p. 408). Currently, a hierarchical approach based on OLS is the most popular method of examining moderator effects. This section will consider moderated multiple regression (MMR) using both OLS and ML. The following data set can be seen to contain a moderator effect:

```
Data mod;
b1 = .7;b1res = sqrt(1 - b1**2);
b2 = .5;b2res = sqrt(1 - b2**2);
n = 500;
no2 = n/2;
Retain err 1995;
Do i = 1 to n;
x = rannor(err);
If (i gt no2) then m = 1;else m = 0;
If (i gt no2) then delta = .25;else delta = 0;
If (m = 0) then y = b1*x + delta + rannor(err)*b1res;
If (m = 1) then y = b2*x + delta + rannor(err)*b2res;
x = 15*x + 100;
y = 10*y + 50;
xm = x*m;
Output;
End;
```

For the first 250 observations, $M = 0$ and the population value of r (ρ) is .7, whereas for the remaining 250 observations, $M = 1$ and ρ is .49. Therefore, M, in fact, moderates the relationship between X and Y. Moreover, the population mean of Y is .25 z-score units larger when $M = 1$. Consider a hierarchy of the following tests, which is basic to the OLS approach:

1. Denote the basic predictor (academic ability in this case) as X and the criterion (academic performance) as Y.

2. Denote the moderator (gender) as M. There are many ways to code categorical variables (Bernstein, 1988; Pedhazur, 1982). For present purposes, use **dummy coding** in which one gender (say males) is defined as 1 and the other is defined as 0.

3. Define the interaction code (I) as the product of X and M. In other words, $I = X$ for males and $I = 0$ for females. The moderator need not be a categorical variable, but, as McClelland and Judd (1993) have shown, tests of a presumed moderator that is normally distributed or even approximately so (as is true of many test scores) has very little power.

4. Compute r_{xy}. The effect, if significant, is that there is a relationship between X and Y, ignoring M, for example, that there is a relationship between academic ability and academic performance, ignoring gender, so r_{xy} is significantly greater than .0.

5. Compute the multiple correlation predicting academic performance from academic ability and gender, $R^2_{XM.Y}$.

6. Test the significance of the difference between $R^2_{XM.Y}$ and r^2_X. The effect, if significant, is that there is a relationship between M and Y, controlling for X, for example, that there is a relationship between gender and academic performance, correcting for academic ability, so $R^2_{XM.Y}$ is significantly greater than r^2_X. Alternatively, one can look at the significance of M's β weight in this three-variable model as the test is equivalent.

7. Compute the multiple correlation predicting academic performance from academic ability, gender, and their interaction, $R^2_{XMI.Y}$.

8. Test the significance of the difference between $R^2_{XMI.Y}$ and $R^2_{XM.Y}$. The effect, if significant, is the one of greatest interest. It is that the relationship between X and Y differs as a function of M, for example, that the relationship between academic ability and performance differs for males and females, so $R^2_{XMI.Y}$ is significantly greater than $R^2_{XM.Y}$.

The SAS steps used to obtain the relevant data were as follows:

```
Proc Corr Data=mod;var x;with y;by m;
Proc Reg Data=mod corr simple;
Model y = x/stb;
Model y = x m/stb;
Model y = x m xm/stb;
```

Proc Corr generated r_{xy} and the univariate data for each group, and Proc Reg performed the MMR analysis proper. Unfortunately, SAS does not directly calculate the F ratio associated with the changes in R^2, but these are not difficult to obtain, especially when the results are copied to a spreadsheet.

The results were that $\overline{Y}$ and r_{xy} were 49.7 and .535 when $M = 0$ and 53.4 and .736 when $M = 1$, respectively. The values of R^2 were 0.3966, 0.4167, and 0.4332 for the three models. Applying Equation 9.10 to the first pair of R^2 values produces an F ratio of 17.13 with 1 and 498 df.

Applying it to the second pair of R^2 values produces an F ratio of 14.44 with 1 and 497 *df*. Both of these are significant well beyond the .01 level. The first effect indicates that the mean difference between 49.7 and 53.4 was significant; the second means that the difference in slopes of the regression lines was significant.

The strategy used in ML regression is also hierarchical, although Proc Calis works best with a different setup. Whereas Proc Reg defined criterion variable Y for all cases but used different rules to define it in the first and last half of the data set, we are going to create two different criteria, Y1 and Y2, as follows:

```
Data mod;
delta = .25;
b1 = .7;b1res = sqrt(1 - b1**2);
b2 = .5;b2res = sqrt(1 - b2**2);
n = 500;
Retain err 1995;
Do i = 1 to n;
x = rannor(err);
y1 = b1*x + delta + rannor(err)*b1res;
y2 = b2*x + rannor(err)*b2res;
x = 15*x + 100;
y1 = 10*y1 + 50;
y2 = 10*y2 + 50;
Output;
Keep x y1 y2;
End;
```

Following are the steps involved in the analysis:

```
Proc Calis ucov aug data = mod;
lineqs y1 = i1 intercep + s1 x + e1,
       y2 = i1 intercep + s1 x + e2;
std e1 e2 = ve1 ve1;
Proc Calis ucov aug data = mod;
lineqs y1 = i1 intercep + s1 x + e1,
       y2 = i2 intercep + s1 x + e2;
std e1 e2 = ve1 ve1;
Proc Calis ucov aug data = mod;
lineqs y1 = i1 intercep + s1 x + e1,
       y2 = i2 intercep + s2 x + e2;
```

```
std e1 e2 = ve1 ve1;
Proc Calis ucov aug data = mod;
lineqs y1 = i1 intercep + s1 x + e1,
       y2 = i2 intercep + s2 x + e2;
std e1 e2 = ve1 ve2;
Run;
```

Note that this particular simulation causes Y1 to differ from Y2 with regard to its intercept, slope, and error terms. Each time `Proc Calis` is called, two regression lines are fit, respectively predicting Y1 and Y2 from X. As noted previously, the options `ucov and aug` are needed to create an intercept term. They respectively denote the **u**ncorrected **cov**ariance matrix ($\sum XY$ and $\sum X^2$) **aug**mented with a vector of 1s.

The following table defines the four models. A C denotes a variable constrained to equality for Y1 and Y2, and an F denotes a variable that is allowed to vary freely. Also presented are the G^2 value for that model, the df for the model, and the difference G^2 between that model and the next more general one in the hierarchy. The difference G^2 tests the specific effects of the constraint. All such difference G^2 values are based on 1 df.

Model	Intercept	Slope	Error	Model G^2	df	Difference G^2
1	C	C	C	52.76	4	28.71
2	F	C	C	24.05	3	12.42
3	F	F	C	11.63	2	9.70
4	F	F	F	1.93	1	

Both the model G^2 and the difference G^2 values are sufficiently large to reject all but the last model in which the intercept, slope, and error vary among groups. Note that another way to fit models is to obtain parameters separately by groups (in SAS, using the by option) and then to fit the subjects as a whole. Recall that you can add the G^2 values for the groups separately to obtain the overall model G^2. The raw-score regression equations are

$$Y_1 = 0.4495 * X + 7.3504 + E_1,$$
$$Y_2 = 0.3325 * X + 16.3961 + E_2$$

and the standardized regression equations are

$$Y_1 = 0.8546 * X + 0.1372 + 0.1379 * E_1,$$
$$Y_2 = 0.6660 * X + 0.3224 + 0.1670 * E_2.$$

One minor problem is that the model implies R^2 values of nearly 1.0 (.98 and .97, respectively). This is because the error terms (E1 and E2) are being treated as if they were predictors. Indeed, if you knew them, prediction would be of that magnitude. However, they are unknowns. Adding in the following provides the R^2 values:

```
Proc Calis Data = mod;
lineqs y1 = s1 x + e1,
       y2 = s2 x + e2;
std e1 e2 = ve1 ve2;
```

The resulting values were 0.462136 and 0.262338 for Y1 and Y2, respectively.

In testing for moderation with real-life problems, do not forget that apparent moderator effects can arise from a quadratic relationship between the predictor and criterion (MacCullum & Mar, 1995). Also be aware of the very limited power there is to detect moderator effects between two continuous, single-peaked predictors (McClelland & Judd, 1993).

▣ ANALYSIS OF VARIANCE

The ANOVA follows the same logic as OLS regression except that the predictors are discrete. In many cases, they will be dummy codes (values that are either 0 or 1), but this is far from universally the case.

Simple Analysis of Variance

We begin this section by using Proc ANOVA to test differences among the three levels of the variable identified as Group in data set Simple of Chapter 5. The additional SAS code needed to run the analysis is as follows:

```
Proc ANOVA Data=Simple;
Class Group;
Model Y = Group;
Means Group/regwq;
Run;
```

The **Class** statement identifies Group as a future independent variable and generates appropriate codes for it. One difference between **Proc ANOVA** and other regression procedures such as **Proc Reg** or **Proc GLM** is that all the predictors have to be categorical. Thus, even though Group takes on the values 1, 2, and 3, these do not stand for three numeric (ordinal or higher) values. Instead, they are simply names, and the necessary **Class** statement makes this explicit.

In general, a categorical variable coded at k levels requires $k-1$ individual variables to describe it. For example, consider the following:

```
If Group = 1 then Group1 = 1;else Group1 = 0;
If Group = 2 then Group2 = 1;else Group2 = 0;
```

Group1 and Group2 are each dummy codes that jointly describe which of the three groups each observation falls into. Note that the statement **If Group = 3 then Group3 = 1;else Group3 = 0;** is redundant because an observation that is coded 0 on Group1 and Group2 is already known to be in Group3. One could use Group1 and Group2 in place of the single variable Group and apply these predictors to **Proc Reg**, but there is no reason to do this because the **Proc ANOVA** format is more convenient and provides additional useful output not directly possible within **Proc Reg**.

The Model statement has the same function as it does in **Proc Reg** and other regression programs in specifying the dependent variable or criterion (Y) and the independent variable(s) or predictor(s), **Group** in this case. The **Means** statement computes means for the designated levels of the predictor(s) and, optionally, specifies one of several possible tests for differences among levels of the predictor. The Ryan–Einot–Gabriel–Welsch multiple range test has been chosen simply for illustrative purposes—different tests are appropriate for different situations.

1. The first information to be output is the class level information, which identifies the predictor, the number of levels it has, and the values it assumes. The number of observations in the data set follows. Check this to make sure you have properly specified the model.

```
Class     Levels     Values
GROUP          3     1 2 3
Number of observations in data set = 15
```

2. Next is the summary table. This section is nearly identical to its Proc Reg counterpart. Note that the Model effect in general contains all of the sources listed in the Model statement. However, in the simple ANOVA, the Model effect and the one listed effect are identical. In general, the Model effect is a pooling of effects of various interest so it is not of great interest.

```
Source                 DF  Sum of Squares  Mean Square  F Value    Pr > F
Model                   2      2.67630077   1.33815039     1.20    0.3344
Error                  12     13.36315998   1.11359667
Corrected Total        14     16.03946075

              R-Square              C.V.     Root MSE            Y Mean
              0.166857          21.41502   1.05527090        4.92771435
```

3. The next data are the individual effect(s) defined in the Model statement. In this case, the Group effect is not significant.

```
Source    DF     ANOVA SS    Mean Square    F Value    Pr > F
GROUP      2   2.67630077     1.33815039       1.20    0.3344
```

4. The means and results of any tests of specific group means appear in the final section. Note that the *df* and the MSE (mean square error) are identical to the values presented in the summary table. The Ryan–Einot–Gabriel–Welsch multiple range test, like most such tests, defines a critical value that is based on the mean square error and number of means compared. In this case, none of the three group means differs from one another so they are grouped together under the symbol A.

```
Ryan-Einot-Gabriel-Welsch Multiple Range Test for Variable: Y
NOTE: This test controls the Type I experimentwise error rate.
Alpha= 0.05  df= 12  MSE=  1.113597
Number of Means             2          3
Critical Range  1.4541707 1.7804927
Means with the same letter are not significantly different.
REGWQ Grouping        Mean     N GROUP
              A      5.5095     5 2
              A
              A      4.7543     5 3
              A
              A      4.5194     5 1
```

Considerations Applicable to Higher Order Designs

A more complete discussion of the analysis of variance is presented in the literature; see for example, Winer, Brown, and Michels (1991). We will limit discussion to some basic distinctions and approaches.

1. One basic distinction was introduced in Chapter 5, namely, that between a fixed and a random effect. Levels of a fixed effect are chosen because one is specifically interested in these levels and does not need to generalize to levels not included in the study. The investigator is interested in estimating the means of the chosen levels. Levels of a random effect are presumably sampled at random, though this is honored more in the breach with regard to subjects who are the most commonly encountered random effect in psychological research. Because they are sampled at random, conclusions generalize to those levels that were not studied but could have been. Interest centers on how much variance there is among levels.

2. A second important consideration is whether two independent variables are **crossed, nested,** or **partially replicated.** In a crossed design, a given level of one variable has the same meaning for all levels of a second variable. A common example of a crossed design is a repeated measure. Let observation X_{ij} denote the observation obtained from the ith level of a variable under investigation and the jth subject. In a repeated measure, $X_{13}, X_{23}, X_{33}, \ldots$ all describe observations obtained from the same individual, so the second subscript has the same meaning for all levels of the first. In general, levels of any two variables are crossed if they are combined in all possible ways. A basic characteristic of a crossed design is that one can obtain the interaction between the two variables as well as their (separate) main effects.

    Conversely, the simple ANOVA provides a simple example of a nested design—subjects X_{13} and X_{23} are arbitrarily related. They have nothing in common except for the act of arbitrarily designating them as the third subject within their respective groups. Subjects $X_{11}, X_{12}, X_{13}, \ldots$ all "belong" to group 1 as they can be considered to belong to the same "nest." A given level of one variable is paired with one and only one level of a second. Sources of variance in a nested design are usually broken down into (a) the main effect of the nesting variable and (b) the variation in the nested variable within levels of the nesting variable. If the nested variable is subjects, its variation is usually pooled into a single term defining experimental error.

Variation in one variable computed within a single level of a second variable is called a **simple** effect. The effect of subjects nested within a given level of a treatment variable is an example of a simple effect. Sometimes, simple effects for a crossed variable may be of interest. For example, if an experimental and a control group are each studied over a series of learning trials, the normal procedure is to obtain the group effect, the trial effect, and the interaction. However, it may be of more interest to obtain the group effect and the simple effect of trials within each of the groups to see whether learning took place in each considered separately.

In between combining all levels of two variables and only pairing one, there are a variety of **partial** replication designs. In such designs, some, but not all, levels are combined. Complex designs may involve combinations of crossed, nested, and partially replicated variables.

This section will consider the traditional approach to the ANOVA. In this approach, the expected sums of squares are defined for each main effect and interaction. For example, consider designs in which there are two independent variables whose levels are crossed and whose subjects are nested within groups. Denote these two independent variables as **A** and **B**. Denote the number of population levels of **A** as A and the number of levels actually chosen as a. In a fixed-effect model, A equals a, and in a random-effect model, A is normally much greater than a. Likewise, let B and b denote the corresponding number of population levels and chosen levels of variable **B**. Although the numbers of population levels in a random effect (A and B) might be small in a given application, assume for present discussion that these are indefinitely large. As a result, $a/A = b/B = .0$ for random effects. Conversely, $a/A = b/B = 1$ in a fixed-effect model. Finally, let there be n subjects in each of the $a \cdot b$ groups.

The expected mean squares (variances) among levels of **A** (σ_A^2), levels of **B** (σ_B^2), the **AB** interaction (σ_{AB}^2), and subjects nested within groups (σ^2) are defined in Winer et al. (1991, p. 303). Their notation is somewhat different from ours. They also consider a more general case in which there may be a finite number of subjects in the population; we assume it is indefinitely large. Given these changes, the expected mean squares for each of these terms, in general, are

$$E(ms_A) = \sigma^2 + n(1 - b/B)\sigma_{AB}^2 + nb\sigma_A^2, \tag{9.11a}$$

$$E(ms_B) = \sigma^2 + n(1 - a/A)\sigma_{AB}^2 + nA\sigma_B^2, \tag{9.11b}$$

$$E(ms_{AB}) = \sigma^2 + n\sigma_{AB}^2. \tag{9.11c}$$

However, note that if **B** is a fixed effect, $n(1 - b/B)\sigma_{AB}^2$ reduces to 0 because $b/B = 1$. Consequently, the expected mean squares for the two main effects are simply

$$E(ms_A) = \sigma^2 + nb\sigma_A^2 \tag{9.12a}$$

and

$$E(ms_B) = \sigma^2 + nb\sigma_A^2. \tag{9.12b}$$

However, if **B** is a random effect, $n(1 - b/B)$ is n because $b/B = 0$, and, therefore,

$$E(ms_A) = \sigma^2 + n\sigma_{AB}^2 + nb\sigma_A^2. \tag{9.13a}$$

Note that this is independent of whether **A** is fixed or random.

Likewise, if **A** is a random effect,

$$E(ms_B) = \sigma^2 + n\sigma_{AB}^2 + nA\sigma_B^2. \tag{9.13b}$$

Again, this is independent of whether **B** is fixed or random.

The expected mean square for the interaction term is always defined by Equation 9.11c, regardless of how **A** and **B** are defined.

Note that σ_A^2 consists of two parts when **B** is fixed: (a) a part that reflects random variation within groups (σ^2) and (b) a part that reflects the effect itself $(nb\sigma_A^2)$. Making **B** a random effect adds a third source of variance, $n\sigma_{AB}^2$. Put in somewhat different terms, **A** is only affected by sampling error due to variation within treatment combinations (usually subjects) and true mean differences when **B** is fixed, but when **B** is random, an additional source of sampling error in levels of **B** that were versus were not chosen appears. When both effects are fixed, the design is known as Model I (Eisenhart, 1947) or a **factorial design**. When both effects are random, the design is known as Model II or a **variance component model**. Finally, when one effect is fixed and the other is random, the design is known as Model III or a **mixed model**. Model I is far more often encountered than either of the other two. However, the repeated-measure design is a special case of Model III in which the random effect is subjects and there is only one observation per combination of subject and condition. Because of this lack of replication, σ^2 is unknown. These different expectations lead to differences in the appropriate choice of error term. There are two basic requirements for an appropriate error term. First, the error term for a given effect must have all the expected sources of variance in that effect except one term specific to the effect itself (b, in the preceding paragraph). The variance within groups (σ^2) meets this criterion for testing all three effects in Model I and for testing the interaction in Model II. In turn, the interaction meets this criterion for testing the main effects in Model II. In Model III, the interaction has the proper components to test the fixed effect and the variance within groups has the proper components to test the random effect.

The second major consideration in choosing an error term is that the F ratio of effect to error must be the ratio of two independently derived random variables, each of which is distributed as χ^2. This is not a problem with Models I and II, but can be a problem with Model III (Scheffé, 1959; Winer et al., 1991). This is especially true when the random effect is also a repeated measure. In fact, subjects are normally considered a random variable to allow generalization beyond the specific subjects chosen and frequently appear as a repeated measure. Conversely, random effects other than subjects are not encountered in psychological research nearly as often. Examples include random selection of schools, classrooms, school districts, and so on, in educational research. The reason a problem arises is because error over subjects is generally not independent over conditions—subjects who obtain high scores in one condition will probably obtain high scores in another. Consider the repeated-measure ANOVA of Chapter 5. The correlations among the five conditions over the three subjects is as follows:

| | Condition | | | | |
Condition	1	2	3	4	5
1	1.00				
2	.69	1.00			
3	.67	−.07	1.00		
4	.95	.42	.87	1.00	
5	.91	.33	.92	1.00	1.00

This should not be a surprising finding—subjects who obtain high scores in one condition probably do so in others, so the negligible correlation between conditions 2 and 3 (−.07) is the exception rather than the rule. However, it violates the assumption of independent error. Moreover, the variances within conditions ranged from .26 to 1.64. Clearly, a variance–covariance matrix with equal variances and no covariances is highly unrealistic. This matrix will be of the form $k\mathbf{I}$, where k is a scalar constant and $\mathbf{I}$ is an identity matrix (a matrix with 1s along the diagonal and 0s off the diagonal). As things turn out, a matrix of this form is not necessary. However, a circular ($\mathbf{H}$) matrix is assumed (Winer et al., 1991). This is a matrix in which $\sigma_{jj} + \sigma_{kk} - 2\sigma_{jk}$ = a constant for all combinations of conditions (values of j and k). Circularity thus defines a form of **homogeneity** in that the variance of the difference between any two levels of the treatment variable is a constant. Although a circular matrix can, and usually does, have nonzero covariances, it can be transformed to one of the form $k\mathbf{I}$ with appropriate matrix operations (in effect, factoring the variables). Box (1954; also see Winer et al., 1991) has proposed a test of circularity.

Winer et al. (1991) further note a special type of $\mathbf{H}$ matrix having **compound symmetry**. This means that (a) all covariances are a constant, σ_{ij}, and (b) all variances equal a second constant plus this constant, $\sigma^2 + \sigma_{ij}$. This is a more restrictive form of the variance component model, but it does allow a common correlation among errors. This common correlation would reflect individual differences across conditions. It only requires two parameters be estimated.

There have been four reactions to the problem that arises from the lack of circularity.

1. One can ignore these restrictive assumptions. Specifically, one can test the main effect in a repeated-measure design against the interaction because the interaction does contain the desired variance components. Although many users of the ANOVA are not aware of the need for circularity, they could argue with at least some merit that if the resulting F ratio is very large, the effects of violating this assumption will not be sufficient to negate the presence of mean differences. This is perhaps the most common situation.

2. One can employ the Geisser–Greenhouse (1958) correction (also see Kirk, 1995). The form used by SAS involves testing F with $(k-1)$ df instead of $(k-1) \cdot (n-1)$ df, where k is the number of levels of the effect (a or b, as the case may be in the preceding example). This is not a major penalty to pay if the experiment has a substantial number of subjects, but it can be extremely conservative in small-sample studies. The test effectively assesses the departure from sphericity

and adjusts the *df* by this number. Huynh and Feldt (1976) have developed a second, more liberal test of this form, which SPSS also incorporates.

3. One can treat the problem as one in discriminant analysis/multivariate analysis of variance (MANOVA–these two procedures being formally equivalent; the major difference is that discriminant analysis emphasizes parameter estimation and the MANOVA stresses the testing of null hypotheses). This option has been available for a long time and will be considered in more detail in the next chapter.

4. One can treat the problem as one of covariance structure analysis and model the variance–covariance matrix. This alternative has been implemented in SAS `Proc Mixed` and is the newest option. In particular, it allows one to model data that clearly do not fit the classical assumptions as outlined previously. It will be considered at the end of the next major section.

Regardless of choice of method, complex ANOVA designs require the user to specify which error terms are to be used to test which effects. In SAS, this involves the `Test` statement, illustrated in the next section.

Analysis of Variance With Unequal *n*

The `Test` statement also plays an important role when the ANOVA is used with unequal numbers of subjects in the various groups, known as the nonorthogonal ANOVA, to which we now turn. We will do so by first presenting an equal-*n* ANOVA and then deleting observations to produce an unequal-*n* ANOVA.

In Chapter 5, a data set was created for purposes of performing a two-way ANOVA. We will now perform that analysis on the SAS data set A for a two-way factorial ANOVA presented in Chapter 5 (data set Twoway). The statements needed to generate the ANOVA are as follows:

```
Proc ANOVA Data=Twoway;
Class Conda Condb;
Model y = Conda|Condb;
Means Conda|Condb;
Quit;
Run;
```

The vertical bar used in the Model statement denotes "generate the main effects and interaction," so Conda|Condb is totally identical to Conda Condb Conda*Condb. The Means statement generates group means and standard deviations. The output parallels the simple ANOVA example. The Model has 8 *df* as it confounds the effects of the two main effects and the interaction. The first useful item of information is a listing of the class variables (Conda and Condb) along with the number of levels and the values defining each level. The total number of observations then appears. The result correctly indicates that there are three levels with values of 1, 2, and 3 for each variable and a total of 45 observations. This is as intended.

The results continue with an ANOVA table for the model as a whole. In this case, the model jointly reflects the Conda, Condb, and Conda*Condb effects with 2, 2, and 4 *df*, respectively, so these results are not of much didactic interest save to note that the eight means should not be assumed equal. However, the *df*, sums of squares (SS), mean squares (MS), *F* ratios, and *p* values for the two main effects and the interaction appear next and are of utmost interest. They indicate that the two main effects are significant, but the interaction is not, that is, the two main effects are additive, as was intended by the simulation.

Source	DF	ANOVA SS	Mean Square	F Value	Pr > F
Conda	2	17.55807098	8.77903549	9.78	0.0004
Condb	2	18.76557797	9.38278899	10.46	0.0003
Conda*Condb	4	2.30422883	0.57605721	0.64	0.6360

The means appear next, but these data also do not add to what was presented in the simple ANOVA.

Proc ANOVA was dictated here because there was an equal number of subjects (5) per combination of Conda and Condb. The log indicated that this was the case as it states:

NOTE: The PROCEDURE ANOVA used 1.92 seconds.

We will now make the analysis unbalanced by eliminating one of the observations. This was accomplished by changing the statement in the data step that originally read Output; to If Conda*Condb*Subj < 45 then output;, which eliminates the last observation. The log indicated that the data set only contained 44 observations and the following message appeared when Proc ANOVA was run:

WARNING: PROC ANOVA has determined that the number of observations in each cell is not equal.
         PROC GLM may be more appropriate.

Changing `Proc ANOVA` to `Proc GLM` can eliminate this message. However, there are important additional considerations (`Proc GLM` requires more execution time than `Proc ANOVA` does, but this difference is negligible on contemporary computers). The form of `Proc GLM` output is not extremely different from that of `Proc ANOVA`. However, `Proc GLM` does provide the options of four different estimates of each sum of squares. Two have been presented previously. Type I sums of squares adjust for previous terms in the model but ignore terms to be added, and reflect the contribution of a predictor at the time it is entered. Type II sums of squares adjust for all appropriate terms in the model, as defined earlier in this chapter. Type III and IV sums of squares are very similar in that they both are estimates of what the sums of squares would be if the design were balanced. A formal discussion of these terms may be found in the SAS/STAT user's manual (SAS Institute, 1989a, pp. 109–125). However, it is probably better to consider the choice in the context of the problem at hand. Note that with equal-n, which is what `Proc ANOVA` assumes, these four quantities will all equal one another so only one needs be computed. In the present case, because the data only lacked one observation from being orthogonal, these sums of squares would be highly similar. A related difference in the nonorthogonal ANOVA relevant to `Proc GLM` is that one can obtain either observed means, using a statement of the form `Means effect(s)`, or least squares means using the statement `Lsmeans effect(s)`. The latter estimates what the means would be if there were equal-n.

The SAS defaults for `Proc GLM` are to present the Type I and Type III sums of squares. If either of the other two are desired, add `/SSN` to the end of the `Model` statement, where N is the desired type. The analysis of covariance example will illustrate the differences among these sums of squares. In addition, both `Proc ANOVA` and `Proc GLM` have a `Test` statement that achieves several possible ends. First, the error term is normally the variance that remains after all the sources of variance specified in the `Model` statement have been computed. In the example we have been using, this source of variance is the pooled within-group variation, as it should be in any factorial design. However, more complex designs may have several error terms. The `Test` statement allows the user to override the default error term and substitute another. In addition, its syntax allows the user to specify which sum of squares is to be used in defining the hypothesis and the error term. The effects need not use the same class of sums of squares. Some additional differences among the various types of sums of squares will be considered in the next section.

The following statement tests effect *A* using effect *B* as an error term with Type II sums of squares for both:

```
Test h=a e=b/htype=2 etype=2;
```

The `Test` statement follows the `Model` statement. This may be illustrated using data set Ho of Chapter 5. This simulated data with one between-group variable (*B*) and one within-group (repeated) measure (*W*). The term swg identifies the subjects within each group. The SAS statements needed to conduct the ANOVA are as follows:

```
Proc ANOVA Data=Ho;
Class B W swg;
Model Y = B|W B(swg);
Test H=B E=B(swg);
Means B|W;
Run;
```

The term `B(swg)` that appears in the `Model` statement denotes a nested effect, variation in total scores within each group, pooled over groups. It is used to test the between-group measure. It reflects individual differences in response to the criterion, and its present value is 9.39. This specification leaves the error term used to test the within-group measure as the residual. This represents the interaction between the within-group measure and subjects for each group, again pooled over groups. The mean square for this error term is .53 on 32 *df*. The fact that it is much smaller than the between-group error term here is a typical outcome because it reflects individual differences in magnitude of the within-group independent variable, which are much smaller than overall individual differences. The summary table contains the `Model` effect, which describes the sum of every effect specified in the `Model` statement. It is of no interest in a balanced design as it is redundant. Equality will not hold in an unbalanced design, but the disparity between the sum of individual effects and the omnibus `Model` effect is still not of major interest.

The *df*, sums of squares, mean squares, *F* ratios, and *p* values for the sources named in the model, B, W, B*W, and B(swg), then appear. However, these *F* ratios and *p* values are derived from the within-group error term so the only results of value that are of interest are W and B*W. The entire set of effects is as follows. Note that each *F* ratio is the mean square divided by .53.

Source	DF	ANOVA SS	Mean Square	F Value	Pr > F
B	1	59.01974164	59.01974164	111.05	0.0001
W	4	19.84391272	4.96097818	9.33	0.0001
B*W	4	18.99397025	4.74849256	8.93	0.0001
B(SWG)	8	75.10595520	9.38824440	17.66	0.0001

Next come what is identified as Tests of Hypotheses Using the ANOVA MS for B(SWG) as an Error Term. This should be used instead of the previously noted value for B. The F ratio is much smaller here as it is tested against the between-group error term, which is the proper term.

Source	DF	ANOVA SS	Mean Square	F Value	Pr > F
B	1	59.01974164	59.01974164	6.29	0.0365

The "bottom line" is that all three effects are significant for this data set with the F ratios for B, W, and B*W equaling 6.29, 9.33, and 8.93. The B effect is based on 1 and 8 df, and the W and B*W effects are based on 4 and 32 df.

▣ ANALYSIS OF COVARIANCE

We have previously noted that the goal behind the ANCOVA is to obtain a measure that is correlated with the criterion but independent of the predictor. In that way, when the criterion is adjusted for the covariate, the error of measurement is reduced, enhancing treatment effects. To illustrate, we will take the ANCOVA data of Chapter 5, perform an ordinary simple ANOVA, and then perform two ANCOVAs using Proc GLM. The covariate (X) is entered before the treatment variable (Group) in the first of these ANCOVAs, and the converse ordering is used in the second.

```
Proc ANOVA Data=Ancov;Class Group;
Model Y = Group;
Proc GLM Data=Ancov;Class Group;
Model Y = X|Group/ss1 ss2 ss3 ss4;
Means Group;
Proc GLM Data=Ancov;Class Group;
```

```
Model Y = Group|X/ss1 ss2 ss3 ss4;
Means Group;
Run;
```

Proc Reg cannot be used directly because Group is a Class variable, and Proc Reg cannot generate the appropriate dummy codes. The user can generate these in the data step, but it is unnecessary. Similarly, Proc GLM cannot process more than one Model statement as can Proc Reg. Several alternatives could have generated equivalent results, such as creating Group1 and Group2 (i.e., as dummy variables) within the data step as described previously.

Perhaps the major difference between the ANOVA and the two ANCO-VAs is that the error MS values were .89, .44, and .44, respectively. The two values for the two ANCOVAs are identical because order of entry does not affect the final term in the model (the residual, which is the error term in this case). The error term drops by roughly 50% because the covariate and covariate by independent variable interaction effects are lumped into an error term in the ANOVA but are separated in the ANCOVA. In the present case, the covariate effect is of substantial magnitude. Within the ANCOVA, the following table of *F* ratios describes the effects of SS type and order of entry for the two main effect terms. The interaction, which was always entered last, had the same value (3.92) in all cases.

| | | *Effect* | |
Entry Order	*SS Type*	*Group*	*X*
Group alone	—	4.77	
X then Group	I	10.50	5.27
Group then X	I	9.54	7.19
X then Group	II	11.05	7.19
Group then X	II	11.05	7.19
X then Group	III	11.05	11.88
Group then X	III	11.05	11.88
X then Group	IV	11.05	11.88
Group then X	IV	11.05	11.88

Consistent with the difference in error terms, the smaller value of *F*, by far, is for Group alone in the simple ANOVA. Note that the order of entry affects only the Type I SS. In this case, both the Group and the covariate effects were smaller in magnitude when they were entered first as opposed to second. In other words, the unadjusted Group effect was smaller than the Group effect controlling for X, and the unadjusted X effect was smaller than the X effect controlling for Group. This implies that the two effects were negatively correlated. The two effects can also be positively correlated. In that case, the unadjusted effects will be stronger. In essence, Type III and IV sums of squares apportion shared estimates. Also note that although the logic of the ANCOVA is to enter the covariate first, the SAS/STAT user's manual (SAS Institute, 1989b, p. 896) uses the reverse ordering. However, SAS also stresses using the Type III or IV sum of squares for which ordering makes no difference. A case could be made for using any of these sums of squares. The major difference is that the procedure we use provides the observed magnitude of the covariate effect. The SAS procedure yields an estimate of the covariate effect adjusting for the treatment effect. The issue is not what is always best, but what best answers the question at hand.

▣ CANONICAL CORRELATION ANALYSIS

In Chapter 5, we created a data set in which there were four observable variables in what was called the X set and three in another that was called the Y set. As shown in Figure 5.3, X_1 and X_2 were related to Y_1 and Y_2 through one latent variable, F_1, and X_3 and X_4 were related to Y_3 through a second latent variable, F_2. We will now proceed to a canonical correlation analysis of these data, which can be obtained using the following command:

```
Proc Cancorr Data=canon all vdep wdep;var x1-x4;with y1-y3;
Run;
```

The options vdep and wdep supplement the basic analysis with regression analyses regressing the X set upon each individual member of the Y set and vice versa. Part of what the all option provides is a **redundancy analysis**, described as follows:

1. The results begin with the simple statistics. Note that the seven observable measures are standardized in the population. This causes various standardized and raw-score results to differ only in sampling error. We will present only the standardized results.

Transformation of the raw data into a different scale is left as an exercise.

```
Means and Standard Deviations
            4 'VAR' Variables
            3 'WITH' Variables
         1000 Observations
Variable              Mean              Std Dev
X1                 0.022842            0.985014
X2                 0.020936            1.023442
X3                 0.022297            0.981104
X4                 0.018067            0.963121
Y1                 0.017837            0.993202
Y2                 0.056196            1.018316
Y3                -0.012684            1.005047
```

2. The correlations among the measures then appear. These are broken down into the Var set (X_1-X_4), the With set (Y_1-Y_3), and the correlations between the two sets.

	X1	X2	X3	X4
X1	1.0000	0.1971	0.0319	0.0358
X2	0.1971	1.0000	0.0562	0.0026
X3	0.0319	0.0562	1.0000	0.3479
X4	0.0358	0.0026	0.3479	1.0000

	Y1	Y2	Y3
Y1	1.0000	0.3325	0.0226
Y2	0.3325	1.0000	0.0427
Y3	0.0226	0.0427	1.0000

	Y1	Y2	Y3
X1	0.1833	0.2745	-0.0121
X2	0.1595	0.2138	-0.0209
X3	-0.0169	0.0423	0.4152
X4	0.0094	0.0143	0.5436

3. The next section contains regression statistics for predicting the Var (X) variables from the With (Y) variables. It is instructive to see where these numbers come from, and that is why values of R^2 relating the X and Y sets to the latent variables were obtained.

Consider the squared multiple correlation (SMC) obtained from using Y_1 to Y_3 to explain X_4 (.30). This is equal, within the approximation imposed by sampling error, to $\sum R^2_{Y.F} r^2_{X.F}$. The $R^2_{Y.F}$ is the SMC between the four Y predictors and a given F, and the $r^2_{X.F}$ term is the squared simple correlation between an individual member of the X set, X_4 in this case, and that given F. Summation proceeds over the two F terms. It can be shown from the analyses performed in Chapter 5 that the two values of $R^2_{Y.F}$ are .60 and .63 and the two values of $r^2_{X.F}$ are .05 and .69. Consequently, the R^2 between the Y set and X_4 is $.60 \cdot .05 + .63 \cdot .69$, or .30, which agrees to the second decimal point with what was obtained. The remaining data in this table are the standard adjustment for number of predictors, the upper and lower 95% confidence intervals, F, and the associated p value. Relevant formulas may be found in such sources as Nunnally and Bernstein (1994, Chapter 5). Note that the unweighted R^2 of .15 is simply the average of the four R^2 values (.09, .06, .17, and .30). The weighted value may be obtained using the variances of the X set members presented before $(.98^2, 1.02^2, .098^2, \text{ and } .96^2)$.

Squared Multiple Correlations and F Tests
3 numerator df   996 denominator df

	R-Squared	Adjusted R-Squared	Approx 95% CI for RSQ Lower CL	Upper CL	F Statistic	Pr > F
X1	0.085500	0.082745	0.053	0.119	31.0398	0.0001
X2	0.055434	0.052589	0.029	0.084	19.4843	0.0001
X3	0.174333	0.171846	0.131	0.217	70.0993	0.0001
X4	0.295582	0.293460	0.247	0.342	139.3111	0.0001

Average R-Squared: Unweighted   = 0.152712
Weighted by Variance = 0.149189

4. The next section contains the correlations among the regression estimates, the β weights (standardized regression coefficients), the b weights (raw regression coefficients, which, in the present case, are almost identical to the standardized coefficients because the raw population data themselves are standardized), the standard errors of b, the values of t testing the null hypothesis that the coefficients are .0, and the p values. In the interest of space, only the b weights will be presented. Note that X_1 and X_2 depend on Y_1 and Y_2, but X_3 and X_4 depend on Y_3, as was intended by the simulation.

	X1	X2	X3	X4
Y1	0.1037	0.0997	−0.0387	0.0001
Y2	0.2411	0.1819	0.0374	−0.0090
Y3	−0.0248	−0.0310	0.4145	0.5440

5. The following section contains semipartial (part), squared semi-partial, partial, and squared partial correlations between the X and Y terms. The semipartial terms remove the effects of the other regressors (Y terms in this case); the partials also remove the effects of the other X terms. The pattern of results is similar to that noted for β weights so they will not be presented.

6. The next section contains the R^2 values and associated regression statistics obtained from using the X set as predictors of Y. Their form parallels what was just presented, so we will limit the presentation to the R^2 values and the β weights. Again, note the relationships among X_1, X_2, Y_1, and Y_2 and among X_3, X_4, and Y_3.

	R-Squared	Adjusted R-Squared	Approx 95% CI for RSQ Lower CL	Upper CL	F Statistic	Pr > F
Y1	0.050487	0.046669	0.025	0.077	13.2263	0.0001
Y2	0.102541	0.098934	0.067	0.138	28.4216	0.0001
Y3	0.355889	0.353300	0.306	0.401	137.4413	0.0001

Average R-Squared: Unweighted   = 0.169639
                   Weighted by Variance = 0.169975

	Y1	Y2	Y3
X1	0.1582	0.2414	−0.0306
X2	0.1302	0.1647	−0.0307
X3	−0.0346	0.0267	0.2598
X4	0.0155	−0.0040	0.4544

7. The results of the canonical correlation analysis proper then appear. The first part consists of the canonical correlations, their values adjusted for the number of predictors, their standard error, and their squared value. Note that there are two canonical correlations of at least modest size (.60 and .34) and one of negligible size (.05). This is because the two data sets have exactly two latent variables in common.

	Canonical Correlation	Adjusted Canonical Correlation	Approx Standard Error	Squared Canonical Correlation
1	0.597841	0.595003	0.020331	0.357413
2	0.341850	0.338885	0.027941	0.116861
3	0.047079	0.036943	0.031568	0.002216

8. The canonical correlations are functions of an eigenanalysis of a complicated product matrix. These eigenvalues, their difference, and the proportion and cumulative proportion of total variance explained appear next. These further explicate what was noted in the preceding section.

	Eigenvalue	Difference	Proportion	Cumulative
1	0.5562	0.4239	0.8052	0.8052
2	0.1323	0.1301	0.1916	0.9968
3	0.0022	.	0.0032	1.0000

9. The likelihood ratios associated with the three eigenvalues are presented along with the associated F ratios, df, and p values. Note that the first two functions are significant; the third is not.

	Likelihood Ratio	Approx F	Num DF	Den DF	Pr > F
1	0.56623512	52.5109	12	2627.523	0.0001
2	0.88118109	21.6324	6	1988	0.0001
3	0.99778352	1.1051	2	995	0.3316

10. Four additional inferential test results follow. Wilks's lambda, Pillai's trace, the Hotelling–Lawley trace, and Roy's largest root all evaluate the null hypothesis that there are no significant canonical relationships in the data. The first three are highly similar but distinct; various authors differ in their preference. They are all suited to detecting both concentrated and diffuse structures in the sense of Chapter 5. Roy's largest root is generally regarded below the other three and is designed for use with a concentrated structure. These tests will be discussed in more detail in the next chapter when we consider discriminant analysis. For now, simply note that they all confirm the presence of relationships between the variables in the X and Y sets.

```
Multivariate Statistics and F Approximations
S=3   M=0    N=495.5
Statistic                     Value        F  Num DF    Den DF  Pr > F
Wilks's Lambda           0.56623512  52.5109      12  2627.523  0.0001
Pillai's Trace           0.47649137  46.9692      12      2985  0.0001
Hotelling-Lawley Trace   0.69075714  57.0834      12      2975  0.0001
Roy's Greatest Root      0.55621058 138.3574       4       995  0.0001
```

11. The sections that follow present canonical coefficients, which are pattern (regression) weights. These begin with the raw (b) weights for the Var (X) set, followed by the raw weights for the With (Y) set and the corresponding standardized (β) coefficients. We will not present the b weights, which are highly similar to the β weights. Note that the first optimal (canonical) function of X_1 to X_4 (V_1— we will follow the same format in identifying canonical variables as we used in identifying observables, which will be different from the way they appear in the SAS printout) is primarily defined by X_3 and X_4. Likewise, the first optimal function of Y_1 to Y_3 (W_1) is primarily defined by Y_3. Conversely, V_2 is primarily defined by X_1 and X_2, and W_2 is primarily defined by Y_1 and Y_2. Note also that V_3 and W_3 can be disregarded because their associated canonical correlation was nonsignificant. Indeed, if your analysis produces many nonsignificant functions, you may wish to use Ncan=xxx to limit the number of functions reported.

```
Standardized Canonical Coefficients for the 'VAR' Variables
              V1              V2              V3
X1        -0.0816          0.7421         -0.2052
X2        -0.0741          0.5299          0.3207
X3         0.4360          0.0475         -0.9269
X4         0.7588          0.0460          0.7181
Standardized Canonical Coefficients for the 'WITH' Variables
              W1              W2              W3
Y1        -0.0546          0.3742          0.9906
Y2        -0.0395          0.8094         -0.6849
Y3         0.9999          0.0291          0.0345
```

12. The canonical structure follows. These are the correlations between observables and the canonical functions. These involve correlations with both a variable's "own" function (e.g., X_1 with V_1) and the "other" function (e.g., X_1 with W_1). Because the two pairs of functions were generated by independent latent variables, the structures are similar to the patterns.

```
Correlations Between the 'VAR' Variables and
Their Canonical Variables
                V1              V2              V3
X1       -0.0552         0.8497         -0.1459
X2       -0.0637         0.6789          0.2301
X3        0.6932         0.1170         -0.6655
X4        0.9074         0.0905          0.3891
Correlations Between the 'WITH' Variables and
Their Canonical Variables
                W1              W2              W3
Y1       -0.0451         0.6440          0.7637
Y2       -0.0150         0.9351         -0.3541
Y3        0.9970         0.0721          0.0277
Correlations Between the 'VAR' Variables and the
Canonical Variables of the 'WITH' Variables
                W1              W2              W3
X1       -0.0330         0.2905         -0.0069
X2       -0.0381         0.2321          0.0108
X3        0.4144         0.0400         -0.0313
X4        0.5425         0.0309          0.0183
Correlations Between the 'WITH' Variables and
the Canonical Variables of the 'VAR' Variables
                V1              V2              V3
Y1       -0.0270         0.2202          0.0360
Y2       -0.0089         0.3197         -0.0167
Y3        0.5961         0.0247          0.0013
```

13. The remaining results relate to **redundancy analysis**. This involves determining the proportions of variance in a given data set that is explained by its own variables and, more crucially, by the other set of variables. The former statistics are a function of eigenanalyses conducted separately upon the X and the Y sets. Thus, the first three eigenvalues of the X set account for 33%, 30%, and 17% of the total. This proportion is then multiplied by the squared canonical correlation to obtain the proportion explained by the other canonical variables. Looking at these statistics is quite important. For example, consider variables in each set that are uncorrelated with the remaining variables in the set but highly corre-

lated with each other. They would produce a very large canonical correlation but would do little to explain the overall variance in the other data set. Note that because there are more X variables (4) than Y variables (3), their total adds to 80% rather than 100%. However, the Y variables are completely explained by the eigenvalues. These results are actually presented twice—once in terms of raw data and then in terms of standardized data; only the standardized data are presented here.

Standardized Variance of the 'VAR' Variables
Explained by

	Their Own Canonical Variables			The Opposite Canonical Variables	
	Proportion	Cumulative Proportion	Canonical R-Squared	Proportion	Cumulative Proportion
1	0.3278	0.3278	0.3574	0.1171	0.1171
2	0.3012	0.6289	0.1169	0.0352	0.1523
3	0.1671	0.7961	0.0022	0.0004	0.1527

Standardized Variance of the 'WITH' Variables
Explained by

	Their Own Canonical Variables			The Opposite Canonical Variables	
	Proportion	Cumulative Proportion	Canonical R-Squared	Proportion	Cumulative Proportion
1	0.3321	0.3321	0.3574	0.1187	0.1187
2	0.4314	0.7635	0.1169	0.0504	0.1691
3	0.2365	1.0000	0.0022	0.0005	0.1696

14. Finally, the SMCs between the Var variables and the successive canonical variables derived from the With variables are presented, followed by the converse. For example, the first row says that X_1 correlates .0011 with the first optimal combination of Y_1 to Y_3, W_1; .0855 with the first two optimal linear combinations, W_1 and W_2, and .0855 with all three (including the nonsignificant W_3). Note that X_1 and X_2 show relatively large increments between W_1 and W_2 because they are only related to W_2. On the other hand, X_3 and X_4 show practically no increment because they are related to W_1 rather than W_2. None of these X variables shows any increment with regard to W_3 because W_3 simply reflects chance. The relationship between Y_1 to Y_3 and V_1 to V_3 may be understood in a similar manner.

M	1	2	3
X1	0.0011	0.0855	0.0855
X2	0.0015	0.0553	0.0554
X3	0.1718	0.1734	0.1743
X4	0.2943	0.2952	0.2956
M	1	2	3
Y1	0.0007	0.0492	0.0505
Y2	0.0001	0.1023	0.1025
Y3	0.3553	0.3559	0.3559

▣ COVARIANCE STRUCTURE ANALYSIS

Basically, although there is a tremendous diversity of covariance structure models, the analysis can be thought of as combining one or more confirmatory factor models and one or more regression models. Somewhat of a novelty is that the criterion in the regression model may be a latent rather than a manifest variable. Second, somewhat of a novel terminology has evolved relative to traditional exploratory factor analysis and multiple regression. However, this is perhaps less of a problem for someone studying this for the first time than for someone who is used to traditional nomenclature.

Figure 9.5 is a path diagram that is more highly structured than the one used to demonstrate canonical correlation analysis (Figure 5.3). Six variables are observable: X_1 to X_4 and Y_1 to Y_2. All of these depend on other terms in the model so they are **endogenous** to the model. One more term is also endogenous, η_1, but it is an unobservable construct (factor). Nine variables are **exogenous** (external) to the model. Two of these are unobservable constructs, ξ_1 and ξ_2. Four are error terms affecting the X variables, δ_1 to δ_4. Two are error terms affecting the Y variables, ε_1 and ε_2. Finally, one is an error term affecting η_1, ζ_1. Path coefficients describe the strengths of relationships. Four of these relate the exogenous constructs to the observable (X) variables: $\lambda_{x_{11}}$, $\lambda_{x_{21}}$, $\lambda_{x_{32}}$, and $\lambda_{x_{42}}$. Two more relate the endogenous construct to the observable (Y) variables: $\lambda_{y_{11}}$ and $\lambda_{y_{21}}$. Two more relate the exogenous constructs to the endogenous construct: γ_{11} and γ_{21}. Finally, there is a possible correlation between the two exogenous constructs, ϕ_{21}. Compare this to Figure 5.3.

The following simulation was created to generate data:

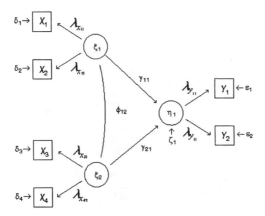

Figure 9.5. Path diagram involving both confirmatory factor analysis and regression models.

```
Data CSA;
Retain err 1999 nobs 1000;
lx11 = .8;delta1 = sqrt(1 - lx11**2);
lx21 = .7;delta2 = sqrt(1 - lx21**2);
lx32 = .9;delta3 = sqrt(1 - lx32**2);
lx42 = .8;delta4 = sqrt(1 - lx42**2);
ly11 = .8;eta1 = sqrt(1 - ly11**2);
ly21 = .7;eta2 = sqrt(1 - ly21**2);
gamma21 = .6;
gamma22 = .6;
zeta1 = sqrt(1 - gamma21**2 - gamma22**2);
Do i = 1 to nobs;
exo1 = rannor(err);
exo2 = rannor(err);
end1 = exo1*gamma21 + exo2*gamma22 + zeta1*rannor(err);
X1 = lx11*exo1 + delta1*rannor(err);
X2 = lx21*exo1 + delta2*rannor(err);
X3 = lx32*exo2 + delta3*rannor(err);
X4 = lx42*exo2 + delta4*rannor(err);
Y1 = ly11*end1 + eta1*rannor(err);
Y2 = ly21*end1 + eta2*rannor(err);
Output;
End;
Keep x1-x4 y1-y2 exo1 exo2 end1;
```

The first three columns of the following table contain the terms of the model as defined both in LISREL notation and in the program and the population parameter and its value as estimated from `Proc Calis`. Note that all of the observed and latent variables are in standard score form in the population. Also note that exo1 and exo2 (exogenous factors ξ_1 and ξ_2) are independent (i.e., $\phi_{21} = 0$) in the population because they are defined by separate calls to the `rannor` function.

Name		Estimate		
In LISREL	In Program	Parameter	Model 1	Model 2
λ_{x11}	lx11	.800	.820	.820
λ_{x21}	lx21	.700	.748	.748
λ_{x32}	lx32	.900	.924	.924
λ_{x42}	lx42	.800	.788	.787
λ_{y11}	ly11	.800	.810	.812
λ_{y21}	ly21	.700	.705	.707
γ_1	gamma21	.600	.613	.608
γ_2	gamma22	.600	.620	.613
ζ_1	zeta1	.529	.508	.505
ϕ_{11}	(defined implicitly)	.000	−.024	.000

Assume an investigator's initial supposition is to specify the model correctly except for allowing ξ_1 and ξ_2 to correlate. The specification is as follows:

```
Proc Calis Cov Data = CSA;var x1-x4 y1-y2;
lineqs x1 = lx11 fex1 + e1,
       x2 = lx21 fex1 + e2,
       x3 = lx32 fex2 + e3,
       x4 = lx42 fex2 + e4,
       y1 = ly11 fen1 + e5,
       y2 = ly21 fen1 + e6,
       fen1 = gamma21 fex1 + gamma22 fex2 + dzeta;
std fex1 fex2 = 2*1.,
dzeta=b,
e1-e6 = ve1-ve6;
cov fex1 fex2 = rho1;
Run;
```

Equations `x1 = lx11 fex1 + e1` to `y2 = ly21 fen1 + e6` define the manifest variables (observables) in terms of the endogenous and exogenous constructs (factors), thus forming the core of the measurement model. Likewise, `fen1 = gamma21 fex1 + gamma22 fex2 + dzeta;` defines the structural model by relating the exogenous constructs and error to the endogenous constructs. The statement `std fex1 fex2 = 2*1., dzeta=b, e1-e6 = ve1-ve6;` fixes the standard deviations of the exogenous constructs at 1.0 (factors are normally so standardized), states that the error in the structural equation needs to be estimated with the variable named b, and further states that the measurement error affecting the observables is to be estimated by the variables named ve1 to ve6. Finally, `cov fex1 fex2 = rho1;` leaves the covariance between the factors to be estimated. Because they are standardized, their covariance is also their correlation.

The fourth column provides the standardized estimates from this first model. As can be seen, these are very close to the true values. The fit of the model was also acceptable in terms of the nonsignificant model G^2 statistic, which was 6.14 on 5 df. However, the estimate of ϕ_{21} (−0.024) was far from statistically significant ($t = -0.626$). A careful investigator would respecify the model by fixing the factor correlation at 0 (some programs such as LISREL would do this automatically). In the present case, respecification is accomplished simply by deleting the statement `cov fex1 fex2 = rho1;`. This increased the model G^2 to 6.53 on 6 df. This marginal increase of .39 was not significant with 1 df. Given this difference G^2, one may assume that ϕ_{21} should be 0, as it is in reality.

Covariance Structure Analysis and Mixed Analysis of Variance Designs

We previously analyzed data set Ho the way perhaps most would—as a classical analysis of variance design with one between-subject and one within-subject effect. However, as we noted, this analysis violates important assumptions about the covariances among error measures. We will now analyze this same data set using the following generalization of Equation 9.1. The notation is ours.

$$\mathbf{y} = \mathbf{X}\mathbf{b}_f + \mathbf{Z}\mathbf{b}_r + \mathbf{e}. \qquad (9.14)$$

The terms $\mathbf{y}$, $\mathbf{X}$, and $\mathbf{b}_f$ are basically the same as in this earlier equation save that a subscript f has been added to denote that this applies only to the fixed effects. The elements of $\mathbf{X}$ can be either continuous or discrete.

The terms Z and b_r are analogous to X and b_f except that they denote the predictors and regression weights for random effects. Furthermore, it is assumed that the variables comprising b_r and e are normally distributed. There are no constraints on the distribution of elements in b_f.

Let G and E be the respective matrices of covariances among the elements of b_r and e. `Proc Mixed` can model various constraints introduced on these two matrices. Specifically, using a `Random` statement but not a `Repeated` statement causes the `TYPE=matrixtype` option (where `matrixtype` refers to the types discussed in the following list) to model G and define E in classical terms as $\sigma^2 I$. Conversely, using a `Repeated` statement without a `Random` statement defines Z as a null matrix (matrix of 0s). Consequently, the random effects drop out of the model, leaving $Y = Xb_f + e$. Here, `TYPE=matrixtype` models E, and no random component appears directly in the model. Using both the `Random` and the `Repeated` statements with separate `TYPE=matrixtype` statements models both G and E. The keyword `GROUP=effect` used with the `Random` and/or `Repeated` statements allows one to evaluate heterogeneity between groups for G and/or E, respectively.

Some of the types of b_r matrices that `Proc Mixed` can model are as follows. SAS Institute (1997) provides a complete list and further mathematical specifications.

1. **Variance components**. This is the default matrix. Each effect has its own variance, but covariances between pairs of conditions are 0. Error is assumed to be the same within levels of a treatment. It corresponds to the classical ANOVA model. It is plausible when the random effect is not a repeated measure, but, as we have seen, it is unlikely to hold when the repeated measure is the random effect of subjects. The model requires one to estimate as many free parameters as there are conditions (main effects plus interactions).

2. **Compound symmetry**, as defined previously. This is somewhat stronger than the assumptions required for a repeated-measure design, but it accommodates the correlated error in subjects as a repeated measure. In addition, only two parameters (a single variance and a single covariance) need be estimated. Problems might emerge if there is considerable heterogeneity of subject variance across conditions, but this is a reasonable choice when this is not the case.

3. **Unstructured**. Both the variances and the covariances are allowed to vary across conditions. This makes no assumptions about the

structure of the error. It requires the maximum possible number of terms to be estimated, $k \cdot (k-1)/2$, where k is the combined number of treatment levels over conditions. This may require too many unknowns to provide an estimable solution.

4. **First-order autoregressive.** This assumes that conditions are ordered in some fashion. There is a variance common to all conditions (σ^2). In addition, the correlation between any pair of adjacent conditions, such as trials 2 and 3, is ρ; the correlation between any pair of conditions separated by two conditions, such as trials 3 and 5, is ρ^2. Consequently, the correlation between any pair of conditions separated by k other conditions is ρ^k. One situation in which this makes sense is a learning study because performance on adjacent trials should be more similar than performance on more widely separated trials. This is one form of what is known as a **simplex** model. It might be suitable for situations in which trials is a repeated measure because the correlations among adjacent trials would be expected to be larger than correlations among more disparate trials, but the assumption that the relationship takes the form of a power law is stronger than most situations assume. It only requires estimation of two free parameters.

5. **Heterogeneous autoregressive.** This is an extension of the first-order autoregressive model in which the condition variances are also allowed to differ. It estimates $k + 1$ free parameters.

6. **First-order factor analytic.** This assumes that a single common factor underlies the conditions. Each covariance is the product of the pattern elements for the two conditions, and each variance is the sum of that condition's squared pattern element and its unique variance. It requires $2 \cdot k$ free parameters—a pattern loading and a uniqueness for each of the k conditions.

7. **General factor-analytic structure.** This is a q-factor extension of the first-order factor-analytic model. It requires $(q + 1) \cdot k$ free parameters—one parameter per factor plus a uniqueness for each condition.

Keep in mind that not all models can be used with all data because there may not be enough data to estimate the unknown parameters. We will analyze the Repeated data set of Chapter 5, first by invoking `Proc ANOVA` as in a classical repeated-measure design:

```
Proc ANOVA Data=Repeated;
Class Subj Cond;
Model Y = Subj Cond;
Means Subj Cond;
```

The mean squares for Subj, Cond, and their interaction (the residual) are 6.74, 2.97, and .61, respectively. The F ratio for Cond was 4.85 on 2 and 8 df, $p < .05$ (an alternate way to obtain these same results is to replace the Model statement with Model Y = Subj|Cond; test h = Cond e=Subj*Cond;—its appearance will be slightly different but it will return the same F ratio).

Next, we add the Proc Mixed code.

```
Proc Mixed Data=Repeated;
Class Cond Subj;
Model Y = Cond;
Repeated/ type=cs subject=Subj r;
Run;
```

Both the Proc and the Class statements parallel their use in Proc ANOVA and other SAS regression procedures. However, instead of including both the fixed condition and the random effects of subjects within a single Model statement, only the fixed effect is specified here. Because we have included a Repeated statement without a Random statement, the options apply to **E**: type=cs models compound symmetry (which holds here in fact); subject=Subj identifies subjects as a repeated measure, and r requests printing of the covariance matrix. This specification reflects present concern with the correlated error of most repeated-measure designs. Testing the particular structure of a random effect is a separate issue.

As can be seen, the Cond*Subj interaction is not mentioned explicitly here or in Proc Anova. The default estimation method of restricted maximum likelihood is used in contrast to two alternatives, (full) maximum likelihood estimation and minimum variance quadratic unbiased estimation. The consequences of using these different methods are not completely clear, and the user might wish to explore differences among them, as we did.

Note that Proc Mixed can be used with continuous predictors as well as in ANOVA designs. These often dictate a different choice of output, which the user can obtain from the SAS Output Delivery System, but we will leave these applications as exercises. Under this specification, the

Subj parameter is the covariance term (σ_{ij}) common to all terms in the covariance matrix, and the **Residual** term is the variance term specific to the diagonal elements, σ^2. Consequently, even though the assumption of compound symmetry was not explicated, it appears implicitly.

1. The output begins with an identification of the variables in the analysis and their levels, as in other ANOVA routines.

2. Because **Proc Mixed** uses an open-form solution, the iteration history appears next. In the present case, the convergence criterion was met at the end of the second iteration.

Iteration	Evaluations	Objective	Criterion
0	1	28.55192408	
1	1	20.54840288	0.00000000

3. The covariance matrix appears next. Note that the difference between the common values of the diagonal terms and the common values of the off-diagonal terms is the condition × subjects error mean square in the conventional repeated-measure analysis $(2.65638859 - 2.04329639 = 0.61309220)$.

Row	COL1	COL2	COL3
1	2.65638859	2.04329639	2.04329639
2	2.04329639	2.65638859	2.04329639
3	2.04329639	2.04329639	2.65638859

4. The covariance parameter estimates follow. Note that the first of these is the covariance and the second is the aforementioned condition × subjects error mean square.

Cov Parm	Subject	Estimate
CS	SUBJ	2.04329639
Residual		0.61309220

5. The next data to appear are various model-fitting results on the overall model. First, there are 15 observations. Multiplying the residual log likelihood (**Res Log Likelihood = –21.3015**) by 2 generates a likelihood ratio chi-square statistic (G^2) test of this overall model. **Akaike's information criterion** and **Schwarz's Bayesian criterion** are descriptive measures of fit discussed in Chapter 8. These are followed by the *df* and the

p value for the inferential test. This test is usually not of major interest because it pools the various effects.

```
        Model Fitting Information for Y
Description                               Value
Observations                            15.0000
Res Log Likelihood                     -21.3015
Akaike's Information Criterion         -23.3015
Schwarz's Bayesian Criterion          -23.7864
-2 Res Log Likelihood                   42.6029
Null Model LRT DF                        1.0000
Null Model LRT P Value                   0.0047
```

6. The analysis concludes with what is of most central interest, the F test on the fixed effect of COND. Note that it is identical to the value obtained with Proc ANOVA because the assumptions made about the $\mathbf{b}_r$ matrix were equivalent. Note also that the F test for the COND effect produces a different p value from the likelihood test for the overall model given in the previous set of results because the latter tests a composite of all the effects rather than this specific effect.

```
          Tests of Fixed Effects
Source    NDF    DDF    Type III F    Pr > F
COND       2      8          4.85     0.0417
```

We next explored several alternative ways to estimate and specify $\mathbf{b}_r$. Changing the estimation method did not have a major effect on the results. However, the specification assumptions were crucial. For example, assuming the $\mathbf{b}_r$ matrix was unstructured produced an F ratio of 25.30 on 2 and 4 df, which is significant well past the .01 level. This makes the rather ad hoc, unparsimonious, and unnecessary assumption (in this case) that every distinct parameter of $\mathbf{b}_r$ takes on a different value. Conversely, assuming variance component and first-order autoregressive models respectively led to F ratios of 1.12 and 2.26 on 2 and 8 df, both of which were nonsignificant. The variance component model produced a small F ratio because it treats the previously noted rather sizable correlations in error across conditions as .0. This effectively erroneously reduces the analysis to a simple ANOVA with a comparable reduction in power. In contrast, the autoregressive model reduced the covariances among conditions as a function of the size of the difference between them. Specifically, the error correlations between adjacent conditions 1 and 2 and 2 and 3 were both estimated to be .67, but the correlation between more widely separated conditions 1 and 3 was estimated to be $.46 = .67^2$. In fact, these correlations are homogeneous in the population. Which, if any, model for $\mathbf{b}_r$ is appropriate depends on the situation. Of course, one should not reason back from a favorable result to choice of model.

Mixed Models and SPSS

SPSS provides some ability to work with mixed models through the General Linear Models options under Analyze, using the Data Editor menu. However, at the time this section was written, SPSS cannot model the sophisticated covariance models that SAS `Proc Mixed` can.

▣ PROBLEM*S*

9.1. Change the weights for the three predictors in data set A to .0, .5, and .6 and repeat the least squares (`Proc Reg`) regression analysis.

9.2. Take the original data set A and add .5 (one third of a z-score unit) to the last 15 observations and examine the residuals.

9.3. Perform a maximum likelihood regression analysis on the data set created in Problem 9.1.

9.4. Redo the tests of moderator effects with least squares (`Proc Reg`) regression analysis. Let (a) b1 = .7, b2 = .6, and delta = .25; (b) b1 = .7, b2 = .5, and delta = .5; and (c) b1 = .7, b2 = .5, and delta = .5.

9.5. Repeat Problem 9.4 using maximum likelihood methods.

9.6. Perform ANOVAs on the data resulting from Problems 5.2 to 5.5 and an ANCOVA on the data of Problem 5.6. Explore the various options for post hoc testing of group mean differences. Report both the uncorrected results and those obtained with the Geisser–Greenhouse correction for the data of Problem 5.4.

9.7. Take the data of Problem 5.3 and delete observations at random to produce an unequal-N data set. Analyze the resulting data.

9.8. Take the data of Problem 5.3 and perform ANOVAs assuming that (a) the first effect is a between-group variable, but the second is a repeated measure, and (b) both effects are repeated measures. Assume that both are fixed effects.

9.9. Take the covariance structure presented in data set CSA and fix the factor correlation at .0. What effect does this have on the change in model G^2? What implications does it have for the model tested?

9.10. Take the data from Problem 5.4 and explore the various options in Proc Mixed.

10

Discriminant Analysis, Classification, and Multivariate Analysis of Variance

CHAPTER OBJECTIVES

Previous chapters considered two traditional multivariate problems: factor analysis (including, by our definition, component factor analysis) and multiple regression. We now turn to a final set of relatively traditional models. Specifically, this chapter is concerned with the following:

1. Discriminant analysis

2. Classification analysis

3. Multivariate analysis of variance

▣ DISCRIMINANT ANALYSIS

Discriminant analysis involves obtaining a linear combination of predictors that best discriminate among groups. Specifically, it involves obtaining a linear combination of predictors that maximizes the ratio of variance between groups to the variance within groups. The criterion groups can fall along a continuum; such as in a market research study that investigates differences among individuals who do not use a product at all, those who use it sometimes, and those who use it often. However, this is by no means necessary—the model can be applied to study differences among people who prefer different brands of a product. In the former case, the measure defining frequency of use is at least ordinal; in the latter it is nominal. Discriminant analysis, unlike most of the other models considered in this chapter, has been available for a relatively long time.

If there are k groups and p predictors, the number of possible dimensions (linear combinations of predictors) along which the groups can vary is the minimum of $k-1$ and p. As a result, the simplest case is when there are only two groups because differences can only exist along one dimension. It is common to separate this case from more complex ones in presenting the statistical theory. However, we will pursue the distinction noted in Chapter 5 between a **concentrated** and a **diffuse** structure of group means in multidimensional space (centroids), which implies more than one potential dimension (the two-group case is a somewhat trivial example of a concentrated structure). We begin by considering the results of a formal discriminant analysis of the data sets identified as Conc and Diff using SAS `Proc Candisc`. Recall that two data sets were defined in Chapter 5. The following is used to perform the actual discriminant analysis on these two sets:

```
Proc Sort Data=Conc;by grp;
Proc Candisc Data=Conc all;class Grp;var x1-x3;
Proc Sort Data=Diff;by grp;
Proc Candisc Data=Diff all;class Grp;var x1-x3;
Run;
```

Results for the Concentrated Data

The output for the concentrated data set (Conc) is as follows:

1. The problem is identified in terms of the number of observations, variables, and classes (groups) along with the various quantities (e.g., df) that are derived from them.

```
Canonical Discriminant Analysis
100 Observations          99 DF Total
  3 Variables             97 DF Within Classes
  3 Classes                2 DF Between Classes
              Class Level Information
      GRP      Frequency      Weight      Proportion
       1             33      33.0000        0.330000
       2             34      34.0000        0.340000
       3             33      33.0000        0.330000
```

2. The individual within-class (also known as the within-group) SSCP matrices consist of values of $\sum xy$ and $\sum x^2$ (sums of cross products and sums of squares based on deviations from **group** means) and appear next. Following are the data for group 1; data for the remaining groups are of the same form and will not be presented.

```
Variable          X1              X2              X3
X1         32.71293802     19.77805694     17.51139932
X2         19.77805694     39.09173356     16.27637429
X3         17.51139932     16.27637429     23.01965963
```

3. The pooled within-class SSCP matrix, which is the sum of the three (in this case) individual SSCP matrices, appears next.

```
Variable          X1              X2              X3
X1        104.4160510     59.5965686      63.9016710
X2         59.5965686     95.9969885      52.7424427
X3         63.9016710     52.7424427      91.5111316
```

4. The total (referred to as the total-sample in SAS) SSCP matrix consists of values of $\sum xy$ and $\sum x^2$, computed as deviations of individual observations about the **grand** mean (i.e., ignoring group membership). The between SSCP matrix consists of values of $\sum xy$ and $\sum x^2$, computed from the deviations of the group means about the grand mean, but is actually computed as the difference between the total and within SSCP matrices. These appear next beginning with the between matrix. They are basically intermediate steps so they will not be presented.

5. The individual within-class, pooled within-class, between, and total-sample matrices are then divided by their respective *df* to

produce variance–covariance matrices. These, too, are basically intermediate steps so they will not be presented.

6. The variance–covariance matrices, in turn, are then used to provide correlation matrices and associated p values. These are important to look at, but do not pay much attention to the individual within-class matrices if they are based on relatively few df. Also note that if there are only two groups, the between-class matrix can only consist of values that are ±1 or (very rarely) 0. In the present case, it is important to note that the between-class correlations are all large, implying that the profile of means over groups is very similar for the three measures—yet another way of saying the structure must be concentrated.

```
GRP = 1
Variable                    X1                 X2                 X3
X1                     1.00000            0.55307            0.63813
                           0.0             0.0008             0.0001
X2                     0.55307            1.00000            0.54258
                        0.0008                0.0             0.0011
X3                     0.63813            0.54258            1.00000
                        0.0001             0.0011                0.0
- - - - - - - - - - - - - - - - - - - - - - - - - - - - - - - - - - - - - -
GRP = 2
Variable                    X1                 X2                 X3
X1                     1.00000            0.59864            0.40512
                           0.0             0.0002             0.0175
X2                     0.59864            1.00000            0.45119
                        0.0002                0.0             0.0074
X3                     0.40512            0.45119            1.00000
                        0.0175             0.0074                0.0
- - - - - - - - - - - - - - - - - - - - - - - - - - - - - - - - - - - - - -
GRP = 3
Variable                    X1                 X2                 X3
X1                     1.00000            0.64398            0.79938
                           0.0             0.0001             0.0001
X2                     0.64398            1.00000            0.67558
                        0.0001                0.0             0.0001
X3                     0.79938            0.67558            1.00000
                        0.0001             0.0001                0.0
Pooled Within-Class Correlation Coefficients / Prob > |R|
Variable                    X1                 X2                 X3
X1                     1.00000            0.59526            0.65372
                           0.0             0.0001             0.0001
X2                     0.59526            1.00000            0.56272
                        0.0001                0.0             0.0001
```

X3	0.65372	0.56272	1.00000
	0.0001	0.0001	0.0

Between-Class Correlation Coefficients / Prob > |R|

Variable	X1	X2	X3
X1	1.00000	0.99885	0.98447
	0.0	0.0306	0.1123
X2	0.99885	1.00000	0.99177
	0.0306	0.0	0.0818
X3	0.98447	0.99177	1.00000
	0.1123	0.0818	0.0

Total-Sample Correlation-Coefficients / Prob > |R|

Variable	X1	X2	X3
X1	1.00000	0.62572	0.69476
	0.0	0.0001	0.0001
X2	0.62572	1.00000	0.59779
	0.0001	0.0	0.0001
X3	0.69476	0.59779	1.00000
	0.0001	0.0001	0.0

7. Next, the simple (univariate) statistics are presented for the total sample and the individual groups.

Total-Sample

Variable	N	Sum	Mean	Variance	Std Dev
X1	100	29.22749	0.29227	1.17944	1.08602
X2	100	36.60347	0.36603	1.03272	1.01623
X3	100	39.83093	0.39831	1.10310	1.05028

GRP = 1

Variable	N	Sum	Mean	Variance	Std Dev
X1	33	−5.58147	−0.16914	1.02228	1.01108
X2	33	1.04614	0.03170	1.22162	1.10527
X3	33	−6.14868	−0.18632	0.71936	0.84815

GRP = 2

Variable	N	Sum	Mean	Variance	Std Dev
X1	34	12.07938	0.35528	0.81885	0.90490
X2	34	14.53028	0.42736	0.78015	0.88326
X3	34	19.53624	0.57460	0.65219	0.80758

GRP = 3

Variable	N	Sum	Mean	Variance	Std Dev
X1	33	22.72958	0.68878	1.39628	1.18164
X2	33	21.02705	0.63718	0.97376	0.98679
X3	33	26.44337	0.80131	1.46778	1.21152

8. The group means are then standardized. Two sets of results are presented. In the first, the total-sample standard deviation is used, and, in the second, the pooled within-class standard deviation is used. These two sets of results are somewhat redundant because the only difference is the value the deviation of the group mean from the grand mean is divided by, and, unless group differences are extremely large, the two values will be numerically similar. Only the pooled within-class standardized class means will be presented here. Note that group 3 obtains the lowest mean score of variable 1 but the highest score on variables 2 and 3. Group 1 falls at the other extreme except for variable 2 where its mean is marginally smaller than group 2's mean. This is indicative of a concentrated structure—the group means on all predictors are ordered the same way. This includes the case in which the ordering is reversed on one or more predictors, because this can be handled by reversing (reflecting) measures appropriately.

```
Variable            1               2               3
X1          -.4447229145   0.0607223850   0.3821604572
X2          -.3360755913   0.0616461356   0.2725613910
X3          -.6019109434   0.1814958598   0.4149152091
```

9. The Mahalanobis distance from group i to group j is defined as $d'W^{-1}d$, where d is a vector of mean differences between the two groups and W^{-1} is the inverse of the pooled within-class variance–covariance matrix. This is a measure of spatial distance (dissimilarity) of the groups. Their squares (D^2) are then presented. Note that group 3 is relatively dissimilar to group 1 $(D^2 = 1.08117)$, but similar to group 2 $(D^2 = .10453)$. The dissimilarity of group 1 and group 2 is intermediate $(D^2 = .61657)$. A somewhat more subtle point is that the square roots of these values (1.04, .32, and .79) are such that the sum of the latter two distances is nearly equal to the first distance. This is evidence that the group differences are concentrated along a single line. If you were to walk three blocks and then walk four blocks on a given street, the total distance would be seven blocks. However, if you changed directions and could walk across fields, the total distance would be the square root of the sums of the two distances squared using the Pythagorean theorem $(\sqrt{3^2 + 4^2} = \sqrt{25} = 5)$.

```
From GRP            1              2              3
        1           0        0.61657        1.08117
        2     0.61657              0        0.10453
        3     1.08117        0.10453              0
```

10. These distances are converted to *F* ratios that test the null hypothesis that the population distances are 0. In the present case, the tests are based on 3 and 95 *df*.

```
From GRP            1              2              3
        1           0        3.37076        5.82385
        2     3.37076              0        0.57149
        3     5.82385        0.57149              0
```

11. The associated *p* values then appear.

```
From GRP            1              2              3
        1      1.0000         0.0217         0.0011
        2      0.0217         1.0000         0.6351
        3      0.0011         0.6351         1.0000
```

12. Univariate test statistics concerned with the significance of differences among the group means follow. These are based on 2 and 97 *df*. Also presented are the associated values of R^2. There are group differences on all three measures, with the largest being on X3 and the smallest on X2. The relationship between *F* and R^2 is given by

$$F = \frac{R^2(N - k + 1)}{(1 - R^2)k}, \tag{10.1}$$

where *k* is the number of predictors and *N* is the number of observations. The *df* are *k* and $N - k + 1$.

```
                    F Statistics, Num DF = 2   Den DF = 97
             Total    Pooled   Between                  RSQ/
Variable      STD       STD       STD   R-Squared     (1-RSQ)        F      Pr > F
X1         1.0860    1.0375    0.4304    0.105757      0.1183   5.7358      0.0044
X2         1.0162    0.9948    0.3060    0.061060      0.0650   3.1540      0.0471
X3         1.0503    0.9713    0.5152    0.162037      0.1934   9.3784      0.0002
Average R-Squared: Unweighted = 0.1096179   Weighted by Variance = 0.1105597
```

13. Discriminant analysis is based on an eigenanalysis of $W^{-1}B$, where W^{-1}, as before, is the inverse of the pooled within-class variance–covariance matrix and *B* is the between-class variance–covariance matrix. Wilks's lambda, Pillai's trace, and the Hotelling–Lawley trace are functions of these matrices plus the total variance–covariance matrix. Each can be converted to an *F* ratio. They test

the null hypothesis that the group vectors are identical. The tests are related to all (three in this case) eigenvalues of $W^{-1}B$ and will be similar but not identical in magnitude. There is dispute over which to use. Roy's largest root, by contrast, is a function of the first eigenvalue alone. It only provides an approximation of F. Moreover, it is most sensitive when there is a concentrated structure, as here. All four measures, however, are omnibus tests of the overall significance of group differences, that is, whether the group vectors differ, which, in fact, they do.

```
Multivariate Statistics and F Approximations
S=2 M=0 N=46.5
Statistic                     Value      F  Num DF  Den DF  Pr > F
Wilks's Lambda            0.82915185  3.1098      6     190  0.0063
Pillai's Trace            0.17189351  3.0089      6     192  0.0078
Hotelling-Lawley Trace    0.20479094  3.2084      6     188  0.0051
Roy's Greatest Root       0.19843751  6.3500      3      96  0.0006
```

14. The individual eigenvalues provide canonical correlations (correlations between linear combinations of the predictors and codes for groups). The relationship between a given eigenvalue (λ) and its associated canonical correlation (R) is given by

$$R = \sqrt{\frac{\lambda}{1 + \lambda}}. \qquad (10.2)$$

Note that the first canonical correlation is relatively large, but the second is very small—again, a characteristic of a concentrated structure.

```
       Adjusted      Approx      Squared
       Canonical     Canonical   Standard    Canonical
       Correlation   Correlation  Error      Correlation
  1    0.406915      0.376876    0.083862    0.165580
  2    0.079456      0.020872    0.099869    0.006313
                  Eigenvalues of INV(E)*H
                  = CanRsq/(1-CanRsq)
       Eigenvalue  Difference  Proportion  Cumulative
  1    0.1984      0.1921      0.9690      0.9690
  2    0.0064        .         0.0310      1.0000
         Test of HO: The canonical correlations in the
            current row and all that follow are zero
```

15. The eigenvalues also provide significance tests for the two functions. Note that the first function is significantly greater than chance, so groups differ along this dimension. Conversely, the second function is not significantly greater than chance so the groups

only differ on the first function (dimension). This, again, is what is meant by a concentrated structure.

```
Likelihood
          Ratio  Approx F  Num DF  Den DF  Pr > F
1  0.82915185    3.1098        6     190  0.0063
2  0.99368668    0.3050        2      96  0.7379
```

16. The structure consists of correlations between the individual variables and the discriminant axis. There are three such structures: (a) total, (b) between, and (c) pooled within. The total consists of correlations between individual scores and the axis, ignoring group mean differences; the between consists of correlations between the group means and the axis, ignoring variation within groups; and the within consists of correlations between individual scores and the axis, adjusting for group mean differences. Because there is only one significant discriminant axis, the data for CAN2 may be ignored both here and in the following discussion. Its printing can be suppressed once the outcome is known by specifying NCAN=1 in the Proc statement, but we will not do so here. The total structure confounds differences between and within groups. However, the remaining two are indexes to help define which variable(s) is (are) most important. In particular, note that variation within groups is most strongly related to X3. Group mean differences are equally discriminating. Like any other structure (correlation) measure, these indexes of predictor importance ignore relationships within other predictors.

```
Total Canonical Structure
                 CAN1                CAN2
X1           0.790268           -0.609822
X2           0.604135           -0.315109
X3           0.988884            0.136385
Between Canonical Structure
                 CAN1                CAN2
X1           0.988838           -0.148997
X2           0.994854           -0.101324
X3           0.999638            0.026921
Pooled Within Canonical Structure
                 CAN1                CAN2
X1           0.763377           -0.642836
X2           0.569518           -0.324165
X3           0.986791            0.148518
```

17. Canonical coefficients are regression weights. The total-sample and within-class coefficients are β weights, and the raw canonical coefficients are b weights. These weights indicate the relationship between a measure and group membership controlling for relationships with the other predictors, as is true for any set of regression weights. The difference between the total-sample and within-class measures is that the former are scaled to produce scores with a standard deviation of 1.0, ignoring group membership, and the latter are scaled to produce scores that have an average standard deviation of 1.0 within groups. Note that by this index X3 is clearly the most discriminating predictor. In contrast, X2 contributes nothing of its own—the reason it has a positive structure is that it is correlated with the other predictors. Again, disregard the data for CAN2.

Total-Sample Standardized Canonical Coefficients

	CAN1	CAN2
X1	0.239176864	-1.272660525
X2	-0.056404422	-0.196220715
X3	0.939118722	1.136925544

Pooled Within-Class Standardized Canonical Coefficients

	CAN1	CAN2
X1	0.228496081	-1.215828059
X2	-0.055215848	-0.192085882
X3	0.868489726	1.051419945

Raw Canonical Coefficients

	CAN1	CAN2
X1	0.220232288	-1.171856402
X2	-0.055503556	-0.193086766
X3	0.894156607	1.082493047

18. The final statistics are the class means on canonical variables. These are the group averages as scaled on the discriminant axis. As can be seen, group 3 is at the positive pole of CAN1 (CAN2 is irrelevant), and group 1 is at the negative pole. These data are scaled so that the average score over all data points is 0.

GRP	CAN1	CAN2
1	-.6058142150	-.0275991969
2	0.1680982730	0.1051590351
3	0.4326220549	-.0807464757

Results for the Diffuse Data

1. We begin the analysis of the diffuse data by presenting the various correlation matrices. Look, in particular, at the between-class structure—although the profile over groups is similar for X1 and X2, both are dissimilar to X3. This is consistent with a diffuse structure.

```
GRP = 1
Variable                X1                X2                X3
X1                 1.00000           0.70669           0.60336
                   0.0               0.0001            0.0002
X2                 0.70669           1.00000           0.46258
                   0.0001            0.0               0.0067
X3                 0.60336           0.46258           1.00000
                   0.0002            0.0067            0.0
-----------------------------------------------------------------
GRP = 2
Variable                X1                X2                X3
X1                 1.00000           0.71215           0.35818
                   0.0               0.0001            0.0375
X2                 0.71215           1.00000           0.42620
                   0.0001            0.0               0.0120
X3                 0.35818           0.42620           1.00000
                   0.0375            0.0120            0.0
-----------------------------------------------------------------
GRP = 3
Variable                X1                X2                X3
X1                 1.00000           0.78899           0.80402
                   0.0               0.0001            0.0001
X2                 0.78899           1.00000           0.67698
                   0.0001            0.0               0.0001
X3                 0.80402           0.67698           1.00000
                   0.0001            0.0001            0.0
Pooled Within-Class Correlation Coefficients / Prob > |R|
Variable                X1                X2                X3
X1                 1.00000           0.73740           0.63469
                   0.0               0.0001            0.0001
X2                 0.73740           1.00000           0.53655
                   0.0001            0.0               0.0001
X3                 0.63469           0.53655           1.00000
                   0.0001            0.0001            0.0
Between-Class Correlation Coefficients / Prob > |R|
Variable                X1                X2                X3
```

X1	1.00000	0.93773	0.14380
	0.0	0.2258	0.9081
X2	0.93773	1.00000	-0.20891
	0.2258	0.0	0.8660
X3	0.14380	-0.20891	1.00000
	0.9081	0.8660	0.0

Total-Sample Correlation Coefficients / Prob > |R|

Variable	X1	X2	X3
X1	1.00000	0.69270	0.59097
	0.0	0.0001	0.0001
X2	0.69270	1.00000	0.41595
	0.0001	0.0	0.0001
X3	0.59097	0.41595	1.00000
	0.0001	0.0001	0.0

2. We will skip over the raw means to look at the pooled within-class standardized class means. Note that, although the profiles for X1 and X2 are similar, in that group 2 is at the high end in contrast with groups 1 and 3, group 3 is at the high end for X3 with group 1 at the low end. This is why the between-class results were as noted (the raw and total standardized data reveal the same results).

Variable	1	2	3
X1	-.0721467451	0.0927155233	-.0233783396
X2	-.2974814438	0.6505645859	-.3727972204
X3	-.4424364502	-.0799528628	0.5248121271

3. Next, consider the D^2 results. Instead of the square roots of these measures (the actual distances) being additive, the D^2 measures themselves are, which is consistent with the Pythagorean theorem for a two-dimensional space as noted previously (as these distances are all significantly greater than 0, the F ratios and associated probabilities will not be presented).

From GRP	1	2	3
1	0	1.55481	1.57707
2	1.55481	0	3.08470
3	1.57707	3.08470	0

4. Another aspect of the findings is that the groups differ on X2 and X3 but not X1.

Variable	Total STD	Pooled STD	Between STD	R-Squared	RSQ/ (1-RSQ)	F	Pr > F
X1	1.0032	1.0110	0.0860	0.004945	0.0050	0.2410	0.7863
X2	1.0855	0.9905	0.5677	0.184165	0.2257	10.9483	0.0001
X3	1.0221	0.9577	0.4657	0.139813	0.1625	7.8831	0.0007

Average R-Squared: Unweighted = 0.1096411 Weighted by Variance = 0.1139635

5. The groups differ in terms of all of the multivariate test statistics:

Multivariate Statistics and F Approximations
S=2 M=0 N=46.5

Statistic	Value	F	Num DF	Den DF	Pr > F
Wilks's Lambda	0.55243212	10.9386	6	190	0.0001
Pillai's Trace	0.50086114	10.6912	6	192	0.0001
Hotelling-Lawley Trace	0.71370692	11.1814	6	188	0.0001
Roy's Greatest Root	0.53256389	17.0420	3	96	0.0001

6. Most important from the present perspective is that the groups clearly differ on both discriminant axes, as the less discriminating axis still produced an *R* of .39.

	Canonical Correlation	Adjusted Canonical Correlation	Approx Standard Error	Squared Canonical Correlation
1	0.589490	0.568899	0.065579	0.347499
2	0.391615	.	0.085090	0.153362

Eigenvalues of INV(E)*H
= CanRsq/(1-CanRsq)

	Eigenvalue	Difference	Proportion	Cumulative
1	0.5326	0.3514	0.7462	0.7462
2	0.1811	.	0.2538	1.0000

Test of H0: The canonical correlations in the
current row and all that follow are zero

	Likelihood Ratio	Approx F	Num DF	Den DF	Pr > F
1	0.55243212	10.9386	6	190	0.0001
2	0.84663751	8.6949	2	96	0.0003

7. Ignoring the total structure because of its confounding of between- and within-class covariation, X3 correlates negatively with axis I and positively with axis II. Both X1 and X2 correlate positively with both axes.

```
Between Canonical Structure
                 CAN1              CAN2
X1           0.689122          0.724645
X2           0.897927          0.440145
X3          -0.618016          0.786165
Canonical Discriminant Analysis
Pooled Within Canonical Structure
                 CAN1              CAN2
X1           0.066571          0.120029
X2           0.584598          0.491345
X3          -0.341422          0.744699
```

8. The pooled within-class standardized canonical coefficients indicate that axis I contrasts X2 with X3, whereas axis II contrasts X1 with X2 and X3. Because the three measures are in comparably sized units, the raw canonical coefficients describe similar relationships, although this is not critical to the analysis.

```
                 CAN1                 CAN2
X1        -0.384628047        -1.073860691
X2         1.292524891         0.727345604
X3        -0.790802051         1.036011171
Raw Canonical Coefficients
                 CAN1                 CAN2
X1        -0.380434008        -1.062151161
X2         1.304880798         0.734298673
X3        -0.825736745         1.081778292
```

9. Finally, the class means on canonical variables indicate that group 2 is located toward the positive pole of axis I, group 3 is located toward the negative, and group 1 is in the middle. In contrast, groups 2 and 3 are slightly toward the positive end of axis II with group 1 located toward the negative pole.

```
GRP               CAN1                 CAN2
1          -.0068728564         -.5972653716
2          0.8684368175         0.2907896768
3          -.8878802282         0.2976638865
```

Quite clearly, the three groups differ on the two axes, but for different reasons.

There is an important assumption underlying this analysis that has thus far been implicit. The assumption is that the population variance–covariance matrices are the same for the several groups; that is, they are **homoscedastic**. To the extent they are not, the pooled statistics presented previously, such as the pooled within-group correlation matrix, are an arbitrary mixture of the separate estimates. Slight disparities are tolerable. However, such outcomes as a large positive correlation between a pair of predictors in one group and a large negative correlation in another (moderation) greatly complicate any interpretation. Box's M test for homoscedasticity is implemented in such programs as SAS' `Proc Discrim`, which it uses for classification analysis. Keep in mind that Box's test is a very powerful one; you should not treat the variance–covariance matrices as heteroscedastic just because it is significant, especially when differences in variability or correlation magnitude are small and the sample size is large. Also keep in mind that a significant value can easily arise from failure to meet the multivariate normality assumption. Also note that heteroscedasticity means that the effects that separate the groups are not simply additive.

▣ CLAΙΙIFICATION ANALYΙIΙ

Discriminant analysis is primarily concerned with determining the dimensions along which groups differ. Classification analysis is more specifically concerned with seeing how well observations are placed into groups. In principle, a single program with suitable options could perform both functions. However, SAS provides two specialized procedures, `Proc Candisc`, which we have seen, and `Proc Discrim`, a classification procedure to which we now turn. We will use the concentrated data for illustration. The syntax is `Proc Discrim Data=Conc all;class Grp;var x1-x3;`.

This is basically equivalent to the syntax of `Proc Candisc` in identifying Grp as the categorical variable representing the classification and x1 to x3 as the predictors. Moreover, using the `all` option provides the same basic discriminant function data as is found using `Proc Candisc`. Although the format of the printout used to be quite different from `Proc Candisc`, recent revisions have made them much more similar. The current version begins with a description of the problem. This is followed by the individual and pooled within-class, between-class, and total sum of square and cross product, covariance, and correlation matrices, simple statistics, and standardized means. The latter are standardized both on the total and within groups. The first item to differ is that the program lists the natural log of the pooled covariance matrix, which is an index of the extent to which the measures vary in multidimensional space. This is followed by the Mahalanobis D^2 data, F ratios, and p values, the univariate tests of group difference, and the multivariate test statistics, for example, Wilks' lambda, again as before.

The core information follows:

1. The first item consists of what are known as **Fisher classification functions**. These are basically a series of regression equations relating the predictors to each group in turn—the higher the score, the more a person is similar to a member of that group.

	1	2	3
CONSTANT	−0.03983	−0.18771	−0.36741
X1	−0.16318	−0.14832	0.12779
X2	0.26230	0.19371	0.21493
X3	−0.23473	0.60098	0.63627

2. The distance of each score from the original set of scores is determined, and the observation is assigned to the closest group (smallest D^2). The accuracy of this process is as follows:

From GRP	1	2	3	Total
1	21	7	5	33
	63.64	21.21	15.15	100.00
2	10	13	11	34
	29.41	38.24	32.35	100.00
3	8	8	17	33
	24.24	24.24	51.52	100.00
Total	39	28	33	100
Percent	39.00	28.00	33.00	100.00
Priors	0.3333	0.3333	0.3333	

3. Summary error data are then provided:

```
            Error Count Estimates for GRP:
                  1          2          3        Total
   Rate        0.3636     0.6176     0.4848     0.4887
   Priors      0.3333     0.3333     0.3333
```

4. By default, the posterior probabilities of membership in each group, $p(j/X)$, where j defines the group and X the observation, are then computed and subjects reassigned. The numbers of observations falling in each category, which are identical to those given previously, are presented along with the average posterior probability.

```
                 Classified into GRP:
     From GRP           1            2            3
         1             21            7            5
                     0.5367       0.3899       0.4340
         2             10           13           11
                     0.4750       0.4103       0.4719
         3              8            8           17
                     0.5136       0.3727       0.5006
     Total             39           28           33
                     0.5161       0.3944       0.4809
     Priors          0.3333       0.3333       0.3333
     Posterior Probability Error Rate Estimates for GRP:
     Estimate           1            2            3        Total
     Stratified      0.3942       0.6701       0.5237     0.5293
     Unstratified    0.3961       0.6687       0.5239     0.5296
     Priors          0.3333       0.3333       0.3333
```

The stratified results take possibly unequal class size into account; the unstratified results do not. It makes very little difference here because the class sizes were nearly identical. There are many different classification strategies one can use with Proc Discrim, and one of its most powerful features is that the classification model can be applied to one set of data and cross-validated upon another. In particular, one can classify observations on the basis of the individual variance–covariance matrices instead of the pooled matrix when there is a high degree of heteroscedasticity.

Discriminant and Classification Analysis in SPSS

To run a discriminant analysis in SPSS, choose Analyze, Classify, Discriminant from the SPSS Data Editor menu.

▣ MULTIVARIATE ANALYSIS OF VARIANCE

We considered a data set, Ho, in Chapter 5, which can be analyzed in terms of one between (B) and one within (W) variable. However, another way to consider these data is to determine if W has the same average elements (centroids) for the two levels of B. In other words, we are effectively doing a simple analysis on B. However, instead of using a scalar criterion as in data set Oneway, the criterion is a vector. This is the multivariate analysis of variance (MANOVA). It is formally equivalent to a discriminant analysis, but it appears in a somewhat different context and there is usually (but not necessarily) more emphasis on inferential issues (Are the centroids significantly different across groups?) than in descriptive ones (What are the discriminant weights that most effectively contrast the groups?). The two, however, are also obverses—independent variables in discriminant analysis are dependent variables in the MANOVA and vice versa. The assumptions underlying the MANOVA are somewhat more general than those underlying the repeated-measure ANOVA, and many statisticians prefer the MANOVA for this reason.

Multivariate Data Analysis in SAS

To perform a MANOVA on data set Ho, it is necessary to restructure the file. Recall that the original data set consisted of two levels of B, five levels of W, and five participants/level of B, identified as swg, and the criterion was identified as Y. The file thus consisted of 50 ($2 \cdot 5 \cdot 5$) observations. Each observation contained values for B, W, swg, and Y. To perform a MANOVA on these data, the criterion is redefined so that it contains the five levels of W. This involves a set with 10 observations (two groups by five participants per group). The five values of B will appear in each observation. Modifying the original data step can do this, but because users may have already defined this data set, we will transform it rather than create it from scratch. This transformation is also needed in other SAS analysis of variance procedures. The steps are as follows:

```
Data Ho;
Retain err 12102 nbet 2 nwithin 5 nwg 5 rw .8;
Retain U 10 a1 1. b1-b4 (-.7 -.5 .1 .04 .0);
```

```
Retain a1b1-a1b5 (-.8 -.3 .1 .3 .0) a2b1-a2b5
(.0 .0 .0 .0 .0);
rwres = sqrt(1 - rw**2);
Array As (*) a1-a2;
Array Bs (*) b1-b5;
Array A1Bs (*) A1B1-A1B5;
Array A2Bs (*) A2B1-A2B5;
b5 = 0;
A2 = -A1;
Do I = 1 to nwithin - 1;
b5 = b5 - Bs(I);
End;
Do I = 1 to nwithin - 1;
A2Bs(i) = -A1Bs(i);
End;
Do i = 1 to nwithin - 1;
a1b5 = a1b5 - A1Bs(i);
a2b5 = a2b5 - A2Bs(i);
End;
Do swg = 1 to nwg;
Do B = 1 to nbet;
Berr = rannor(err);
Werr = rannor(err);
Do W = 1 to nwithin;
werri = rw*werr + rwres*rannor(err);
If B = 1 then AB = A1Bs(W);else AB = A2Bs(W);
Y = U + As(B) + Bs(W) + AB + werri + Berr;
Keep B W swg Y;
Output;
* Put A1-A2;
* Put B1-B5;
* Put A1B1-A1b5;
* Put A2B1-A2B5;
End;
End;
End;
Data Ho1 Ho2 Ho3 Ho4 Ho5;
Set Ho;
Id = 10*B + swg;
If (W ne 1) then Go to A;
```

```
Y1 = Y;
Output Ho1;
A: If (W ne 2) then Go to B;
Y2 = Y;
Output Ho2;
B: If (W ne 3) then Go to C;
Y3 = Y;
Output Ho3;
C: If (W ne 4) then Go to D;
Y4 = Y;
Output Ho4;
D: If (W ne 5) then Go to E;
Y5 = Y;
Output Ho5;
E:Z = 1;
Data Ho1;set Ho1;Keep ID B swg Y1;
Data Ho2;set Ho2;Keep ID Y2;
Data Ho3;set Ho3;keep ID Y3;
Data Ho4;set Ho4;Keep ID Y4;
Data Ho5;set Ho5;Keep ID Y5;
Proc Sort Data=Ho1;by ID;
Proc Sort Data=Ho2;by ID;
Proc Sort Data=Ho3;by ID;
Proc Sort Data=Ho4;by ID;
Proc Sort Data=Ho5;by ID;
Data HoAll;Merge Ho1 Ho2 Ho3 Ho4 Ho5;by ID;
```

The first data step (Data Ho) creates various values that are common to each dependent variable. Then, the next data step (Data Ho1 Ho2 Ho3 Ho4 Ho5) creates the five specific dependent variables. The contents of each initially contain the data from data set Ho because of the Set operation in the second line. The first of these contains values of Y when W = 1, the second contains values of Y when W = 2, and so on. The variable ID is created so that these can later be merged—each individual subject will have a unique value. Notation such as B: represents statement labels that the program jumps to whenever the particular condition is not met, and the final statement simply defines a dummy operation because a label must have some operation that follows it. Next, each of the data sets is called up and its contents limited to the variable ID and Y except for Ho1, where the values of B and swg are also kept. Each is then sorted.

The Merge operation then places the values of Y1 to Y5 together for each participant in data set HoAll.

The ANOVA ca be formatted as follows to perform the MANOVA:

```
Proc ANOVA Data=HoAll;
Class B swg;
Model Y1-Y5 = B;
MANOVA h=b/Printh Printe Summary;
```

Note that, unlike the scalar ANOVAs performed in Chapter 9, several dependent variables appear to the left of the equal sign in the Model statement. However, this will not by itself provide a MANOVA—it will simply perform a series of ANOVA on each variable, using B as the predictor in this case. The MANOVA statement itself invokes the procedure. Like the Test statement, you need to provide a hypothesis variable and an error variable unless, as here, you use the default error term. Printh prints what was identified previously as the between-group SSCP matrix, and Printe prints the within-group SSCP matrix. Summary provides analyses on the individual dependent variables.

The output begins like the other ANOVAs we have considered by identifying the value of the predictors. The summary tables for the individual criteria appear next. They indicate that the two B groups did not differ significantly on Y1 or Y3, but they did differ on Y2, Y4, and Y5. The error (within-group) SSCP follows.

1. The first output of major interest is identified as the Partial Correlation Coefficients From the Error SS&CP Matrix / Prob > Irl. These are simply the pooled within-group correlations among the five criterion measures, as follows:

DF = 8	Y1	Y2	Y3	Y4	Y5
Y1	1.000000	0.826106	0.867324	0.808965	0.901288
	0.0001	0.0061	0.0025	0.0083	0.0009
Y2	0.826106	1.000000	0.817081	0.649784	0.794074
	0.0061	0.0001	0.0072	0.0582	0.0106
Y3	0.867324	0.817081	1.000000	0.879034	0.729094
	0.0025	0.0072	0.0001	0.0018	0.0258
Y4	0.808965	0.649784	0.879034	1.000000	0.521329
	0.0083	0.0582	0.0018	0.0001	0.1501
Y5	0.901288	0.794074	0.729094	0.521329	1.000000
	0.0009	0.0106	0.0258	0.1501	0.0001

2. This is followed by the between-group SSCP matrix. The corresponding correlation matrix is not presented, but, in this case, it could only consist of values of ±1 because there are only two groups. Next are the eigenvalues of what this program identifies as E Inverse * H, but which were previously identified as $W^{-1}B$. The eigenvectors and eigenvalues are referred to as characteristic roots and characteristic vectors, respectively. Both the former and the latter terminologies are in common usage. Because there are but two groups, only one eigenvalue can have a nonzero value, so we will limit presentation to that value.

```
Characteristic  Percent                Characteristic Vector V'EV=L
Root
                                Yi        Y2        Y3        Y4        Y5
82.3385658     100.00  -1.39326672 0.11281694 -0.40367967 0.91170643 0.98827997
```

This means that the most discriminating linear combination contrasts Y2, Y4, and Y5 with Y1 and Y3. Despite successfully differentiating the groups, Y2 is actually given the smallest weight. The remainder of the analysis consists of the multivariate test statistics previously considered (e.g., Wilks' lambda). All were significant, denoting that the vector consisting of Y1 to Y5 differed in the two B groups.

Multivariate Data Analysis in SPSS

To conduct a multivariate data analysis in SSPS, choose Analyze, General Linear Models, Multivariate from the SPSS Data Editor menu.

▣ PROBLEMƒ

10.1. Take the concentrated and diffuse data sets you generated in response to Problem 5.9 and perform discriminant analyses upon them.

10.2. Using these same two data sets, perform classification analyses. In particular, vary the option that controls the prior probabilities and note how this affects your final outcome.

10.3. Use the repeated-measure data you generated in Problem 5.9 and perform a MANOVA.

10.4. Modify both the concentrated and the diffuse data sets so that group 1 has 50% of the subjects and the remaining two groups each have 25%. Run a classification analysis with each data set under the following conditions: (a) with equal prior probabilities (the default in many programs) and (b) with the prior probabilities proportional to the sample sizes. What differences are there in the two sets of results?

10.5. Modify both the concentrated and the diffuse data sets so that the variance of group 1 is doubled on all three measures. Compare the results of performing a classification analysis with pooled versus separate variance–covariance matrices.

11

Analysis of Categorical Data

CHAPTER OBJECTIVES

This chapter concludes the discussion of analyzing categorical data. It represents more recently developed models than the previous chapter. Specifically, we will discuss the following:

1. Logistic regression and predictor–criterion models, with particular emphasis on SAS `Proc Catmod`

2. Other forms of categorical modeling, most specifically log linear

3. Problems associated with applying standard factor analysis to categorical data such as items

4. Applications based on item response theory

▣ CATEGORICAL MODELS

Logistic Regression

Logistic regression is one of a growing series of statistical models for categorical data. Chapter 6 introduced the logistic function as

$$p = \frac{e^{da(X-b)}}{1 + e^{da(X-b)}}. \tag{6.1}$$

It is not very difficult to solve for X in this equation. The result is

$$X = \frac{1}{da} \ln\left[\frac{p}{1-p}\right] + b. \tag{11.1a}$$

Define $\ln[p/(1-p)]$ as the **log-odds ratio** (lor) (log-odds ratios are also known as **logits**). In particular, note that $\text{lor}(p) = 0$ when $p = .5$. This point on the function is known as the **threshold** of X.

Restating Equation 11.1a in terms of the log-odds ratio,

$$X = \frac{1}{da} \text{lor}(p) + b. \tag{11.1b}$$

Note that X is a linear function of $\text{lor}(p)$ with slope $1/da$ and intercept b. This linearity is why many of the most popular categorical models analyze $\text{lor}(p)$ instead of p itself. Logistic regression is thus a form of generalized linear model in which a series of predictors are used in conjunction with a categorical outcome. Note that the model relates the predictors linearly to $\text{lor}(X)$ but not to X itself. The lor (logit) is a **link function**.

Normally, logistic regression and other categorical modeling parameters are estimated by the method of maximum likelihood (ML). SAS' `Proc Catmod` is one of several tools used to this end. `Proc Catmod` is a rather complex program and, in some ways, its syntax conflicts with what you have learned from the syntax of more commonly used programs for continuous data such as `Proc GLM`. SAS has other tools for specialized situations, such as `Proc Logistic` and `Proc Probit` for fitting logistic, cumulative normal, and Gompertz curves (all of which are forms of growth curves). However, these will not be considered here, as the intent is to familiarize the reader with `Proc Catmod`, which is the most general of the SAS procedures.

Categorical Modeling and `Proc Catmod`

Keep in mind that `Proc Catmod`, by default, models the data in terms of their internal sorting values. In other words, suppose the dichotomous variable $Y = f(X)$ and the relationship is positive. Because 0 sorts ahead of 1, the analysis will model the probability that the response is a 0 so X will have negative regression weights. You may not wish to model the response category 0. You can force SAS to analyze a data set named "all" and model the higher of the two levels of Y using the following statements.

Also note that by limiting outcomes to "0" and "1" we are dealing with
binary logistic regression, an important special case of the more general
model in which outcomes may take on more than two values.

```
Proc Sort Data=all;by descending y;
Proc Catmod Data=all order=data;
[additional Proc Catmod statements]
```

The keyword `Order=Data` says that the data will be modeled in terms
of the first variable encountered in the data set, and the `Sort` command
guarantees that it will be a 1 or, in general, the higher of the two values.

Logistic regression will first be illustrated using the psychometric function data generated in Chapter 6, Pmetrica. These data were sorted using
the `by descending` keyword so that we model the probability of a "1"
rather than a "0." The `Catmod` statements that model the lor of $X = Y$
are as follows:

```
Proc Catmod Data=Pmetrica order=data;
Direct x;
Model y=x/pred=prob;
Run;
```

1. The `Direct` statement says to incorporate the numeric value of X
    rather than generate a series of dummy predictors. Two parameters
    are to be estimated: (a) a single slope for X and (b) an intercept.
    Were there no `Direct` statement, a slope would be estimated for
    each value of X. Note that `Proc Catmod` treats variables as classi-
    fication variables unless a `Direct` statement dictates that they be
    treated as continuous. In contrast, `Proc GLM` and other OLS regres-
    sion algorithms treat variables as continuous unless a `Class` vari-
    able dictates otherwise (`Proc ANOVA` treats all independent vari-
    ables as classificatory). This is because predictors in `Proc Catmod`
    are assumed, by default, to be categorical, whereas predictors in the
    more "classic" procedures such as `Proc Reg` are assumed, by de-
    fault, to be continuous.

    The output begins with a general description of the database.
    This identifies the criterion as "Y," says that it has two levels, says
    that each observation is equally weighted, identifies the data set as
    `Pmetrica`, and reports that there are 7000 observations. The only
    novel point is the **population**. This defines the number of cells used
    for analysis and, by default, is equal to the number of distinct lev-

els of the independent variable, seven in this case. However, it can be overridden. One purpose of overriding the default populations is to equate the number of cells in a hierarchical analysis, as will be illustrated later. Another is to make each individual observation a population. Defining a unique identifier for each observation (e.g., "ID") and using this in a population statement (e.g., Population ID) can do this. This is left as an exercise.

```
CATMOD PROCEDURE
Response: Y                  Response Levels (R)  =     2
Weight Variable: None        Populations (S)      =     7
Data Set: PMETRICA           Total Frequency (N)  =  7000
Frequency Missing: 0         Observations (Obs)   =  7000
```

2. The POPULATION PROFILES are the number of observations per sample (1000 each).

```
POPULATION PROFILES
                Sample
Sample   X       Size

   1    170      1000
   2    180      1000
   3    190      1000
   4    200      1000
   5    210      1000
   6    220      1000
   7    230      1000
```

3. The RESPONSE PROFILES indicate which response number in the data is modeled. The results of the Sort cause "1" responses to be modeled, as intended.

```
RESPONSE PROFILES
Response  Y

   1      1
   2      0
```

4. By default, the iteration history describing the changing values of the parameter estimates is presented under the heading of

MAXIMUM-LIKELIHOOD ANALYSIS. These data are not extremely important as long as the process converges, which it did (note that each parameter estimate changes less in magnitude from step [defined in the algorithm as Sub Iteration] to step). The noiter keyword is commonly used to suppress this somewhat uninformative set of results.

MAXIMUM-LIKELIHOOD ANALYSIS

Iteration	Sub Iteration	-2 Log Likelihood	Convergence Criterion	Parameter Estimates 1	Parameter Estimates 2
0	0	9704.0605	1.0000	0	0
1	0	5756.067	0.4068	-12.5789	0.0655
2	0	5275.4076	0.0835	-18.4524	0.0967
3	0	5217.0009	0.0111	-21.3059	0.1119
4	0	5215.5359	0.000281	-21.8479	0.1148
5	0	5215.5347	2.3262E-7	-21.8640	0.1149
6	0	5215.5347	1.69E-13	-21.8641	0.1149

5. The MAXIMUM-LIKELIHOOD ANALYSIS-OF-VARIANCE TABLE contains the G^2 and p values for the intercept, the effect (X or slope), and the residual (the residual df is the number of populations -1). In the present case, both parameter estimates clearly differ from .0, and the residual is nonsignificant. This means that the mean probabilities for the various values of X fit a logistic function as defined in terms of Equations 5–7 and 11.1.

MAXIMUM-LIKELIHOOD ANALYSIS-OF-VARIANCE TABLE

Source	DF	Chi-Square	Prob
INTERCEPT	1	1868.66	0.0000
X	1	1907.51	0.0000
LIKELIHOOD RATIO	5	5.28	0.3831

6. The ANALYSIS OF MAXIMUM-LIKELIHOOD ESTIMATES contains the parameter estimates and their standard errors and repeats the G^2 and p values. These data indicate that the best fitting function is $Y = p(X) = .1149X - 21.8641$. The redundancy of the G^2 values will hold in any analysis in which each effect has 1 df, as it does here. However, if X were treated as a series of seven categorical values, the total effect, based on 6 df, would have been reported as

a single entity in the previous section. As long as there are several parameter estimates within an effect, it is possible that individual estimates are nonsignificant yet the effect as a whole may be significant or vice versa.

```
ANALYSIS OF MAXIMUM-LIKELIHOOD ESTIMATES
                                   Standard     Chi-
Effect       Parameter   Estimate    Error    Square    Prob

-------------------------------------------------------------

INTERCEPT         1      -21.8641    0.5058   1868.66   0.0000
X                 2        0.1149    0.00263  1907.51   0.0000
```

7. The MAXIMUM-LIKELIHOOD PREDICTED VALUES FOR RESPONSE FUNCTIONS AND PROBABILITIES are an important, but optional part of the results generated by the pred=prob keyword as they provide residual statistics. Consider the first population $(X = 170)$. As noted in Chapter 6, the probability of an affirmative response is .087, so the probability of a negative response is the complement or .913. The observed standard error is

$$\sqrt{\frac{pq}{N}} = \sqrt{\frac{p(1-p)}{N}},$$

which equals

$$\sqrt{\frac{.087 \cdot .913}{1000}} = .00891241$$

in this example. These appear under Observed in the second and third lines, respectively. The function value (-2.3508278) is the logit of the probability of an affirmative response. The corresponding residual values are obtained from the parameter estimates by inserting the value of 170 into the equation. Proper residual analysis implies looking into reasons that led to an exceptionally large discrepancy between observed and predicted values or exceptionally large standard errors. Pred=freq gives analogous residual statistics based on frequencies rather than probabilities.

MAXIMUM-LIKELIHOOD PREDICTED VALUES FOR RESPONSE FUNCTIONS AND PROBABILITIES

Sample	X	Y	Function Number	Observed Function	Observed Standard Error	Predicted Function	Predicted Standard Error	Residual
1	170		1	-2.3508278	0.11220313	-2.3364256	0.06721661	-0.0144022
		1	P1	0.087	0.00891241	0.08815081	0.00540289	-0.0011508
		0	P2	0.913	0.00891241	0.91184919	0.00540289	0.00115081
2	180		1	-1.1526795	0.07404361	-1.1877413	0.04676339	0.03506178
		1	P1	0.24	0.01350555	0.23366315	0.00837367	0.00633685
		0	P2	0.76	0.01350555	0.76633685	0.00837367	-0.0063369
3	190		1	-0.0120001	0.06324669	-0.039057	0.0351996	0.02705685
		1	P1	0.497	0.0158111	0.49023699	0.00879655	0.00676301
		0	P2	0.503	0.0158111	0.50976301	0.00879655	-0.006763
4	200		1	1.03559932	0.07191528	1.1096273	0.04092266	-0.074028
		1	P1	0.738	0.01390525	0.75205962	0.00763068	-0.0140596
		0	P2	0.262	0.01390525	0.24794038	0.00763068	0.01405962
5	210		1	2.16432687	0.10403636	2.25831159	0.05910814	-0.0939847
		1	P1	0.897	0.00961202	0.90536507	0.00506434	-0.0083651
		0	P2	0.103	0.00961202	0.09463493	0.00506434	0.00836507
6	220		1	3.66356165	0.20254787	3.40699588	0.08183114	0.25656577
		1	P1	0.975	0.0049371	0.96792246	0.00254074	0.00707754
		0	P2	0.025	0.0049371	0.03207754	0.00254074	-0.0070775
7	230		1	4.95482051	0.37929433	4.55568017	0.10621842	0.39914035
		1	P1	0.993	0.00263647	0.98960191	0.00109298	0.00339809
		0	P2	0.007	0.00263647	0.01039809	0.00109298	-0.0033981

Another example of logistic regression may be obtained by dichotomizing the criterion measures (Y) of the data generated earlier in this chapter. Do this by using the previously generated criterion measures as latent variables and establishing a **threshold** or **cutoff** such that the dichotomized (observable) measure equals 1 if the latent variable exceeds this cutoff and equals 0 otherwise. Because the latent measures are standard normal ($\mu = 0$, $\sigma = 1$), ordinary normal curve statistics can be used to infer the expected criterion distribution. The existing multiple regression spreadsheet can be modified as follows.

We have formed two dichotomous variables, DY1 and DY2, each of which is based on Y. DY1 is 1 if Y exceeds the value of cut1 (set at .0) and 0 otherwise. DY2 is 1 if Y exceeds the value of cut2 (set at .5) and 0 otherwise. Following are the `Proc Catmod` statements:

```
Data A;
Retain nobs 1000 b1-b3 (.4 .5 .6) err 3213 cut1-cut2 (0 .5);
Bres = sqrt(1 - b1**2 - b2**2 - b3**2);
Array Xs (*) X1-X3;
Do I = 1 to nobs;
Do J = 1 to 3;
Xs(J) = rannor(err);
End;
Yp = b1*x1 + b2*x2 + b3*x3;
Y = Yp + bres*rannor(err);
X1 = 50 + X1*10;
X2 = 50 + X2*10;
X3 = 50 + X3*10;
If Y gt cut1 then DY1 = 1;else DY1 = 0.;
If Y gt cut2 then DY2 = 1;else DY2 = 0.;
Output;
End;
Proc Catmod Data=A;
Direct x1-x3;
Model dy1 = x1-x3/noiter noprofile;
Proc Catmod Data=A;
Direct x1-x3;
Model dy2 = x1-x3/noiter noprofile;
Run;
```

The **Direct** statement treats each of the variable(s) (X1 to X3 in this case) as continuous variables instead of as categories. **Noiter** suppresses an iteration-by-iteration listing of such information as the parameter estimates.

The residual G^2 using DY1 as the criterion is a nonsignificant 711.7 on 996 df. This is consistent with the idea that the three predictors exhaust all systematic prediction (which, in fact, they do). However, this is not definitive in real problems because the large number of error allows degrees of freedom significant sources of variation to "hide" in some of these degrees of freedom.

The parameter estimates and associated G^2 values are as follows:

Effect	Parameter	Estimate	Standard Error	Chi-Square	Prob
INTERCEPT	1	−26.8175	1.7362	238.57	0.0000
X1	2	0.1320	0.0118	124.33	0.0000
X2	3	0.1762	0.0139	161.23	0.0000
X3	4	0.2276	0.0159	203.66	0.0000

Using DY2 as the criterion led to a residual of 596.59. This is also nonsignificant. The parameter estimates are as follows:

Effect	Parameter	Estimate	Standard Error	Chi-Square	Prob
INTERCEPT	1	−30.3087	2.0173	225.73	0.0000
X1	2	0.1392	0.0131	112.13	0.0000
X2	3	0.1941	0.0158	151.58	0.0000
X3	4	0.2327	0.0174	178.19	0.0000

Note that the intercept has changed substantially, but the slope estimates have not.

Despite the fact that the term **logistic regression** implies the application of continuous predictors to a dichotomous criterion, the logic is no different from the discrete predictor case. The upshot is a model that can be described as Criterion = f(Predictors) and programs such as Proc Catmod "do not know the difference." For this reason, the term **predictor–criterion model** is more general.

Hierarchical Models

Instead of entering X1, X2, and X3 all at the same time, it is often desirable to enter them successively. In particular, consider the following five Proc Catmod requests:

```
Proc Catmod Data=A;
Population X1-X3;
Direct x1-x3;
Model dy1 = /noint noiter noprofile;
Proc Catmod Data=A;
Population X1-X3;
Direct x1-x3;
Model dy1 = /noiter noprofile;
Proc Catmod Data=A;
Population X1-X3;
Direct x1-x3;
Model dy1 = x1/noiter noprofile;
Proc Catmod Data=A;
Population X1-X3;
Direct x1-x3;
Model dy1 = x1-x2/noiter noprofile;
Proc Catmod Data=A;
Population X1-X3;
```

```
Direct x1-x3;
Model dy1 = x1-x3/noiter noprofile;
Run;
```

The first of these, model dy1 = /noint..., lists no predictors and, in addition, specifies that the intercept term is to be omitted. This produces a model in which the probability of an affirmative response is .5 (logit = 0), independent of X (i.e., is a horizontal line). The next model statement also lists no predictors, but, by omitting noint, it estimates the most likely outcome probability with a nonzero intercept. Following this, X1, X2, and X3 are entered, in turn. The Population statements force the same numbers of cells in all cases. Had this not been done, there would only have been two populations to compare (0 and 1) in the first two models (because all 1000 X1 values were different, Population X1, or any of the individual predictors, would have produced identical outcomes as to numbers of populations).

The results were as follows:

Effect	Model			Difference		
	G^2	df	p	G^2	df	p
None	1386.29	1000	.00	—	—	—
Response	1385.51	999	.00	.78	1	.38
X1	1298.10	998	.00	87.41	1	.00
X1 + X2	1125.45	997	.00	172.65	1	.00
X1, X2, and X3	711.70	996	1.00	413.75	1	.00

The values under *Model* are the likelihood ratios noted previously, with their respective *df* and *p* values. Note that because the first model (None) estimates no parameters, it preserves the original 1000 observations as *df*, and each successive model uses 1 *df*. It assumes that the probability of an affirmative response is .5 and that this value is independent of X1, X2, or X3. The fact that its G^2 is very large means this model can be rejected. The next step is to allow the probability of an affirmative response to be other than .5 but independent of the predictors. This model, too, can be rejected. Moreover, the difference G^2 is quite small, so there is no evidence that the probability of an affirmative response is other than .5 (which, in fact, it was). Entering X1 does significantly reduce G^2 but still leaves a significant residual. This is also true for X2. Entering X3 further also significantly reduces G^2 but leaves a nonsignificant residual as well. Although further prediction might be contained in the large number of residual *df* (which we know not to be the case but would ordinarily be ignorant of), there is no evidence where this might be. Consequently, the decision is to accept the main effects of X1 to X3 and the best guess one would make not knowing how the data were generated is that these are sufficient. The model can be stated as

$$\text{logit}(p) = 26.8175 - 0.1320X_1 - 0.1762X_2 - 0.2276X_3.$$

The signs are negative because *p* is defined as the proportion of events coded 0 rather than events coded 1.

It is preferable to have a reason for choosing a particular order in evaluating a hierarchy, for the same reason it was in the previous chapter's discussion of moderated multiple regression. If choice of order is arbitrary, evaluate a couple of different orders to see how much difference it makes, but this is left as an exercise here.

Log-Linear Models

A great many, though not all, models used to analyze categorical data are log linear in the sense that Equation 11.1 is—the log probability of outcomes is a linear function of some property of the predictor. However, in a more narrow sense, log-linear models are often used to denote the decomposition of multiway contingency tables as an extension of the test for nonindependence commonly used with two-dimensional tables. For example, consider a table that contains all combinations of four binary dimensions, that is, consists of 16 cells organized in a $2 \times 2 \times 2 \times 2$ manner. One might ask whether the first and second of these dimensions are related, the first and third, and so on. Olzak and Wickens (1983) reported a prototypical study in which subjects made two detection judgments concurrently in the sense that a faint stimulus could be presented to the right eye, left eye, both, or neither eyes, and subjects responded by stating which eye or eyes were presented with stimuli. For simplicity, we will assume that these responses, like the stimuli, were binary in the sense that subjects made two responses denoting whether they thought that stimuli were or were not presented to the left and right eye, respectively. This is a simplification of the experiment as subjects could express degrees of confidence about each judgment. There is a long history related to the formal study of what is known as **perceptual independence** in cognitive psychology. This begins with Thurstone's (1927/1994) law of comparative judgment and involves the theory of signal detection (Macmillan & Creelman, 1991; Swets, 1986a, 1986b), followed by Garner and Morton's (1969) work and the more recent work of Ashby and Townsend (1986; Ashby & Maddox, 1990; Ashby & Perrin, 1988) on **general recognition**. In particular, Wickens and Olzak (Wickens, 1989; Wickens & Olzak, 1989) cast the problem in terms of log-linear modeling. What is presented here combines elements of both Wickens and Olzak and Garner and Morton.

Consider Figure 11.1, which illustrates perceptual independence in a situation similar to that presented in Olzak and Wickens (1983). Assume one process involves the presentation of a stimulus, A. This stimulus affects a perceptual stage, $p(A)$, but so does error, $e(A)$. The latter causes judgments of A made at a decision stage, $D(A : a)$, to be partially inaccurate. This decision stage produces an overt response, a. For present purposes, assume that A and a are both observable binary events, specifically presence versus absence and judged presence versus absence, but that the information at $p(A)$ and $D(A : a)$ and error affecting the process, $e(A)$, are continuous variables. In addition, assume that $e(A)$ is normally distributed. Next, assume a second process involving stimulus B. It contains analogous terms $p(B)$, $D(B : b)$, and $e(B)$. Despite its formal similarity to the first process, nothing in the one influences anything in the other—it is as if two unrelated individuals were making two totally different judgments. Response a is influenced by stimulus A, and response b is influenced by stimulus B. The resulting data form a 2 (levels of A) × 2 (levels of a) × 2 (levels of B) × 2 (levels of b) **contingency table**.

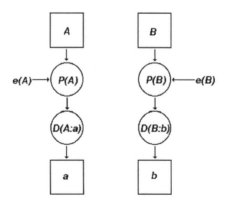

Figure 11.1. Model for perceptual independence.

The following program generates perceptually independent data:

```
Data Ind;
Title 'Complete Independence Model';
Retain n 4000 err 1995;
Do i = 1 to n;
If (mod(i, 2) = 1) then sa = 1;else sa = 0;
If (mod(i, 4) le 1) then sb = 1;else sb = 0;
xa = sa + rannor(err) - .5;
xb = sb + rannor(err) - .5;
If (xa ge .0) then ra = 1;else ra = 0;
If (xb ge .0) then rb = 1;else rb = 0;
Drop n i err;
Output;
End;
Run;
```

The terms SA, RA, SB, and RB correspond to the two binary stimulus and response processes noted previously. Because this is an experiment, it was possible for the experimenter to control the frequency with which SA and SB were presented, and, following standard practice, the two levels of each appeared equally often and independently of one another. The XA and XB terms represent the intervening latent processes, and the calls to rannor(err) represent the random error. In this simulation, the probabilities of "yes" responses (XA and XB ≥ 0) and of "no" responses (XA and XB < 0) are .5 in the population, matching the stimulus probabilities, but the simulation can easily be modified to introduce what is known as a **response bias**. The simulation gave rise to the following table of joint frequencies:

SA	SB	RA	RB	F(X)
0	0	0	0	454
0	0	0	1	226
0	0	1	0	217
0	0	1	1	103
0	1	0	0	200
0	1	0	1	488
0	1	1	0	94
0	1	1	1	218
1	0	0	0	212

1	0	0	1	86
1	0	1	0	477
1	0	1	1	225
1	1	0	0	94
1	1	0	1	230
1	1	1	0	204
1	1	1	1	472

This is obviously a very difficult table to understand, which is precisely why some model is needed to analyze its results. If you look carefully, you can tell that frequencies tend to be higher when a response is correct, so that SA = RA and/or SB = RB, than when it is incorrect, but it is difficult to do more than make this inference.

Proc Catmod is invoked as follows:

```
Proc Catmod Data=Ind;
%Include 'D:\Research\Gar_Mor\Loglin.Txt';
Run;
```

The statement that begins with %Include means "insert the text found in the relevant file." This strategy was used because the entire program simulates various alternatives to perceptual independence, but the code used to run each instance of Proc Catmod is the same. Using %Include avoids the need to insert several identical copies of this shared text. This shared text is as follows:

```
MODEL SA*SB*RA*RB=_RESPONSE_/corrb covb freq oneway predict prob xpx;
LOGLIN SA|SB;
RUN;
LOGLIN SA|SB RA;
RUN;
LOGLIN SA|SB RA RB;
RUN;
LOGLIN SA|SB RA RB SA*RA;
RUN;
LOGLIN SA|SB RA RB SA*RA SB*RB;
RUN;
LOGLIN SA|SB RA RB SA*RA SB*RB SA*RB;
RUN;
LOGLIN SA|SB RA RB SA*RA SB*RB SA*RB SB*RA;
RUN;
LOGLIN SA|SB RA RB SA*RA SB*RB SA*RB SB*RA RA*RB(SA*SB);
RUN;
LOGLIN SA|SB RA RB SA*RA SB*RB SA*RB SB*RA RA*RB(SA*SB)
SA*SB*RA;
LOGLIN SA|SB RA RB SA*RA SB*RB SA*RB SB*RA RA*RB(SA*SB)
```

```
SA*SB*RA SA*SB*RB;
Quit;
Run;
```

The statement MODEL SA*SB*RA*RB=_RESPONSE_ defines the contingency table as four-way combinations. The statements following the slash describe various options (there is no "all" to obtain all possible output as in other SAS procedures, which will be explained in the context of the printout). The keyword ML could have been added to specify ML estimation, but that is the default. Each combination of a LOGLIN and a RUN; statement defines the variables to be estimated in a particular model, so that the first says "fit the model to SA, SB, and their two-way association; ignore everything else." This adjusts only for the experimentally defined contingencies. The succeeding LOGLIN statements have the following respective meanings:

Term	Interpretation
RA	Tendency for *a* to deviate from 50/50
RB	Tendency for *b* to deviate from 50/50
SA*RA	Association between *A* and *a*, i.e., accuracy in judging *A*
SB*RB	Association between *B* and *b*, i.e., accuracy in judging *B*
SA*RB	Tendency for *A* to influence responses to *b*
SB*RA	Tendency for *B* to influence responses to *a*
RA*RB(SA*SB)	Tendency for *a* and *b* to be the same within stimulus combinations
SA*SB*RA	Effect of *B* on the accuracy in judging *A*
SA*SB*RB	Effect of *A* on the accuracy in judging *B*

In this specification, the RA effect is the natural logarithm of the ratio of the probability of a "no" response to a "yes" response, which is why the term log-odds ratio (lor) was introduced, correcting for the probability of SA and SB and the joint probability. The two-way associations of SA*RA and SB*RB are the log-odds ratio of being correct relative to being incorrect. Note SA and RA effects were always entered before SB and RB effects. That choice was arbitrary, and it means that effects involving event *B* are corrected for event *A* but not vice versa. It is also possible to enter both events at the same stage, in which case effects such as RA and

RB are corrected for each other. Unless there is a good theoretical reason for dictating order, alternative ordering should always be explored, because it takes very little additional time. Hopefully, such relatively minor differences should not affect the outcome materially. At the same time, always enter simpler effects such as RA before more complex ones such as SA*RA. Entry of SA*SB before RA and RB is an exception because it was built into the experimental design.

1. As with any model analyzed by Proc Catmod, the output begins with some general information that helps determine if the model is properly specified. The response analyzed is the four-way contingency table of SA*SB*RA*RB noted previously. Because all are binary, there are a total of 16 response levels. Observations are not weighted. There is only one population being analyzed (if, for example, separate contingency tables were formed by gender for comparison, there would be two). Ind is the name of the SAS data set. The total number of observations is 4000. None was missing, so the total number of (valid) observations is also 4000.

```
CATMOD PROCEDURE
Response: SA*SB*RA*RB      Response Levels (R)= 16
Weight Variable: None      Populations (S)= 1
Data Set: IND              Total Frequency (N)= 4000
Frequency Missing: 0       Observations (Obs)= 4000
```

2. Next to be presented are the one-way (marginal) frequencies for the four variables. Note that each of the two stimulus categories are split exactly in half, but the two response categories each differ from equality due to sampling error.

```
ONE-WAY FREQUENCIES
Variable   Value   Frequency
-----------------------------
SA          0        2000
            1        2000
SB          0        2000
            1        2000
RA          0        1990
            1        2010
RB          0        1952
            1        2048
```

3. The printout continues with the number of samples (one in this case) and the sample size.

Sample	Sample Size
1	4000

4. The _RESPONSE_ command, which defines the categorical analysis as log linear, normally involves the conjunction of several variables. RESPONSE PROFILES are the specific combinations. These help identify various items such as the parameter estimates given later. For example, response 1 consists of level 0 for all variables.

Response	SA	SB	RA	RB
1	0	0	0	0
2	0	0	0	1
3	0	0	1	0
4	0	0	1	1
5	0	1	0	0
6	0	1	0	1
7	0	1	1	0
8	0	1	1	1
9	1	0	0	0
10	1	0	0	1
11	1	0	1	0
12	1	0	1	1
13	1	1	0	0
14	1	1	0	1
15	1	1	1	0
16	1	1	1	1

5. RESPONSE FREQUENCIES are the number of times each combination occurred. If you use the _RESPONSE_ information that identifies each of these 16 cells, you can collapse them in various ways to obtain 2×2 contingency tables (e.g., SA by RA). Examine these. You should find that all but SA by RA and SB by RB produce independence in the sense that the $X_{11} \cdot X_{22}$ will equal $X_{21} \cdot X_{12}$ for any of these matrices, X.

			Response	Number				
Sample	1	2	3	4	5	6	7	8
1	454	226	217	103	200	488	94	218

			Response	Number				
Sample	9	10	11	12	13	14	15	16
1	212	86	477	225	94	230	204	472

6. RESPONSE PROBABILITIES are the frequencies divided by the total number of observations (4000 in this case).

			Response	Number				
Sample	1	2	3	4	5	6	7	8
1	0.1135	0.0565	0.05425	0.02575	0.05	0.122	0.0235	0.0545

			Response	Number				
Sample	9	10	11	12	13	14	15	16
1	0.053	0.0215	0.11925	0.05625	0.0235	0.0575	0.051	0.118

7. The _RESPONSE_ MATRIX describes the effects in terms of the response profiles. For example, the profile for the first eight members of the _RESPONSE_ matrix was at level 0 of SA and the last eight were at level 1. The effect in the first column, SA, contrasts these. The first set is assigned a value of +1 and the last set a value of –1 because the last element in the profile is always coded –1. In the same way, you can verify that the second codes SB. An association, SA*SB in this case, is simply the product of its respective components. If you have constructed orthogonal contrasts in the ANOVA, you have used this same principle.

	1	2	3
1	1	1	1
2	1	1	1
3	1	1	1
4	1	1	1
5	1	–1	–1

```
 6   1  -1  -1
 7   1  -1  -1
 8   1  -1  -1
 9  -1   1  -1
10  -1   1  -1
11  -1   1  -1
12  -1   1  -1
13  -1  -1   1
14  -1  -1   1
15  -1  -1   1
16  -1  -1   1
```

8. The next item is the iteration history. In the present case, the convergence criterion was satisfied at the second step. `-2 Log Likelihood` is used in forming the test statistic.

Iteration	Sub Iteration	-2 Log Likelihood	Convergence Criterion	Parameter Estimates 1	2	3
0	0	22180.71	1.0000	0	0	0
1	0	22180.71	0	1.388E-17	-7.88E-18	7.883E-18

9. The `MAXIMUM-LIKELIHOOD ANALYSIS-OF-VARIANCE TABLE`, as in the logistic regression problem, contains the G^2 values for the specific effects and for the model. The reason that the effect G^2 values are all 0 is because the stimulus combinations appeared with equal frequency by design. The large value of the residual means that additional relationships are needed to account for the contingencies.

Source	DF	Chi-Square	Prob
SA	1	0.00	1.0000
SB	1	0.00	1.0000
SA*SB	1	0.00	1.0000
LIKELIHOOD RATIO	12	1182.14	0.0000

10. Again, as in the logistic regression problem, the `ANALYSIS OF MAXIMUM-LIKELIHOOD ESTIMATES` is presented and contains the parameter numbers, which relate back to the response profile; the parameter estimates; the standard errors; the G^2 values; and the associated probability.

Effect	Parameter	Estimate	Standard Error	Chi-Square	Prob
SA	1	1.39E-17	0.0158	0.00	1.0000
SB	2	-788E-20	0.0158	0.00	1.0000
SA*SB	3	7.88E-18	0.0158	0.00	1.0000

11. The **COVARIANCE MATRIX OF THE MAXIMUM-LIKELIHOOD ESTIMATES** describes the relationships among the estimates.

	1	2	3
1	0.00025000	0.00000000	0.00000000
2	0.00000000	0.00025000	0.00000000
3	0.00000000	0.00000000	0.00025000

12. The **CORRELATION MATRIX OF THE MAXIMUM-LIKELIHOOD ESTIMATES** is derived from the preceding covariance matrix.

	1	2	3
1	1.0000000	0.0000000	0.0000000
2	0.0000000	1.0000000	0.0000000
3	0.0000000	0.0000000	1.0000000

13. **MAXIMUM-LIKELIHOOD PREDICTED VALUES FOR RESPONSE FUNCTIONS AND PROBABILITIES** basically contrasts observed and predicted values. Consider first the first of the 15 lines beneath the caption that begins with 1 under **Sample. Function Number 1** means that it is the first element of the **_RESPONSE_** profile whose observed **Response Probability** was .1135 (see item 6). In **Proc Catmod**, probabilities are converted to logits taken relative to the last element whose probability was .118, and ln(.1135/.118) = -.0388818. A standard error is defined accordingly. This particular model predicts equal outcomes in all 16 cells so the predicted function is .0. These data can be used to obtain the model G^2 using Equation 7.4a. The residual is simply the difference between the two. There is no entry for the 16th element because it is being compared to itself so its logit will always be 0 by definition. Following these data are the results expressed as proba-

bilities (the keyword `pred=freq` will output frequencies instead of probabilities).

						-------Observed-------		------Predicted------		
					Function		Standard		Standard	
Sample	SA	SB	RA	RB	Number	Function	Error	Function	Error	Residual
1					1	-0.0388818	0.0657365	0	0.04472136	-0.0388818
					2	-0.736444	0.08089143	0	0.04472136	-0.736444
					3	-0.7770816	0.08201792	0	0.04472136	-0.7770816
					4	-1.52225	0.10875377	0	0.04472136	-1.52225
					5	-0.8586616	0.08437206	0	0.04472136	-0.8586616
					6	0.03333642	0.06455869	0	0.04472136	0.03333642
					7	-1.6136842	0.11294663	0	0.04472136	-1.6136842
					8	-0.7724839	0.08188895	0	0.04472136	-0.7724839
					9	-0.8003927	0.08267784	0	0.04472136	-0.8003927
					10	-1.7026317	0.11724569	0	0.04472136	-1.7026317
					11	0.01053751	0.06492365	0	0.04472136	0.01053751
					12	-0.7408786	0.08101289	0	0.04472136	-0.7408786
					13	-1.6136842	0.11294663	0	0	-1.6136842
					14	-0.7188997	0.08041437	0	0	-0.7188997
					15	-0.838859	0.08378905	0	0	-0.838859
	0	0	0	0	P1	0.1135	0.00501542	0.0625	0.00171163	0.051
	0	0	0	1	P2	0.0565	0.00365061	0.0625	0.00171163	-0.006
	0	0	1	0	P3	0.05425	0.00358144	0.0625	0.00171163	-0.00825
	0	0	1	1	P4	0.02575	0.00250434	0.0625	0.00171163	-0.03675
	0	1	0	0	P5	0.05	0.00344601	0.0625	0.00171163	-0.0125
	0	1	0	1	P6	0.122	0.00517484	0.0625	0.00171163	0.0595
	0	1	1	0	P7	0.0235	0.00239519	0.0625	0.00171163	-0.039
	0	1	1	1	P8	0.0545	0.00358921	0.0625	0.00171163	-0.008
	1	0	0	0	P9	0.053	0.00354228	0.0625	0.00171163	-0.0095
	1	0	0	1	P10	0.0215	0.00229335	0.0625	0.00171163	-0.041
	1	0	1	0	P11	0.11925	0.00512419	0.0625	0.00171163	0.05675
	1	0	1	1	P12	0.05625	0.003643	0.0625	0.00171163	-0.00625
	1	1	0	0	P13	0.0235	0.00239519	0.0625	0.00171163	-0.039
	1	1	0	1	P14	0.0575	0.00368082	0.0625	0.00171163	-0.005
	1	1	1	0	P15	0.051	0.00347847	0.0625	0.00171163	-0.0115
	1	1	1	1	P16	0.118	0.00510088	0.0625	0.00171163	0.0555

We will now turn to a consideration of the remaining effects. Consider the following table of model and difference effects (the stimuli effect is the composite of SA, SB, and SA*SB; all difference effects are based on 1 *df* except RA*RB(SA*SB) which is based on 4):

		Model			*Difference*	
Effect	*df*	G^2	*p*		G^2	*p*
Stimuli	12	1182.14				
RA	11	1182.04	.00		.10	.75
RB	10	1179.74	.00		2.30	.13
SA*RA	9	609.53	.00		570.21	.00

SB*RB	8	3.88	.87	605.65	.00
SA*RB	7	3.31	.85	.57	.45
SB*RA	6	1.97	.92	1.34	.25
RA*RB(SA*SB)	2	1.16	.56	.81	.37
SA*SB*RA	1	.51	.48	.65	.42
SA*SB*RB	0	.00	—	.51	.48

Note that only two effects are significant, SA*RA and SB*RB. These each describe the effect of a stimulus on its proper response. RA and RB are independent of one another and of "improper" stimuli, which is what one would expect when perceptual independence is present. An important question is how to define the magnitudes of the two significant effects. One could do so by considering the magnitudes at the very end (the saturated model, which, by definition, has 0 df). However, the effects at this point are influenced by a variety of terms that are strictly due to chance, something true in OLS as well as ML. Note that the more effects that are present, the lower effect magnitudes tend to be because any shared variance is omitted, as is true with any form of regression weight. Perhaps a better possibility is to consider effect magnitudes in the model containing the last effect retained (SB*RB). The parameter estimates of this model are as follows:

Effect	Estimate	Standard Error
SA	.00	.02
SB	.01	.02
SA*SB	.00	.02
RA	-.01	.02
RB	-.03	.02
SA*RA	.39	.02
SB*RB	.41	.02

All but SA*RA and SB*RB can be treated as 0. In this model, we further conclude that the two effects are of equal magnitude.

Now consider a model in which perceptual independence does not hold. In particular, assume that error affecting A and error affecting B

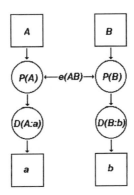

Figure 11.2.   Correlated error affecting *P*(*A*) and *P*(*B*).

are correlated as in Figure 11.2. If the random noise (EC) is positive, it erroneously causes both XA and XB to be large, so that the two responses tend to jointly assume values of "1", more often than they would under independence. Conversely, if the random noise is negative, it erroneously causes both XA and XB to be too small, so that the two responses tend to jointly assume values of "0."

The program that simulates this process is as follows:

```
Data Err_cor2;
Title 'Error Correlation Model 2 (Ashby & Townsend)';
Retain n 4000 err 1995;
Do i = 1 to n;
If (mod(i, 2) = 1) then sa = 1;else sa = 0;
If (mod(i, 4) le 1) then sb = 1;else sb = 0;
ec = rannor(err);
xa = sa + sqrt(.5)*ec + sqrt(.5)*rannor(err) - .5;
xb = sb + sqrt(.5)*ec + sqrt(.5)*rannor(err) - .5;
If (xa ge .0) then ra = 1;else ra = 0;
If (xb ge .0) then rb = 1;else rb = 0;
Drop n i err ec;
Output;
End;
```

The preceding `Proc Catmod` statements would be appended here, as well as in the models discussed later. In this model, the error correlation is positive, which simulates what is known as an "assimilation effect" in the sense that stimuli or stimulus attributes are both overjudged or both underjudged. Simulating the converse "contrast effect," which perhaps occurs more often, only involves a change in one statement, but is left as an exercise. The following describes the results:

1. Following are the response frequencies:

Sample	1	2	3	4	5	6	7	8
1	555	127	160	158	276	424	32	268
Sample	9	10	11	12	13	14	15	16
1	270	32	427	271	158	154	138	550

2. Here are the hierarchical effects. Note that, in addition to the accuracies (SA*RA and SB*RB), the term representing the association between responses within stimulus combinations, RA*RB(SA*SB), also contributes to the model.

		Model			*Difference*	
Effect	*df*	G^2	*p*		G^2	*p*
Stimuli	12	1653.63	.00			
RA	11	1653.61	.00		.02	.89
RB	10	1653.36	.00		.25	.62
SA*RA	9	1048.10	.00		605.26	.00
SB*RB	8	376.16	.00		671.94	.00
SA*RB	7	375.08	.00		1.08	.30
SB*RA	6	374.16	.00		.92	.34
RA*RB(SA*SB)	2	.15	.93		374.01	.00
SA*SB*RA	1	.08	.78		.07	.79
SA*SB*RB	0	.00			.08	.78

3. The total effect of RA*RB(SA*SB) actually consists of four parameters, one per stimulus combination. Their magnitudes are 0.3645, 0.4284, 0.4155, and .3529. All are significantly greater than 0 and, in fact, differ purely by chance.

A third form of perceptual nonindependence is when the response to one stimulus is determined in part or in whole by an improper stimulus. In the extreme case, there is a single perceptual process leading to one response, RB in this case, being judged solely by the other stimulus.

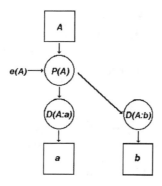

Figure 11.3. Single perceptual process causing RB to be determined by SA rather than SB.

Figure 11.3 illustrates this. Note that even though the same stimulus is being judged the decisional processes are separate so the responses need not be identical.

Following is the program that generates data consistent with this model:

```
Data Sing_PP;
Title 'Single Perceptual Process Model';
Retain n 4000 err 1995;
Do i = 1 to n;
If (mod(i, 2) = 1) then sa = 1;else sa = 0;
If (mod(i, 4) le 1) then sb = 1;else sb = 0;
xa = sa + rannor(err) - .5;
xb = sa + rannor(err) - .5;
If (xa ge .0) then ra = 1;else ra = 0;
If (xb ge .0) then rb = 1;else rb = 0;
Drop n i err;
Output;
End;
```

1. Following are the response frequencies:

Sample	1	2	3	4	5	6	7	8
1	454	226	217	103	491	197	216	96
Sample	9	10	11	12	13	14	15	16
1	99	199	226	476	94	230	204	472

2. Here are the hierarchical effects. The main points to note are that the accuracy in judging B (SB*RB) is by chance, but the influence of SA on RB is quite strong. There is a suggestion that SA influences the accuracy of SB, but that just misses significance ($p = .052$) and, in fact, is a Type I error.

Effect	df	Model G²	p	Difference G²	p
Stimuli	12	1161.99	.00		
RA	11	1161.89	.00	.10	.75
RB	10	1161.89	.00	.00	1.00
SA*RA	9	591.67	.00	570.22	.00
SB*RB	8	591.59	.00	.08	.78
SA*RB	7	7.18	.41	584.41	.00
SB*RA	6	5.84	.44	1.34	.25
RA*RB(SA*SB)	2	4.61	.10	1.23	.87
SA*SB*RA	1	3.77	.05	.84	.36
SA*SB*RB	0	.00		3.77	.05

3. The magnitudes of the effects of SA on RA and SB, respectively, are 0.3919 and 0.3972.

This is illustrative of only some of the possible influences.

Categorical Data Analysis in SPSS

There are several options for performing a categorical data analysis in SPSS. All may be found under the `Analyze` option in the SPSS Data Editor menu. From there, choose either (a) `Regression`, `Binary Logistic`, `Multinomial Logisitic`, `Ordinal`, `Probit` or (b) `Loglinear` with its various options.

▣ FACTOR ANALYSIS OF CATEGORICAL DATA

One of the most common mistakes made in data analysis is to factor items or other categorical data and ignore the difference between such data and the continuous data that are implicit in the model. Chapter 5 contained data set One—which generates single-factor data and which was further studied in Chapter 6. Now view these six items as standard normal latent data that give rise to manifest item responses. In particular, first assume that the decision rule is to have the response be a 0 if the latent observation is less than or equal to 0 and 1 if the latent observation is greater than 0. This corresponds to a test where all items are of equivalent, intermediate difficulty. Next, assume that the decision rule is to have the response for three of the items be a 0 if the latent observation is less than or equal to –1 and 1 if the latent observation is greater than –1. Call this the same difficulty set. Conversely, the decision rule for the remaining three items is to have the response for the remaining three items be a 0 if the latent observation is less than or equal to +1 and 1 if the latent observation is greater than +1. This corresponds to a test consisting of three easy items and three difficult items. Call this the different difficulty set.

There are two things to note. The first is that the average correlations will be lower in both data sets because dichotomizing loses information in the bivariate case, though it can spuriously increase the rate of Type I errors with two predictors (Maxwell & Delaney, 1993). Second, although correlations will tend to be homogeneous in the same difficulty set, they will tend to form two clusters in the different difficulty set—an easy item will correlate more highly with another easy item than with a difficult item and vice versa. These results can look just like those produced by latent two-factor data. Both considerations will tend to diffuse the factor structure in the sense of reducing the magnitude of the first eigenvalue. Depending on the criterion used to determine the number of factors, this can lead to spurious results. In particular, factors that emerge from differences in the statistical distribution of variables are historically known as difficulty factors, so named because they represent factors based on differences in difficulty level rather than content when found in the literature on skills. One interpretation is that they arise from nonlinearity (McDonald, 1985). Regardless of the area they appear in, they are spurious.

The following program was used to create data:

```
Data One;
Title "Single-Factor Homogeneous Correlations";
Retain b .7 nobs 1000 nitems 6 err 1998;
Bres = sqrt(1 - b**2);
Array Y (*) s1-s6;
```

```
Array P (*) sa1-sa6;
Array D (*) di1-di6;
Do I = 1 to nobs;
F = rannor(err);
Do j = 1 to nitems;
Y(j) = b*F + bres*rannor(err);
If Y(j) ge 0 then p(j) = 1;else p(j) = 0;
If j le 3 then cutoff = 1;else cutoff = -1;
If Y(j) ge cutoff then d(j) = 1;else d(j) = 0;
End;
Keep s1-s6 sa1-sa6 di1-di6;
Output;
End;
```

Note the use of the `Array` statements to facilitate forming the latent data set (s1–s6), the same difficulty observations (sa1–sa6), and the different difficulty observations (di1–di6).

The latent, same difficulty, and different difficulty data sets were then factored using both principal components and maximum likelihood, the latter under the hypothesis of a single factor. The code is as follows:

```
Proc Factor All Data=One;var s1-s6;
Proc Factor All Data=One method=ml n=1;var s1-s6;
Proc Factor All Data=One;var sa1-sa6;
Proc Factor All Data=One method=ml n=1;var sa1-sa6;
Proc Factor All Data=One;var di1-di6;
Proc Factor All Data=One method=ml n=1;var di1-di6;
```

The correlations within the latent data set are as follows:

	s1	s2	s3	s4	s5	s6
s1	1.000	.502	.515	.489	.468	.482
s2	.502	1.000	.487	.475	.467	.516
s3	.515	.487	1.000	.494	.482	.515
s4	.489	.475	.494	1.000	.462	.503
s5	.468	.467	.482	.462	1.000	.469
s6	.482	.516	.515	.503	.469	1.000

Likewise, the correlations within the same difficulty data set are as follows:

	sa1	sa2	sa3	sa4	sa5	sa6
sa1	1.000	.342	.358	.352	.326	.356
sa2	.342	1.000	.296	.299	.304	.322
sa3	.358	.296	1.000	.314	.336	.318
sa4	.352	.299	.314	1.000	.334	.296
sa5	.326	.304	.336	.334	1.000	.314
sa6	.356	.322	.318	.296	.314	1.000

Finally, the correlations within the different data set are as follows:

	di1	di2	di3	di4	di5	di6
di1	1.000	.252	.353	.183	.152	.177
di2	.252	1.000	.288	.122	.150	.168
di3	.353	.288	1.000	.162	.161	.193
di4	.183	.122	.162	1.000	.311	.318
di5	.152	.150	.161	.311	1.000	.282
di6	.177	.168	.193	.318	.282	1.000

As can be seen, the latent correlations ranged from .462 to .516, and the same difficulty correlations ranged from .296 to .358. The correlations ranged from .252 to .353 for items of equivalent difficulty and from .122 to .193 for items of different difficulty in the different data set. These are as expected.

Figure 11.4 is a scree plot. The data thus far considered are identified as Pearson correlations; the data denoted "polychoric" will be discussed shortly. Note that both the same difficulty and the different difficulty data sets (filled and open circles) are more diffuse than the latent data (filled squares), though only the different difficulty data generated two eigenvalues greater than 1, which would lead one to accept a second factor by the oft-used Kaiser–Guttman criterion. In fact, it is possible to show that the same difficulty data could also give rise to spurious factors with a greater number of variables (Bernstein & Teng, 1989).

The ML factoring gave rise to G^2 values against the null hypothesis of exactly one factor. These were 5.60, 4.60, and 103.46 for the latent, same difficulty, and different difficulty data, respectively. The first two values are not significant, but the third is ($p < .001$, all $df = 9$). Not surprisingly, the two clusters of correlations are nearly unrelated.

One suggested approach to the problems associated with factoring categorical data is to obtain polychoric correlations (known as tetrachoric correlations when the data are dichotomies, as here). These are intended

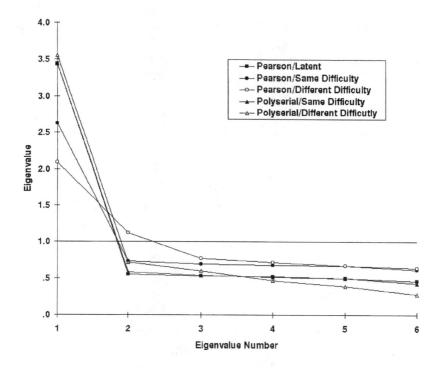

Figure 11.4. Scree plots for latent and dichotomized data.

to correct for the information lost in categorization as they estimate what the correlations between the latent variables would be. There are somewhat different models used to generate such data in the context of factor analysis. In the present case, the correlations were generated as follows:

```
Proc Freq Data=One;Tables (sa1-sa6)*(sa1-sa6) (di1-
di6)*(di1-di6) /PLCORR;
```

The PLCORR option generates the desired values along with a number of other statistics that are not of present interest. The PRELIS option in LISREL uses a very similar algorithm. The correlations were then organized into correlation matrices (see Chapter 4) and made part of two data sets, SPolys and DPolys, for the same and different data, respectively, as follows:

```
Data SPolys (type=corr);
Input _type_ $ _name_ $ Poly1-Poly6;
Cards;
```

```
N                1000   1000   1000   1000   1000   1000
Corr Poly1  1.000   .512    .533    .526    .490    .530
Corr Poly2   .512  1.000   .449    .453    .460    .484
Corr Poly3   .533    .449  1.000   .473    .504    .479
Corr Poly4   .526    .453    .473  1.000   .501    .449
Corr Poly5   .490    .460    .504    .501  1.000   .474
Corr Poly6   .530    .484    .479    .449    .474  1.000
Data DPolys (type=corr);
Input _type_ $ _name_ $ Poly1-Poly6;
Cards;
N                1000   1000   1000   1000   1000   1000
Corr Poly1  1.000   .456    .591    .654    .439    .554
Corr Poly2   .456  1.000   .503    .335    .435    .511
Corr Poly3   .591    .503  1.000   .479    .459    .619
Corr Poly4   .654    .335    .479  1.000   .545    .554
Corr Poly5   .439    .435    .459    .545  1.000   .502
Corr Poly6   .554    .511    .619    .554    .502  1.000
```

These are known as `Type=corr` data sets (see the **SAS Procedures Guide**, SAS Institute, 1990) and can be input directly to such routines as `Proc Factor`. Further note that the correlations within SPolys ranged from .449 to .553, but the correlations within DPolys ranged from .335 to .654. In fact, the following additional analyses were conducted.

```
Proc Factor All Data=SPolys;var Poly1-Poly6;
Proc Factor All Data=SPolys method=ml n=1;var Poly1-Poly6;
Proc Factor All Data=DPolys;var Poly1-Poly6;
Proc Factor All Data=DPolys method=ml n=1;var Poly1-Poly6;
Run;
```

Both sets of polyserial correlations were as concentrated as the latent data, which is one objective of forming them in the first place. However, the G^2 values from the ML tests were 15.32 and 198.96. The first approached significance ($p < .08$), whereas the second was clearly significant ($p < .001$), as is consistent with its greater range of correlations. Thus, polyserial correlations did not rectify the problems posed by using categorical data by this inferential criterion.

▣ ITEM REƧPONƧE THEORY APPLICATIONƧ

As with any statistical application, fitting an item response theory (IRT) model to data involves determining which parameters are needed. Alternate theories require different parameters. Use data from the multivariate correlational simulations in Chapter 5 as an example. One might wish to begin by testing the proposition that the items were equally discriminating though not necessarily equally difficult and that the probability of answering the item by guessing is .0 (i.e., the test uses a short-answer format). Most current IRT models of such data assume that the relationship between the latent variable, θ, and the probability of responding correctly to the item is a logistic function, where the relationship between the two is called an item operating characteristic (IOC) or item trace. Consequently, the model of interest is known as a one-parameter logistic (1PL) model, also called a Rasch model. Early use of the term "Rasch model" also implied a particular form of parameter estimation (Thissen, 1988), a practice that is no longer observed. It is not as restrictive as a parallel model in which items are also equally difficult. However, it is more restrictive than a two-parameter logistic model in which items also could differ in discrimination. For some recent applications of this model, see Hoijtink, Rooks, and Wilmink, (1999), Reise and Widaman (1999), and Steinberg and Thissen (1996). Andrich and Styles (1998) and Noel (1999) present interesting uses of item response theory having a very different form than considered here.

Specifying a 1PL or other model involves telling the program various things, such as the number of items, the format of the data, and, of course, the model to be tested. This varies considerably from program to program. The one to be used here, Thissen's (1987) Multilog, is far from the easiest to use, but it is extremely general as to the variety of models it can test. One major problem is that it is a DOS program, which requires considerable typing skills. However, it is also in the process of being revised for a Windows environment so that it should be user friendlier in the future. The example to be presented uses the 1PL data based on the six items generated in Chapter 5. Data can be input to Multilog either in the form of counts for the various patterns of correct and incorrect responses or as individual observations. Patterns were used in this example. There were six items so there are $2^6 = 64$ patterns.

Two files are input to Multilog, a program file having the extension "TEN" and a data file having the extension "DAT." The "TEN" file either can be created directly or can be generated by a program in the Multilog package called Inforlog. The commands to Inforlog include such things as the number of items, the model to be used, the number of groups, the number of patterns, and the format of the data to be analyzed. In the present case, the data were in the form of a frequency distribution for the 64 patterns, listed, in part, as follows:

```
000000 0 133
000001 0  13
    . . .
111101 0  41
111110 0  49
111111 0 119
```

The output is as follows:

1. Because most versions of Multilog are closely related to the original version, which was written in FORTRAN for a mainframe computer, the first column of output contains information that is meaningful to the line printer used by the mainframe. For example, a "1" in the first column starts a new page. The last line contains the user-written title for the analysis.

```
1 M U L T I L O G
0 FOR MULTIPLE CATEGORICAL ITEM RESPONSE DATA
0 VERSION 5.11
01PL ON SIMULATED DATA
```

2. The next output contains input parameters and should be checked against what was intended for accuracy. N denotes the total number of combinations of frequencies for correct and incorrect responses, which, as noted, were 64. L denotes the total number of items to be estimated. These, in turn, are divided into item statistics (L1 = 6 items) and group statistics (L2 = 1 group). Optionally, an ID field can identify patterns or individual responses. NCHAR is the number of characters used for the identifier. Because an identifier was not used here, it assumes the value of 0. Finally, MCODE denotes the maximum number of codes used; it is not of general interest.

```
0DATA PARAMETERS-
         N        L       L1       L2    NCHAR    MCODE
        64        7        6        1        0        2
```

3. The succeeding output contains various parameters controlling estimation such as the maximum number of iterations. They are of minimal interest to users.

```
OESTIMATION PARAMETERS-
      NCYC     NFRC     NSEG      NP      MAXIT
        25        0        1       7         4
OI/O CONTROL-
      LOTS       I1      I2   RESTRT
         0        1       2        0
OCONVERGENCE CONTROL-
      CRTI     CRTC     STEP      RK      RM   ACCMAX
     .0001    .0010    .5000   .9000  1.0000    .0000
OMISSING VALUE CODE FOR CONTINUOUS DATA= 9.0000
OSWITCHES-
      PUNI     PUNS    PRNTS   READI   READS   FIT   SCORE   KMID   SDIZE   READC
         F        F        F       T       T     T       T      T       F       F
  0   MARG      RWT   INCORE   PRIOR
         T        T        T       F
```

4. These are followed by the initial estimates, which likewise are usually of little interest. Some problems require either the iterations or the initial estimates to be modified in order to achieve convergence, though.

```
1ITEM SUMMARY AT START
01PL ON SIMULATED DATA
OITEM    1       2 GRADED CATEGORIES
OA = P(  7) =    1.000
 B( 1) = P(  1) =      .000
OITEM    2       2 GRADED CATEGORIES
OA = P(  7) =    1.000
 B( 1) = P(  2) =      .000
OITEM    3       2 GRADED CATEGORIES
OA = P(  7) =    1.000
 B( 1) = P(  3) =      .000
OITEM    4       2 GRADED CATEGORIES
OA = P(  7) =    1.000
 B( 1) = P(  4) =      .000
OITEM    5       2 GRADED CATEGORIES
OA = P(  7) =    1.000
 B( 1) = P(  5) =      .000
OITEM    6       2 GRADED CATEGORIES
OA = P(  7) =    1.000
 B( 1) = P(  6) =      .000
OITEM    7 GRP1 GAUSSIAN CONTINUOUS-
OBETA = P(499) =   -1.000, MU = P(  8) =      .000, SIGMA =
P(498) =    1.000
OIN-CORE CATEGORICAL DATA STORAGE AVAILABLE FOR N=  5000,
5000 WORDS
OTH-GROUP ASSIGNMENTS-
```

5. Some more estimation parameters follow, again of little interest.

```
QUAD. PT.  1 LOCATION= -4.500 DENSITY=  1.00000
QUAD. PT.  2 LOCATION= -3.500 DENSITY=  1.00000
QUAD. PT.  3 LOCATION= -2.500 DENSITY=  1.00000
QUAD. PT.  4 LOCATION= -1.500 DENSITY=  1.00000
QUAD. PT.  5 LOCATION=  -.500 DENSITY=  1.00000
QUAD. PT.  6 LOCATION=   .500 DENSITY=  1.00000
QUAD. PT.  7 LOCATION=  1.500 DENSITY=  1.00000
QUAD. PT.  8 LOCATION=  2.500 DENSITY=  1.00000
QUAD. PT.  9 LOCATION=  3.500 DENSITY=  1.00000
QUAD. PT. 10 LOCATION=  4.500 DENSITY=  1.00000
```

6. The codes and formats used for the various categories are indicated. Make sure these are correct.

```
11PL ON SIMULATED DATA
OREADING DATA...
OKEY-
OCODE  CATEGORY
   0      111111
   1      222222
0
OFORMAT FOR DATA-
 (6A1,1X,F1.0,1X,F3.0)
OFIRST OBSERVATION AS READ-
 ITEMS 000000
 NORML       .000
 WT/CR    133.00
```

7. Various statistics are presented when convergence occurs.

```
+FINISHED CYCLE 12
 MAXIMUM INTERCYCLE PARAMETER CHANGE=   .00045 P(  7)
 1ITEM SUMMARY
 O1PL ON SIMULATED DATA
```

8. Statistics then appear for each item based on the maximum likelihood estimation process. In particular, each estimate is numbered, followed by the value of the estimate and its standard error first for *a* (discrimination) and then for *b* (difficulty) in this model (other parameter estimates appear in other models). For example, the discrimination parameter for item 1 is parameter estimate 7, its value

is 1.709, and its standard error is .058. Because this is a 1PL model, these three values will remain the same for all items. However, the difficulty parameter for item 1, B(1) in the printout, which is parameter estimate 1, P(1) in the printout, has an estimated value of −.427 and a standard error of .059, and these differ among items.

```
OITEM   1        2 GRADED CATEGORIES
0A  = P(  7) =    1.709
                 (   .058)
 B( 1) = P(  1) =   -.427
                 (   .059)
OITEM   2        2 GRADED CATEGORIES
0A  = P(  7) =    1.709
                 (   .058)
 B( 1) = P(  2) =   -.324
                 (   .057)
OITEM   3        2 GRADED CATEGORIES
0A  = P(  7) =    1.709
                 (   .058)
 B( 1) = P(  3) =   -.062
                 (   .059)
OITEM   4        2 GRADED CATEGORIES
0A  = P(  7) =    1.709
                 (   .058)
 B( 1) = P(  4) =    .062
                 (   .058)
OITEM   5        2 GRADED CATEGORIES
0A =  P(  7) =    1.709
                 (   .058)
 B( 1) = P(  5) =    .426
                 (   .057)
OITEM   6        2 GRADED CATEGORIES
0A  = P(  7) =    1.709
                 (   .058)
 B( 1) = P(  6) =    .520
                 (   .058)
```

9. Every model has a Gaussian, continuous group parameter. This can be important when several groups are being compared, as the estimates compare the groups on the latent variable θ. However, the group parameter can be ignored when, as here, there is only one group.

```
OITEM   7 GRP1 GAUSSIAN CONTINUOUS-
OBETA = P(499) = -1.000, MU = P(  8) =    .000, SIGMA =
P(498) =  1.000
                      (    .000)                  (    .000)
(    .000)
```

10. The data continue with (a) the observed frequencies and (b) the expected frequencies. These are followed by (c) the standardized residual, (d) the expected a posteriori (EAP) estimate of θ and its standard error, and (e) the pattern. In this case, pattern 111111 (six incorrect responses) appeared 133 times in the data. The model predicted 133.7 occurrences. The standardized difference between the two is -.06. People achieving this pattern have an expected value of θ equal to 1.35 ± .64. At the user's discretion, the maximum likelihood estimate (mode of the individual estimates of θ) can be obtained in place of the EAP or mean of the estimates. As with any 1PL model, patterns containing the same number of correct and incorrect responses produce the same estimate of θ. For example, patterns 111112, 111121, 111211, etc. all produce an estimate of -.76.

```
11PL ON SIMULATED DATA
0OBSERVED(EXPECTED)      STD.  :    EAP (S.D.)  :  PATTERN
                         RES.  :                :
     133.0(   133.7)    -.06  :  -1.35 (  .64) :  111111
      13.0(     9.4)    1.19  :   -.76 (  .51) :  111112
       8.0(    11.0)    -.90  :   -.76 (  .51) :  111121
       1.0(     1.7)    -.54  :   -.39 (  .45) :  111122
      19.0(    20.5)    -.33  :   -.76 (  .51) :  111211
       4.0(     3.2)     .46  :   -.39 (  .45) :  111212
       6.0(     3.7)    1.17  :   -.39 (  .45) :  111221
       1.0(     1.1)    -.10  :    .03 (  .53) :  111222
      25.0(    25.3)    -.07  :   -.76 (  .51) :  112111
       3.0(     3.9)    -.47  :   -.39 (  .45) :  112112
       1.0(     4.6)   -1.68  :   -.39 (  .45) :  112121
       1.0(     1.4)    -.31  :    .03 (  .53) :  112122
       6.0(     8.6)    -.89  :   -.39 (  .45) :  112211
       3.0(     2.5)     .28  :    .03 (  .53) :  112212
       4.0(     3.0)     .59  :    .03 (  .53) :  112221
       3.0(     1.9)     .82  :    .44 (  .44) :  112222
      42.0(    39.6)     .38  :   -.76 (  .51) :  121111
       6.0(     6.1)    -.06  :   -.39 (  .45) :  121112
       8.0(     7.2)     .29  :   -.39 (  .45) :  121121
```

2.0(	2.1)	−.09	:	.03 (	.53) :	121122
10.0(	13.5)	−.94	:	−.39 (	.45) :	121211
7.0(	4.0)	1.51	:	.03 (	.53) :	121212
4.0(	4.7)	−.31	:	.03 (	.53) :	121221
3.0(	2.9)	.04	:	.44 (	.44) :	121222
22.0(	16.6)	1.31	:	−.39 (	.45) :	122111
3.0(	4.9)	−.87	:	.03 (	.53) :	122112
5.0(	5.8)	−.32	:	.03 (	.53) :	122121
5.0(	3.6)	.72	:	.44 (	.44) :	122122
11.0(	10.8)	.07	:	.03 (	.53) :	122211
5.0(	6.8)	−.68	:	.44 (	.44) :	122212
13.0(	7.9)	1.79	:	.44 (	.44) :	122221
5.0(	9.2)	−1.39	:	.80 (	.52) :	122222
53.0(	47.2)	.84	:	−.76 (	.51) :	211111
4.0(	7.3)	−1.23	:	−.39 (	.45) :	211112
8.0(	8.6)	−.20	:	−.39 (	.45) :	211121
1.0(	2.5)	−.97	:	.03 (	.53) :	211122
13.0(	16.0)	−.76	:	−.39 (	.45) :	211211
1.0(	4.7)	−1.72	:	.03 (	.53) :	211212
7.0(	5.6)	.61	:	.03 (	.53) :	211221
5.0(	3.5)	.80	:	.44 (	.44) :	211222
21.0(	19.8)	.26	:	−.39 (	.45) :	212111
8.0(	5.9)	.88	:	.03 (	.53) :	212112
10.0(	6.9)	1.19	:	.03 (	.53) :	212121
3.0(	4.3)	−.64	:	.44 (	.44) :	212122
13.0(	12.9)	.04	:	.03 (	.53) :	212211
5.0(	8.1)	−1.08	:	.44 (	.44) :	212212
5.0(	9.5)	−1.45	:	.44 (	.44) :	212221
22.0(	11.0)	3.31	:	.80 (	.52) :	212222
29.0(	31.0)	−.36	:	−.39 (	.45) :	221111
8.0(	9.2)	−.39	:	.03 (	.53) :	221112
17.0(	10.8)	1.90	:	.03 (	.53) :	221121
3.0(	6.8)	−1.45	:	.44 (	.44) :	221122
26.0(	20.1)	1.32	:	.03 (	.53) :	221211
14.0(	12.6)	.39	:	.44 (	.44) :	221212
12.0(	14.8)	−.73	:	.44 (	.44) :	221221
15.0(	17.2)	−.54	:	.80 (	.52) :	221222
23.0(	24.9)	−.37	:	.03 (	.53) :	222111
17.0(	15.6)	.35	:	.44 (	.44) :	222112
10.0(	18.3)	−1.94	:	.44 (	.44) :	222121
25.0(	21.3)	.80	:	.80 (	.52) :	222122
31.0(	34.2)	−.54	:	.44 (	.44) :	222211
41.0(	39.7)	.20	:	.80 (	.52) :	222212

```
   49.0(    46.6)     .35  :    .80 (  .52) :   222221
  119.0(   121.8)    -.25  :   1.39 (  .63) :   222222
```

11. Finally, the model G^2, which equals –2 times the log likelihood of the model, is presented. Note that this value can be obtained by applying Equation 6.5 to the observed and expected frequencies.

```
ONEGATIVE TWICE THE LOGLIKELIHOOD=        61.1

(CHI-SQUARE FOR SEVERAL TIMES MORE EXAMINEES THAN CELLS)
```

🔳 PROBLEMS

11.1. Perform logistic regressions on the data generated in response to Problems 6.1 to 6.3.

11.2. Analyze the data of Problem 6.1 again, this time evaluating the effects of the three predictors hierarchically.

11.3. Analyze the data of Problems 6.7 to 6.9 as log-linear models. Note that such analyses play an important role in current signal detection theory. For a more complete discussion of the relationship between signal detection theory and categorical modeling, see Macmillan and Creelman (1991).

11.4. Take the data of Problem 6.7 and consider only the relationships among SA, SB, and RA. Analyze the data both as a log-linear and as a predictor–criterion model. How do the two sets of parameters relate to one another?

11.5. Obtain real questionnaire data and factor them. Is there a tendency for items with similar distributions (i.e., means and standard deviations) to be defined by the same factors? What effect does converting to polychoric correlations have upon the outcome?

11.6. Analyze the item response data using a program of your own choosing and the data of Chapter 5.

Eigenanalysis of a Gramian Matrix

OBJECTIVE

The goal of this appendix is to show you the basic "physiology" of the process of eigenanalysis, the extraction of eigenvalues and eigenvectors. As you have seen in several chapters, this extraction is basic to multivariate analysis. This is most directly true in component analysis, which is little more than an eigenanalysis per se.

You do not need to perform eigenanalyses by hand. The process is built in to all of the statistics packages, virtually by definition, and it is done in a manner that is far more efficient than the methods we will demonstrate. The goal of this appendix is simply to give you a feel for what underlies your analyses so that they will not be empty operations.

▣ PRINCIPLEſ UNDERLYING EIGENANALYſIſ

Basic Properties

Eigenvectors and eigenvalues can be extracted from any square matrix, Gramian or not. The most important general properties of note are the following:

1. By definition, an eigenvector ($\mathbf{v}$) of a matrix ($\mathbf{Y}$) is a vector that fulfills the relationship $\mathbf{Yv} = \lambda\mathbf{v}$, where λ is its associated eigenvalue. $\mathbf{Y}$ must be a square matrix (data array). Even though it is tempting to cancel $\mathbf{v}$ from both sides of this equation, do not do so—cancellation is not appropriate because the result, $\mathbf{Y} = \lambda$, is not true. To understand this unusual equation, note that a vector, by definition, has a length and a direction. An arbitrarily chosen vector changes both if it is premultiplied by $\mathbf{Y}$ (i.e., if $\mathbf{Y}$ appears to its left, as in the equation). However, as the equation shows, the eigenvector changes only its length and not its direction. Thus, the equation means that $\mathbf{Y}$ "acts like" a scalar in the presence of the eigenvector.

2. It is assumed that $\mathbf{v}$ is not a null vector consisting only of 0s.

3. Successive eigenvalues will be of diminishing magnitude (i.e., if $i < j$, then $\lambda_i > \lambda_j$).

4. Although it is poor word usage to some mathematicians, we will refer to the "ith eigenvector" as the eigenvector associated with λ_I so that the first eigenvector is the eigenvector associated with the largest eigenvalue.

5. If $\mathbf{v}$ is an eigenvector, any scalar multiple, $k\mathbf{v}$, will also be an eigenvector.

6. As a result, eigenvectors are not uniquely defined as to length. However, a vector is normalized if its length is 1.0, so that the sum of its squared elements is 1.0. Unless we state otherwise, assume that eigenvectors have been normalized.

7. There will be as many eigenvalues and eigenvectors in toto as there are rows and columns of $\mathbf{Y}$ (the **order** of $\mathbf{Y}$).

8. There will be as many nonzero eigenvalues as there are linearly independent rows (or columns) of $\mathbf{X}$. This number is known as the **rank** of $\mathbf{X}$, which will equal the **rank** of $\mathbf{Y}$ in this case.

9. The sum of the eigenvalues, $\sum \lambda$, will equal the sum of the diagonal elements or **trace** of $\mathbf{Y}$. As with all matrix operations, the diagonal refers to the elements running from top left to bottom right (strictly speaking, this is the **negative** diagonal). Negative diagonal elements have the property that their row position will equal their column position so they will be of the form y_{ii}, the first subscript identifying the row and the second identifying the column.

10. The product of the eigenvalues equals the determinant of **Y**, $|Y|$. If $|Y|$ is 0, at least one column of **Y** is a linear combination of the remaining columns. The rank of **Y** will be less than its order, and the matrix is then said to be **singular**. Conversely, if **Y** is of **full** rank (its order equals its rank), $|Y|$ will differ from 0 and **Y** is termed **nonsingular**. The rank of a matrix can never exceed its order.

Subtracting λ from each diagonal element of **Y** reduces its rank by 1.

Gramian Matrices

Although eigenanalysis can be applied to any square matrix, it is most common to apply it to a **Gramian matrix**. By definition, a matrix **Y** is Gramian if it can be expressed as the product of a second matrix, **X**, and the transpose of the second matrix, **X′**, so that **X′X = Y**. All Gramian matrices are symmetric in that element y_{ij} = element y_{ji}, though not all symmetric matrices are Gramian. In practice, however, the larger the values of the diagonal elements of any symmetric matrix, the more likely it will have Gramian properties even if it was not explicitly generated as a product.

Eigenvalues and eigenvectors of a Gramian matrix have the following additional properties:

1. None of the eigenvalues will be negative, generally allowing them to be interpreted as variances.

2. All pairs of eigenvectors, v_i and v_j, are **geometrically** orthogonal in that $\sum v_i v_j = 0$ in summation language and $v'_i v_j = 0$ in matrix language. Note that this **inner** product is a scalar and should be contrasted with the **outer** product. The latter is of the form $v_i v'_j$ and produces a matrix. The matrix language expression applied to the matrix **V** that contains all eigenvectors is **V′V = 0**. Note the convention of using lowercase for vectors and uppercase for matrices. Always place the eigenvectors in columns. Geometric orthogonality allows the eigenvectors to be thought of as right-angled axes because they will have a cosine of 0. Note also that the cosine of the angle between any two vectors mathematically equals $v'_i v_j / \sqrt{(v'_i v_i)(v'_j v_j)}$ and is mathematically equivalent to a correlation in statistics.

3. We will define the factor loadings from the **principal component** method of factor analysis as the normalized eigenvectors multiplied by the scalar $\sqrt{\lambda}$. We will denote each such set of loadings as **b**. This conforms to the scaling in such programs as SAS' `Proc Factor` and SPSS' `Factor` (which might be thought of as performing principal component factor analysis). Programs such as SAS' `Princomp`, which performs component analyses, obtain rather different (but related) quantities.

4. Eigenvectors are **statistically** orthogonal in that $\mathbf{V'YV} = 0$. In component analysis, this means that factor scores derived from eigenvectors will be uncorrelated.

5. Because the eigenvalues of a Gramian matrix are nonzero and add to the sum of the diagonal elements of **Y**, which frequently consists of variances, one can literally think of eigenanalysis as a form of "analysis of variance."

6. Because the eigenvectors cannot be less than 0, the $|Y|$ cannot be negative.

Outcomes of Matrix Multiplication with Gramian Matrices

If $\mathbf{X'X} = \mathbf{Y}$, one can let $\mathbf{W} = \mathbf{X'}$, which implies $\mathbf{WW'} = \mathbf{Y}$. In other words, it does not matter whether one premultiplies or postmultiplies a matrix by its transpose to obtain **Y**. Similarly, note that if **Y** can be expressed as $\mathbf{X'X}$, an infinite number of matrices can serve as **X**. To see this, consider a **transformation** matrix **T** that has the property that $\mathbf{T'T} = \mathbf{I}$, which is known as **orthonormality**. In the 2×2 case,

$$\mathbf{T} = \begin{bmatrix} \cos(\theta) & -\sin(\theta) \\ \sin(\theta) & \cos(\theta) \end{bmatrix}.$$

The point is that $\mathbf{X'T'TX} = \mathbf{X'X} = \mathbf{Y}$. Consequently, $(\mathbf{TX})'(\mathbf{TX}) = \mathbf{Y}$, so **TX** will produce the same result as **X**. This is the principle that underlies the **rotation of axes**.

Two of the infinite number of **X** matrices are especially important. One is a matrix with 0s below its diagonal. This is known as a **triangular** matrix and the process of determining it is called **Cholesky** decomposition. This appears in the multivariate literature in what is known as the **square**

root method of factor analysis, where it is basically obsolete, and as an algorithm to extract successive partial correlations. This **X** matrix is sometimes called the square root of **Y**. The second is a matrix of components. Its extraction occurs in the **method of principal components**. It is basic to nearly all forms of factor analysis.

STEPS IN CONDUCTING AN EIGENANALYSIS

1. Start with an arbitrary input vector $\mathbf{v}_i$. A unit vector consisting of 1s will do.

2. Compute $\mathbf{v}_o = \mathbf{Y}\mathbf{v}_i$.

3. Let $\mathbf{v}_d$ denote the difference between $\mathbf{v}_I$ and $\mathbf{v}_n$.

4. If $\mathbf{v}_I$ is very close to $\mathbf{v}_n$, say $\sqrt{\mathbf{v}'_d\mathbf{v}_d}$ (the length of $\mathbf{v}_d$) < .001, then go to step 8.

5. Normalize $\mathbf{v}_o$ to produce $\mathbf{v}_n$.

6. Replace the current $\mathbf{v}_I$ with $\mathbf{v}_n$.

7. Go to step 2.

8. The first eigenvalue, λ_1, is the length of $\mathbf{v}_o = \sqrt{\mathbf{v}'_o\mathbf{v}_o}$.

9. The first eigenvector, $\mathbf{v}_i$, is $\mathbf{v}_n$.

10. Form the reduced (residual) **Y** matrix $\mathbf{Y}_r = \mathbf{Y} - \lambda_1\mathbf{v}_1\mathbf{v}'_1$. Note that the latter employs the **outer** product of $\mathbf{v}_1$.

11. Repeat steps 1 to 10 a total of k times, where k is the order of **Y**, to extract the full set of eigenvalues and eigenvectors.

USING EXCEL TO PERFORM AN EIGENANALYSIS

It is fairly cumbersome to perform all of the steps in Excel so we will only extract the first eigenvalue, its associated eigenvector, and the reduced matrix, leaving the remaining eigenvalues and eigenvectors as an exercise. For this example, the elements of **Y** were chosen to be

$$
\begin{array}{cccc}
1.00 & .80 & .30 & .20 \\
.80 & 1.00 & .30 & .10 \\
.30 & .30 & 1.00 & .40 \\
.20 & .10 & .40 & 1.00
\end{array}
$$

A unit vector was employed as the arbitrary input vector.

1. Format the sheet to three-decimal place accuracy.

2. Place Y in cells A1:D4.

3. Place the arbitrary input vector v_I in cells F1:F4.

4. Select cells H1:H4 at the same time.

5. Enter =MMULT(A1:D4,F1:F4). This is the matrix multiplication function.

6. Press ⎡Shift⎤, ⎡Ctrl⎤, and ⎡Enter⎤ at the same time. This creates what is known as an **array** in Excel—cells H1 to H4 become a unit with values appropriate to each cell generated at the same time. In this case, the contents will be v_o. The equation will appear in brackets.

7. Enter =SQRT(SUMSQ(H1:H4)) in cell H5. This calculates the length of v_o so it will eventually contain λ_1.

8. Enter =H1/H$5 in cell J1. This is the first element of v_n.

9. Drag the contents of J1 through to row 4 to create the remaining elements of v_n.

10. Copy the equation in H5 to cell J5. It should produce a value of 1.0 to denote that the v_n is indeed normalized.

11. Enter =F1-J1 in cell L1. This is the first element of v_d.

12. Drag the contents of L1 through to row 4 to create the remaining elements of v_d.

13. Enter =$H$5*MMULT(J1:J4,TRANSPOSE(J1:J4)) in cells A6:D9.

14. Press $\boxed{\text{Shift}}$, $\boxed{\text{Ctrl}}$, and $\boxed{\text{Enter}}$ at the same time to create an array.

15. Enter **=A1-A6** in cell A11 to create the first element of the residual matrix. This does not have to be entered as an array formula.

16. Drag the equation in cell A11 through to column D.

17. Drag cells A11:D11 through row 15 to complete the residual matrix.

18. You are now ready to iterate. Begin by selecting J1:J4 and performing a copy ($\boxed{\text{Ctrl}}$-C).

19. Do a Paste Special, Values ($\boxed{\text{Alt}}$-E,S,V) to cells F1:F4. Note, do **not** do a Regular Paste ($\boxed{\text{Ctrl}}$-V or $\boxed{\text{Alt}}$-E,P). If you mistakenly perform a Regular Paste (you can tell because many cells will show **#DIV/0!** error), perform an Undo ($\boxed{\text{Alt}}$-E,U or $\boxed{\text{Ctrl}}$-Z) immediately. When a Paste Special is performed, note how the elements in L1:L5 become closer to 0. Keep iterating until the cells contain 0s.

To start over, simply place the original set of 1s (or any other numbers) in cells F1:F4.

USING SAS TO PERFORM AN EIGENANALYSIS

Here is an SAS version, again only going as far as the first residual matrix. To use it, you must have the IML (Interactive Matrix Language) module installed.

```
Proc IML;
R={1 .8 .4, .8 1 .2, .4 .2 1}; *Correlation Matrix;
R1=R;                     *Copy of R for later;
VI={1, 1, 1};             *Initial (arbitrary) vector;
Do ii=1 to 3;             *Compute eigenvectors & eigenvalues;
  Factor=ii;
  If(Factor > 1) then R=R-VOL#(VON*VON`);
  Diffl=10;               *Initialize eventual difference
between normalized input & output;
```

```
ITR=0;                        *Initialize number of iterations;
 Do while(Diffl>.0001);       *Iterate until normalized input vector
equals normalized output vector;
  ITR=ITR+1;
  VIL=sqrt(VI`*VI);           *Obtain length of input vector;
  VIN=(1/VIL)#VI;             *Normalize input vector;
  VO=R*VI;                    *Compute output vector;
  VOL=sqrt(VO`*VO);           *Obtain length of output vector;
  VON=(1/VOL)#VO;             *Normalize output vector;
  Diff=VIN-VON;               *Compute vector of differences
between normalized input & output;
  Diffl=sqrt(Diff`*Diff);     *Compute length of difference vector -
this goes to 0 on convergence;
  VI=VON;                     *Replace previous input vector with
normalized output vector;
 END;                         *End while loop;
 EV=EV||VON;                  *Add eigenvector to eigenvector table;
 Lambda=Lambda||VOL;          *Add eigenvalue to eigenvalue table;
 B=B||sqrt(VOL)#VON;          *B is factor pattern (see later) &
is eigenvector times square root of eigenvalue;
END;                          *End computation of eigenvalues &
eigenvectors;
Rest=B*B`;                    *As a check, BB` should equal R;
Print 'Results from IML iterative Eigen Analysis';
Print EV Lambda B Rest;       *Print results;
Call Eigen(Evals,Evects,R1);  *Use SAS eigenvalue routine to do the
same thing - recall R1=R;
Print 'Results from SAS eigenvalue routine';
Print Evals Evects;
Quit;
Run;
```

As a check, execute Call Eigen(Evals,Evects,R1); while in IML and with Reset Print on. This will provide all the eigenvalues and eigenvectors using a single function call.

Following is an SPSS equivalent:

USING SPSS TO PERFORM AN EIGENANALYSIS

```
matrix.
compute r = {1,.8,.4;.8,1,.2;.4,.2,1}.
compute r1=r.
compute vi = {1;1;1}.
compute von = {0;0;0}.
loop factor=1 to 3.
  do if (factor gt 1).
    compute r=r-vol*(von*t(von)).
  end if.
  compute diffl=10.
  compute itr=0.
  loop if (diffl gt .0001).
    compute itr=itr+1.
    compute vil=sqrt(t(vi)*vi).
    compute vin=(1/vil)*vi.
    compute vo=r*vi.
    compute vol=sqrt(t(vo)*vo).
    compute von=(1/vol)*vo.
    compute diff=vin-von.
    compute diffl=sqrt(t(diff)*diff).
    compute vi=von.
  end loop.
print factor.
print vol.
print von.
end loop.
call eigen(r1,evect,evals).
print evals.
print evect.
end matrix.
```

References

Akaike, H. (1974). A new look at the statistical model identification. *IEEE Transactions on Automatic Control, 19,* 716–723.

Akaike, H. (1987). Factor analysis and AIC. *Psychometrika, 52,* 317–332.

Anderberg, M. R. (1973). *Cluster analysis for applications.* New York: Academic Press.

Anderson, T. W., & Rubin, H. (1956). Statistical inference in factor analysis. *Proceedings of the Third Berkeley Symposium on Mathematical Statistics and Probability, 5,* 111–150.

Andrich, D., & Styles, I. M. (1998). The structural relationship between attitude and behavior statements from the unfolding perspective. *Psychological Methods, 3,* 454–469.

Ashby, F. G., & Maddox, W. T. (1990). Integrating information from separable dimensions. *Journal of Experimental Psychology: Human Perception and Performance, 16,* 698–612.

Ashby, F. G., & Perrin, N. A. (1988). Toward a unified theory of similarity and recognition. *Psychological Review, 95,* 124–130.

Ashby, F. G., & Townsend, J. T. (1986). Varieties of perceptual independence. *Psychological Review, 93,* 154–179.

Bartlett, M. S. (1937). The statistical conception of mental factors. *British Journal of Psychology, 28,* 97–104.

Bartlett, M. S. (1950). A further note on tests of significance in factor analysis. *British Journal of Mathematical and Statistical Psychology, 4,* 1–2.

Bentler, P. M. (1985). *EQS/PC, Version 2.0.* Los Angeles: BMDP Statistical Software.

Bentler, P. M. (1986). *Lagrange multiplier and Wald tests for EQS and EQS/PC.* Los Angeles: BMDP Statistical Software.

Bentler, P. M., & Bonnett, D. G. (1980). Significance tests and goodness of fit tests in the analysis of covariance structures. *Psychological Bulletin, 88,* 508–606.

Bernstein, I. H. (1988). *Applied multivariate analysis.* New York: Springer-Verlag.

Bernstein, I. H., & Havig, P. (1998). *Computer literacy: Getting the most from your PC.* Thousand Oaks, CA: Sage.

Bernstein, I. H., & Teng, G. (1989). Factoring items and factoring scales are different: Spurious evidence for multidimensionality due to item categorization. *Psychological Bulletin, 105,* 467–477.

Bernstein, I. H., Jaremko, M. E., & Hinkley, B. S. (1994). On the utility of the SCL-90-R with low-back pain patients. *Spine, 19,* 42–48.

Bickel, P. J., Hammel, E. A., & O'Connell, J. W. (1975). Sex bias in graduate admissions: Data from Berkeley. *Science, 187,* 398–404.

Blashfeld, R. K., & Aldenderfer, M. S. (1978). The literature on cluster analysis. *Multivariate Behavioral Research, 13,* 271–295.

Bock, R. D. (1972). Estimating item parameters and latent ability when the responses are scored in two or more nominal categories. *Psychometrika, 37,* 29–51.

Bollen, K. A. (1986). Sample size and Bentler and Bonnett's nonnormed fit index. *Psychometrika, 51,* 375–377.

Bollen, K. A. (1988). A new incremental fit index for general structural equation models. *Sociological Methods and Research, 17,* 303–316.

Box, G. E. (1954). Some theorems on quadrative forms applied in the study of analysis of variance problems. I: Effect of inequality of variance in the one-way classification. *Annals of Mathematical Statistics, 25,* 290–302.

Bozdogan, H. (1987). Model selection and Akaike's information criterion (AIC): The general theory and its analytical extensions. *Psychometrika, 52,* 345–370.

Browne, M. W. (1982). Covariance structures. In D. M. Hawkins (Ed.), *Topics in applied multivariate analysis.* Cambridge, UK: Cambridge University Press.

Buse, A. (1982). The likelihood ratio, Wald, and Lagrange multiplier tests: An expository note. *American Statistician, 36,* 153–157.

Darlington, R. B. (1900). *Regression and linear models.* New York: McGraw-Hill.

Davison, M. L. (1983). *Multidimensional scaling.* New York: John Wiley.

Davison, M. L. (1985). Multidimensional scaling vs. components analysis of test intercorrelations. *Psychological Bulletin, 97,* 94–105.

Eisenhart, C. (1947). The assumptions underlying the analysis of variance. *Biometrics, 3,* 1–21.

Everitt, B. S. (1980). *Cluster analysis* (2nd ed.). London: Heineman.

Garner, W. R., & Morton, J. (1969). Perceptual independence: Definitions, models, and experimental paradigms. *Psychological Bulletin, 72,* 233–259.

Geisser, S., & Greenhouse, S. W. (1958). An extension of Box's results on the use of the *F* distribution in multivariate analysis. *Annals of Mathematical Statistics, 29,* 885–891.

Gorsuch, R. L. (1983). *Factor analysis* (2nd ed.). Hillsdale, NJ: Lawrence Erlbaum.

Greenwald, A. G., Pratkanis, A. R., Lieppe, M. R., & Baumgardner, M. H. (1986). Under what conditions does theory abstract research progress? *Psychological Review, 93,* 216–229.

Guttman, L. (1954). Some necessary conditions for common factor analysis. *Psychometrika, 19,* 149–161.

Hambleton, R. K., & Swaminathan, H. (1985). *Item response theory.* Boston: Kluwer-Nijhoff.

Hambleton, R. K., Swaminathan, H., & Rogers, L. (1991). *Fundamentals of item response theory.* Newbury Park, CA: Sage.

Harman, H. H. (1976). *Modern factor analysis* (3rd ed. rev.). Chicago: University of Chicago Press.

Hays, W. L. (1988). *Statistics.* New York: Holt, Rinehart & Winston.

Hintzman, D. L. (1980). Simpson's paradox and the analysis of memory retrieval. *Psychological Review, 87,* 398–410.

Hoelter, J. W. (1983). The analysis of covariance structures: Goodness-of-fit indices. *Sociological Methods and Research, 11,* 325–344.

Hoijtink, H., Rooks, G., & Wilmink, F. W. (1999). Confirmatory factor analysis of items with a dichotomous response format using the multidimensional Rasch model. *Psychological Methods, 4,* 300–314.

Huynh, H., & Feldt, L. S. (1976). Estimation of the Box correction for degrees of freedom from sample data in randomized block and split-plot designs. *Journal of Educational Statistics, 1,* 69–82.

James, L. R., Mulaik, S. A., & Brett, J. M. (1982). *Causal analysis.* Beverly Hills, CA: Sage.

Jöreskog, K. G. (1969). A general approach to confirmatory maximum likelihood factor analysis. *Psychometrika, 34,* 183–202.

Jöreskog, K. G., & Sörbom, D. (1993). *LISREL 8: Structural equation modeling with the SIMPLIS command language*. Chicago: Scientific Software International.

Kaiser, H. J. (1960). The application of electronic computers to factor analysis. *Educational and Psychological Measurement, 20,* 141–151.

Kaiser, H. J. (1970). A second generation little jiffy. *Psychometrika, 35,* 401–417.

Kirk, R. E. (1995). *Experimental design: Procedures for the behavioral sciences*. Pacific Grove, CA: Brooks/Cole.

Krieckhaus, E. E., & Eriksen, C. W. (1960). A study of awareness and its effects on learning and generalization. *Journal of Personality, 28,* 503–517.

Kruskal, J. B., & Wish, M. (1978). *Multidimensional scaling*. Beverly Hills, CA: Sage.

Lee, S. Y. (1985). On testing functional constraints in structural equation models. *Biometrika, 72,* 125–131.

MacCallum, R. C. (1986). Specification searches in covariance structure modeling. *Psychological Bulletin, 100,* 107–120.

MacCallum, R. C., & Mar, C. M. (1995). Distinguishing between moderator and quadratic effects in multiple regression. *Psychological Bulletin, 118,* 405–421.

Macmillan, N. A., & Creelman, C. D. (1991). *Detection theory: A user's guide*. Cambridge, UK: Cambridge University Press.

Maxwell, S. E., & Delaney, H. D. (1993). Bivariate median splits and spurious statistical significance. *Psychological Bulletin, 113,* 181–190.

McClelland, G. H., & Judd, C. M. (1993). Statistical difficulties of detecting interactions and moderator effects. *Psychological Bulletin, 114,* 376–390.

McDonald, R. (1985). *Factor analysis and related methods*. Hillsdale, NJ: Lawrence Erlbaum.

McDonald, R. P. (1989). An index of goodness-of-fit based on noncentrality. *Journal of Classification, 6,* 97–103.

Mulaik, S. A. (1972). *The foundations of factor analysis*. New York: McGraw-Hill.

Mulaik, S. A., James, L. R., Van Alstine, J., Bennett, N., et al. (1989). Evaluation of goodness-of-fit indices for structural equation models. *Psychological Bulletin, 105,* 430–445. (Note: This appears in the SAS printout where it was cited simply as "Mulaik, 1989.")

Noel, Y. (1999). Recovering unimodal latent patterns of change by unfolding analysis: Application to smoking cessation. *Psychological Methods, 4,* 173–191.

Nunnally, J. C., & Bernstein, I. H. (1994). *Psychometric theory* (3rd ed.). New York: McGraw-Hill.

Olzak, L. A., & Wickens, T. D. (1983). The interpretation of detection data through direct multi-variate frequency analysis. *Psychological Bulletin, 93,* 574–585.

Paik, M. (1985). A graphical representation of a three-way contingency table. *American Statistician, 39,* 53–54.

Pedhazur, E. J. (1982). *Multiple regression in behavioral research* (2nd ed.). New York: Holt, Rinehart & Winston.

Reise, S. P., & Widaman, K. F. (1999). Assessing the fit of measurement models at the individual level: A comparison of item response theory and covariance structure approaches. *Psychological Methods, 4,* 3–21.

Sarle, W. S. (1983). *Cubic clustering criterion* (SAS Tech. Rep. No. A-108). Cary, NC: SAS Institute, Inc.

SAS Institute, Inc. (1989a). *SAS/STAT® User's Guide,* Version 6, Fourth Edition, Volume 1, Cary, NC: SAS Institute, Inc.

SAS Institute, Inc. (1989b). *SAS/STAT® User's Guide,* Version 6, Fourth Edition, Volume 2, Cary, NC: SAS Institute, Inc.

SAS Institute, Inc. (1990). *SAS® procedures guide, Version 6* (3rd ed.). Cary, NC: SAS Institute, Inc.

SAS Institute, Inc. (1997). *SAS/STAT software: Changes and enhancements through Release 6.12.* Cary, NC: SAS Institute, Inc.

Scheffé, H. A. (1959). *The analysis of variance.* New York: John Wiley.

Schwarz, G. (1978). Estimating the dimension of a model. *Annals of Statistics, 6,* 461–464.

Sclove, L. S. (1987). Applications of model-selection criteria to some problems of multivariate analysis. *Psychometrika, 542,* 333–343.

Simpson, E. H. (1951). The interpretation of interaction in contingency tables. *Journal of the Royal Statistical Society, Ser. B, 13,* 238–241.

Snook, S. C., & Gorsuch, R. L. (1989). Component analysis versus common factor analysis: A Monte Carlo study. *Psychological Bulletin, 106,* 148–154.

Spearman, C. (1904). General intelligence: Objectively determined and measured. *American Journal of Psychology, 15,* 201–293.

Steinberg, L., & Thissen, D. (1996). Uses of item response theory and the testlet concept in the measurement of psychopathology. *Psychological Methods, 1,* 81–97.

Stigler, S. M. (1988). *The history of statistics: The measurement of uncertainty before 1900.* Cambridge, MA: Belknap.

Swets, J. A. (1986a). Indices of discrimination or diagnostic accuracy: Their ROCs and implied models. *Psychological Bulletin, 99,* 100–117.

Swets, J. A. (1986b). Form of empirical ROCs in discrimination and diagnostic tasks: Implications for theory and measurement of performance. *Psychological Bulletin, 100,* 181–198.

Thissen, D. (1988). *Multilog.* Mooresville, IN: Scientific Software.

Thurstone, L. L. (1927/1994). A law of comparative judgment. *Psychological Review, 101,* 266–270.

Thurstone, L. L. (1935). *Vectors of the mind.* Chicago: University of Chicago Press.

Thurstone, L. L. (1947). *Multiple factor analysis.* Chicago: University of Chicago Press.

Tucker, L. R., & Lewis, C. (1972). A reliability coefficient for maximum likelihood factor analysis. *Psychometrika, 38,* 1–10.

Tversky, A., & Gati, I. (1982). Similarity, separability, and the triangle inequality. *Psychological Review, 89,* 123–154.

Velicer, W. F. (1976). The relation between factor score estimates, image scores, and principal component scores. *Educational and Psychological Measurement, 36,* 149–159.

Von Békésy, G. (1960). *Experiments in Hearing.* New York: McGraw-Hill.

Weisberg, S. (1985). *Applied linear regression* (2nd ed.). New York: John Wiley.

White, H. (1980). A heteroskedasticity-consistent covariance matrix estimator and a direct test for heteroskedasticity. *Econometrika, 48,* 817–838.

Wickens, T. D. (1989). *Multiway contingency tables analysis for the social sciences.* Hillsdale, NJ: Lawrence Erlbaum.

Wickens, T. D., & Olzak, L. A. (1989). The statistical analysis of concurrent detection ratings. *Perception & Psychophysics, 45,* 514–528.

Wilkinson, L. (1979). Tests of significance in stepwise regression. *Psychological Bulletin, 86,* 168–174.

Wilson, E. B., & Hilferty, M. M. (1931). The distribution of chi-square. *Proceedings of the National Academy of Sciences, 17,* 694.

Winer, B. J., Brown, D. R., & Michels, K. M. (1991). *Statistical principles in experimental design* (3rd ed.). New York: McGraw-Hill.

Yates, A. (1987). *Multivariate exploratory data analysis.* Albany: State University of New York Press.

Zwick, W. R., & Velicer, W. E. (1982). Factors influencing four rules for determining the number of components to retain. *Multivariate Behavioral Research, 17,* 253–269.

Zwick, W. R., & Velicer, W. E. (1986). Comparison of five rules to determine the number of components to retain. *Psychological Bulletin, 99,* 432–442.

Author Index

Subject Index